206
Topics in Current Chemistry

Editorial Board:
A. de Meijere · K. N. Houk · H. Kessler
J.-M. Lehn · S. V. Ley · S. L. Schreiber · J. Thiem
B. M. Trost · F. Vögtle · H. Yamamoto

Springer-Verlag Berlin Heidelberg GmbH

Modern Solvents in Organic Synthesis

Volume Editor: Paul Knochel

With contributions by
J. Augé, B. Betzemeier, B. Cornils, D. P. Curran,
P. Knochel, W. Leitner, B. Linclau, A. Loupy,
A. Lubineau, J. J. Maul, P. J. Ostrowski, D. Sinou,
G. A. Ublacker

Springer

This series presents critical reviews of the present position and future trends in modern chemical research. It is addressed to all research and industrial chemists who wish to keep abreast of advances in the topics covered.

As a rule, contributions are specially commissioned. The editors and publishers will, however, always be pleased to receive suggestions and supplementary information. Papers are accepted for "Topics in Current Chemistry" in English.

In references Topics in Current Chemistry is abbreviated Top. Curr. Chem. and is cited as a journal.

Springer WWW home page: http://www.springer.de
Visit the TCC home page at http://www.springer.de/

ISBN 978-3-662-15636-0   ISBN 978-3-540-48664-0 (eBook)
DOI 10.1007/978-3-540-48664-0

Library of Congress Catalog Card Number 74-644622

This work is subject to copyright. All rights are reserved, whether the whole or part of the material is concerned, specifically the rights of translation, reprinting, reuse of illustrations, recitation, broadcasting, reproduction on microfilms or in any other ways, and storage in data banks. Duplication of this publication or parts thereof is only permitted under the provisions of the German Copyright Law of September 9, 1965, in its current version, and permission for use must always be obtained from Springer-Verlag. Violations are liable for prosecution under the German Copyright Law.

© Springer-Verlag Berlin Heidelberg 1999
Originally published by Springer-Verlag Berlin Heidelberg New York in 1999.
Softcover reprint of the hardcover 1st edition 1999

The use of general descriptive names, registered names, trademarks, etc. in this publication does not imply, even in the absence of a specific statement, that such names are exempt from the relevant protective laws and regulations and therefore free for general use.

Cover design: Friedhelm Steinen-Broo, Barcelona; MEDIO, Berlin
Typesetting: Fotosatz-Service Köhler GmbH, 97084 Würzburg

SPIN: 10669000   02/3020 – 5 4 3 2 1 0 – Printed on acid-free paper

Volume Editor

Prof. Paul Knochel
Ludwig Maximilians-Universität München
Institut für Organische Chemie
Butenanderstr. 5–13
D-81377 München, Germany
E-mail: Paul.Knochel@cup.uni-muenchen.de

Editorial Board

Prof. Dr. Armin de Meijere
Institut für Organische Chemie
der Georg-August-Universität
Tammannstraße 2
D-37077 Göttingen, Germany
E-mail: ameijer1@uni-goettingen.de

Prof. Dr. Horst Kessler
Institut für Organische Chemie
TU München
Lichtenbergstraße 4
D-85747 Garching, Germany
E-mail: kessler@ch.tum.de

Prof. Steven V. Ley
University Chemical Laboratory
Lensfield Road
Cambridge CB2 1EW, Great Britain
E-mail: svl1000@cus.cam.ac.uk

Prof. Dr. Joachim Thiem
Institut für Organische Chemie
Universität Hamburg
Martin-Luther-King-Platz 6
D-20146 Hamburg, Germany
E-mail: thiem@chemie.uni-hamburg.de

Prof. Dr. Fritz Vögtle
Kekulé-Institut für Organische Chemie
und Biochemie der Universität Bonn
Gerhard-Domagk-Straße 1
D-53121 Bonn, Germany
E-mail: voegtle@uni-bonn.de

Prof. K. N. Houk
Department of Chemistry and Biochemistry
University of California
405 Higard Avenue
Los Angeles, CA 90024-1589, USA
E-mail: houk@chem.ucla.edu

Prof. Jean-Marie Lehn
Institut de Chimie
Université de Strasbourg
1 rue Blaise Pascal, B. P. Z 296/R8
F-67008 Strasbourg Cedex, France
E-mail: lehn@chimie.u-strasbg.fr

Prof. Stuart L. Schreiber
Chemical Laboratories
Harvard University
12 Oxford Street
Cambridge, MA 02138-2902, USA
E-mail: sls@slsiris.harvard.edu

Prof. Barry M. Trost
Department of Chemistry
Stanford University
Stanford, CA 94305-5080, USA
E-mail: bmtrost@leland.stanford.edu

Prof. Hisashi Yamamoto
School of Engineering
Nagoya University
Chikusa, Nagoya 464-01, Japan
E-mail: j45988a@nucc.cc.nagoya-u.ac.jp

Topics in Current Chemistry
Now Also Available Electronically

For all customers with a standing order for Topics in Current Chemistry we offer the electronic form via LINK free of charge. You will receive a password for free access to the full articles. Please register at:

http://link.springer.de/series/tcc/reg_form.htm

If you do not have a standing order you can nevertheless browse through the table of contents of the volumes and the abstracts of each article at:

http://link.springer.de/series/tcc

There you will also find information about the

- Editorial Board
- Aims and Scope
- Instructions for Authors

Preface

In recent years the choice of a given solvent for performing a reaction has become increasingly important. More and more, selective reagents are used for chemical transformations and the choice of the solvent may be determining for reaching high reaction rates and high selectivities. The toxicity and recycling considerations have also greatly influenced the nature of the solvents used for industrial reactions. Thus, the development of reactions in water is not only important on the laboratory scale but also for industrial applications. The performance of metal-catalyzed reactions in water for example has led to several new hydrogenation or hydroformylation procedures with important industrial applications. The various aspects of organic chemistry in water will be presented in this book. Recently, novel reaction media such as perfluorinated solvents or supercritical carbon dioxide has proven to have unique advantages leading to more practical and more efficient reactions. Especially with perfluorinated solvents, new biphasic catalyses and novel approaches to perform organic reactions have been developed. These aspects will be examined in detail in this volume.

Finally, the performance of reactions in the absence of solvents will show practical alternatives for many reactions.

More than ever before, the choice of the solvent or the solvent system is essential for realizing many chemical transformations with the highest efficiency. This book tries to cover the more recent and important new solvents or solvent systems for both academic and industrial applications.

Munich, June 1999   Paul Knochel

Contents

Contents of Volume 192

Organofluorine Chemistry
Fluorinated Alkenes and Reactive Intermediates

Volume Editor: R. D. Chambers
ISBN 3-540-63171-2

Contents of Volume 193

Organofluorine Chemistry
Techniques and Synthons

Volume Editor: R. D. Chambers
ISBN 3-540-63170-4

Water as Solvent in Organic Synthesis

André Lubineau[1] · Jacques Augé[2]

[1] Laboratoire de Chimie Organique Multifonctionnelle, bat 420, Université de Paris-Sud, F-91405 Orsay, France. *E-mail: lubin@icmo.u-psud.fr*
[2] Université de Cergy-Pontoise, 5 mail Gay-Lussac, Neuville-sur-Oise, F-95031 Cergy-Pontoise, France. *E-mail: auge@u-cergy.fr*

Organic reactions using water as solvent are reviewed with a focus on pericyclic reactions, carbonyl additions, stoichiometric organometallic reactions, oxidations and reductions which show an unusual outcome in terms of reactivity and selectivity compared with those performed in organic solvent. The advantages of using water as a solvent are discussed and related to the hydrophobic effects and the hydrogen-bonding ability of water with a special emphasis on its very high cohesive energy density which strongly favors organic reactions having a negative activation volume.

Keywords. Water, Solvents, Organic synthesis, Reactivity, Friendly processes.

Topics in Current Chemistry, Vol. 206
© Springer-Verlag Berlin Heidelberg 1999

1
Introduction

Water as solvent in organic synthesis means first that water must at least partially solubilize the reagents prone to react, and secondly that water cannot be a
reactive species in the process. In fact chemical transformations in living
systems occur chiefly in an aqueous environment. Nevertheless, in organic synthesis, water was rediscovered as a solvent only in the 1980s [1] and largely
popularized in the 1990s [2]. Among the main advantages of using water as the
solvent are the following: (i) water is cheap and not toxic, (ii) smooth conditions
occur frequently in water-promoted reactions, leading to improvements in
terms of yield and selectivity, (iii) the tedious protection-deprotection steps can
be avoided in particular cases, such as carbohydrate, nucleoside or peptide
chemistry, (iv) water can facilitate ligand exchange in transition-metal catalyzed
reactions, and (v) water-soluble catalysts can be reused after filtration, decantation or extraction of the water-insoluble products.

It is certainly illusive to pretend that there is a common explanation for the
exact role of water as the solvent but, nevertheless, it is important to have an
overview of the unique properties of water to understand at least the outstanding effects of reactions in neat water. More difficult is the understanding of the
reactions in mixed solvents, especially when water is used in small amounts. A
remarkable feature of water-promoted reactions is that the reactants only need
to be sparingly soluble in water and most of the time the effects of water occur
under biphasic conditions. If the reactants are not soluble enough, miscible co-
solvents can be used as well as surfactants or hydrophilic phase-transfer agents,
e.g. carbohydrate [3], carboxylate [4] or sulfonate [5] group, on the hydrophilic
reactant or ligand.

This chapter is devoted to reactions using water as the solvent with special
emphasis on pericyclic reactions, carbonyl additions, organometallic reactions,
oxidations and reductions and is restricted to those which show an unusual outcome when performed in water or in an aqueous medium. Enzymatic reactions
are beyond the scope of this review.

2
The Unique Properties of Liquid Water and Aqueous Solutions

It is widely believed that the unique properties of water are responsible for various physicochemical phenomena such as the aggregation of surfactants, the stability of biological membranes, the folding of nucleic acids and proteins, the binding of enzymes to substrates and more generally complex molecular associations in molecular recognition [6].

Among the unique physicochemical properties of liquid water are the following: (i) the small size of the molecule, (ii) a high cohesive pressure (550 cal/ml), (iii) a large heat capacity, (iv) a large surface tension (72 dynes/cm), (v) a low compressibility, (vi) a decrease of viscosity with pressure, and (vii) a strong and anomalous dependence of the thermal expansion coefficient leading to a density maximum at 4 °C. Even more surprising are the properties of aqueous solutions of non-polar solutes owing to hydrophobic hydration and hydrophobic interactions [7]. The physical origin of these interactions is still controversial [8]. Several models which emphasize the order and disorder of the hydrogen-bond network have been proposed, i.e. (i) the "iceberg" model [9], (ii) the "flickering clusters" model [10], (iii) the random network model of a dynamic equilibrium between some bonded and non-bonded water molecules [11], (iv) the continuum model based on the concept of strained and bent hydrogen bonds [12], and (v) the percolation model in which liquid water is treated as a large macroscopic space-filling hydrogen-bond network [13].

In the two-state model liquid, water is represented as an equilibrium between "structured water" constituted by ice-type clusters having low entropy and low density, and "unstructured water" having higher entropy and density in which each molecule has many neighbors. The dissolution of a non-polar solute in water, which is a thermodynamically unfavored process ($\Delta G_{tr} > 0$), brings about an enhancement of the water structure by augmenting the order of water molecules around the solute ($\Delta S_{tr} < 0$) and strengthening the hydrogen-bond pattern ($\Delta H_{tr} < 0$) at low temperatures. This is the well-known enthalpy-entropy compensation effect. X-ray studies on clathrate hydrate crystals of many non-polar compounds provide evidence of such a water reorganization [14]. The local environment around a non-polar solute is favorable for the formation of hydrogen bonds between the neighboring water molecules [15]. This water reorganization originates the positive heat capacity change ($\Delta C_p > 0$) observed when hydrophobic solutes are dissolved in water [16]. Only the water molecules in the first hydration shell are responsible for the heat capacity change which is proportional to the non-polar accessible surface area of the solute, the coefficient averaging 2 J K^{-1} mol^{-1} Å^{-2} approximately [17]. By contrast the perturbation due to polar and small ionic solutes causes a negative heat capacity change ($\Delta C_p < 0$) owing to an increase in the average length and angle of the water-water hydrogen bonds in the first hydration shell [18]. Based on measurements of liquid hydrocarbon-water surface tensions, a macroscopic measure for the hydrophobic Gibbs energy per unit surface area was evaluated [19] to average 310 J mol^{-1} Å^{-2}, which is somehow higher than the microscopic Gibbs energy correlations ranging from 70 to 130 J mol^{-1} Å^{-2} [8]. The properties of liquid water strongly change at high temperature or pressure. Unlike the effect of pressure [20], the effect of temperature has been extensively studied. When increasing the temperature, water loses its capacity to maintain hydrogen bonds upon intrusion of a non-polar solute; the solvation enthalpy increases gradually and becomes positive at high temperatures. The breakage of hydrogen bonds thus leads to an increase in solvation entropy which also becomes positive. Dramatic changes in the physicochemical properties of water occur when the temperature increases even more. For example, as the temperature rises from 25 to 300 °C, the density of water decreases from 0.997 to 0.713, its dielectric constant decreases from 78.85 to 19.66, its cohesive pressure decreases from 550 to 210, and its pK_a decreases from 14 to 11.30. This means that water can act as an acid-base bicatalyst, which could have ecological applications in recycle, regeneration, disposal and detoxification of chemicals [21].

Table 1. Cohesive energy density (ced), E_T parameter and dielectric constant ε at 25 °C for a range of solvents

	ced (cal cm^{-3})[a]	E_T (kcal mol^{-1})[b]	ε[b]
Water	550.2	63.1	78.5
Formamide	376.4	56.6	109.5
Ethylene glycol	213.2	56.3	37.7
Methanol	208.8	55.5	32.6
Dimethyl sulfoxide	168.6	45.0	48.9
Ethanol	161.3	51.9	24.3
Nitromethane	158.8	46.3	38.6
1-Propanol	144	50.7	20.1
Acetonitrile	139.2	46.0	37.5
Dimethylformamide	139.2	43.8	36.7
2-Propanol	132.3	48.6	18.3
1-Butanol	114.5	50.2	17.1
tert-Butanol	110.3	43.9	12.2
Dioxane	94.7	36.0	2.2
Acetone	94.3	42.2	20.7
Tetrahydrofuran	86.9	37.4	7.4
Chloroform	85.4	39.1	4.7
Toluene	79.4	33.9	2.4
Diethyl ether	59.9	34.6	4.2
Hexane	52.4	30.9	1.9

[a] Ref 24;   [b] Ref 23.

Table 1 gives a classification of organic solvents based on the decreasing values of the cohesive energy density (ced) at 25 °C. The cohesive energy density is readily obtained from the experimental heats of vaporization ΔH_{vap} via the relationship:

$$ced = \Delta U_{vap}/V = (\Delta H_{vap} - RT)/V$$

The cohesive energy density is expressed in terms of pressure, whence the expression cohesive pressure, but any confusion with the internal pressure of the solvent must be avoided: for example, the internal pressure of water, unlike other solvents, increases with increasing temperature until reaching a maximum at 150 °C, whereas the cohesive pressure decreases regularly with increasing temperature [22]. The cohesive energy density of water, which is much higher than for all organic solvents, reflects the unique organization of water molecules through the hydrogen-bond network.

The E_T parameter, which is usually considered as an appropriate indication of solvent polarity, is an empirical parameter based on energy transitions (E_T) corresponding to a charge transfer bond in ethyl-1-methoxycarbonyl-4-pyridinium iodide. This parameter is incontestably more accurate for evaluating the polarity of the solvent than the dielectric constant [23]. Table 1 shows that water is both the more structured and the more polar solvent and this might have implications on the chemical reactivity. However, it must be kept in mind that water behaves as a fluctuating structure which can be modified by the formation (breakage) of more hydrogen bonds, stronger (slighter) hydrogen bonds, or a reduction (increase) of free OH bonds.

3
Origin of the Reactivity in Water

In the Hughes–Ingold theory, solvent effects are rationalized by studying the Gibbs energy of solvation of the reactants and of the transition states. Solvation effects include the solute–solvent interactions and the reorganization of solvent around the solutes. An enthalpically dominated rate enhancement usually rises from a large decrease in the transition-state enthalpy. Thus the well-known SN_1 solvolysis reactions are accelerated in highly polar solvents, such as water, by strong interactions between the carbonium ions and the solvent in the transition state [23]. With the highest E_T parameter, water is the solvent of choice for reactions which go through a more polar transition state.

Interestingly, Dack considered the volume of activation and predicted that solvents accelerated the rate of a reaction when lowering the value of the volume of the reaction by electrostriction [24]. This is the case when the transition state is more polar than the initial state, but this cannot explain why non-polar reactions such as Diels-Alder reactions are strongly accelerated in water compared to other solvents [1]. What occurs when two hydrophobic molecules, susceptible to reaction, are put together in aqueous solution? Due to hydrophobic interactions, they have a tendency to aggregate but this association is not sufficient to explain the rate enhancement. The hydrophobic hydration, which is unfavorable and proportional to the hydrophobic surface area (vide supra), must be lowered by a decrease in the volume of the reactants, which occurs in reaction with a negative volume of activation. In such a reaction, the hydrophobic surface area decreases during the activation process, leading to a less unfavorable hydrophobic Gibbs energy (Fig. 1).

Thus a kinetically controlled reaction between two apolar molecules for which $\Delta V^{\neq}$ is negative must be accelerated in water [25]. The origin of such an

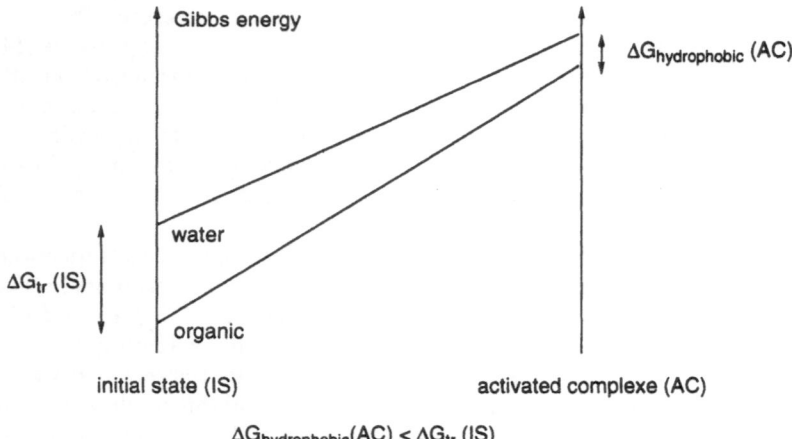

$$\Delta G_{hydrophobic}(AC) < \Delta G_{tr} (IS)$$

Fig. 1. Hypothetical Gibbs energy of the initial state and the activated complex in organic and water solvent: decreasing the hydrophobic hydration with the decreasing of the hydrophobic surface area

acceleration comes from the hydrophobic effects, which are a consequence of the hydrogen-bond network of water. Cohesive energy density is probably the best parameter to account for this type of acceleration. However, if one of the reactants is a hydrogen donor or acceptor, a charge development in the transition state may occur leading to stabilization of the activated complex versus the initial state. This second factor comes from an enhanced hydrogen-bonding interaction. Both contributions could be active in the same reaction, which means a greater destabilization of the hydrophobic reactants in the initial state than in the transition state, and a greater stabilization of a more polar transition state.

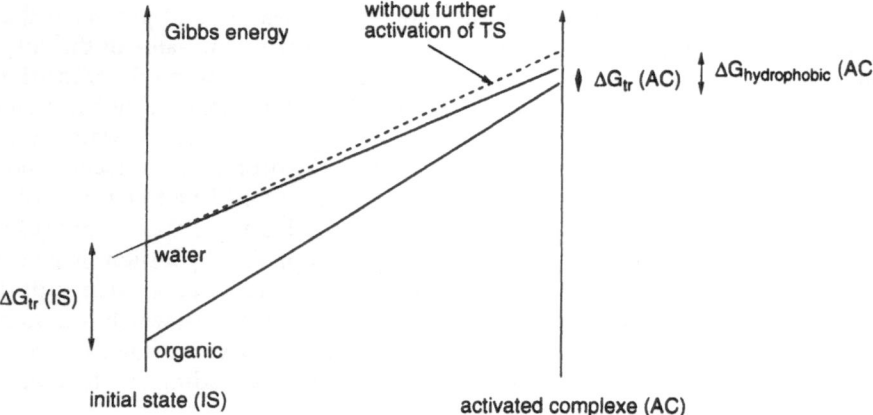

Fig. 2. Gibbs energy of transfer from organic to water solvent for the initial state and the activated complex

The Gibbs energy of transfer for the activated complex from organic or water solvent $\Delta G_{tr}(AC)$ is thus infinitely lower (Fig. 2) and can even be negative in particular cases. The rate acceleration in water is expressed by the negative term $\Delta\Delta G^{\neq} = \Delta G_{tr}(AC) - \Delta G_{tr}(IS)$. The absolute value of this term is all the highest since $\Delta G_{tr}(AC)$ is negative and $\Delta G_{tr}(IS)$ is strongly positive. The respective influence of the hydrophobic effect and hydrogen-bonding contributions was tentatively determined based on Monte Carlo simulations and molecular orbital calculations [26].

Water-tolerant Lewis acids, which can coordinate the reactants, catalyze reactions even in water, but their accelerative effect is less pronounced than in the reactions in organic media, probably by preventing the second factor (hydrogen-bonding enhancement in the transition state) to operate with the same efficiency [27].

4
Pericyclic Reactions

4.1
Diels-Alder Reactions

4.1.1
The Effect of Water and Additives on Chemical Reactivity

The influence of water as solvent on the rate of Diels-Alder reactions is likely the best example of the close relationship between the unique properties of liquid water and its strongly accelerative effect. This is probably due to the large negative value of the activation volume of Diels-Alder reactions; this value (ca. 30 cm^3 mol^{-1}) is even somewhat more negative in water versus organic solvents [28].

Although performed in water by Diels and Alder themselves [29] and in aqueous microemulsion [30], the Diels-Alder reaction was known as a reaction insensitive to solvents, until Breslow observed a dramatic accelerating effect in the aqueous reaction of cyclopentadiene with methyl vinyl ketone [1]. He suggested that hydrophobic packing of the reactants is likely to be responsible for the rate enhancement of Diels-Alder reactions [31]. The implication of the hydrophobic effect is supported by effects of additives. A prohydrophobic (salting-out) agent, such as lithium chloride, which increases the energy cost of cavitation by electrostriction, accelerates the reaction even more. Salting-in additives such as guanidinium chloride, which decreases the hydrophobic hydration by acting as a bridge between water molecules and apolar solutes [32], lead to a small decrease in the rate. Moreover, β-cyclodextrin, which promotes a favorable complexation of hydrophobic substrates, accelerates the cycloaddition; by contrast, α-cyclodextrin, with a small size of the hydrophobic cavity preventing both partners being complexed together, slows down the reaction (Scheme 1).

The implication of hydrophobic effects rather than polar effects was further demonstrated by (i) a deviation from linearity when correlating the Gibbs

solvent	kinetics 10^5 k (M^{-1}s^{-1})	selectivity endo/exo ratio
isooctane	5.94[a]	
methanol	75.5[a]	8.5[c]
formamide	318[b]	8.9[b]
ethylene glycol	480[b]	10.4[b]
water	4400[a]	25[d]
water (LiCl 4.86 M)	10800[a]	28[d]
water ((NH$_2$)$_3$CCl 4.86 M)	4300[a]	22[d]
β-cyclodextrin (10 mM)	10900[a]	
α-cyclodextrin (10 mM)	2610[a]	

[a] Ref 1   [b] Ref 37   [c] value in ethanol (Ref 50)   [d] Ref 51

Scheme 1. Kinetics and selectivity of the cycloaddition between cyclopentadiene and methyl vinyl ketone

activation energy with the solvent polarity as expressed by the E_T parameter [33], and (ii) a linear correlation of the Gibbs activation energy versus solvophobic parameters Sp, which originate from standard free energies of transfer of alkanes from the gas phase to a given solvent [34]. Such a sensitivity of Diels-Alder reactions to solvent hydrophobicity depends however on the nature of the reagents [35] and is more pronounced for the reactions with a more negative activation volume [36].

Formamide and ethylene glycol are other structured solvents (high cohesive energy density) and were often considered as "water-like" solvents. In these solvents the Diels-Alder reaction is accelerated, but to a smaller extent than in water (Scheme 1). Solvophobic binding of diene and dienophile is likely responsible for such an acceleration and β-cyclodextrin which is also able to bind both partners together in these solvents induces further acceleration. However urea and guanidinium ion which are normally antihydrophobic, and thus decrease the rate of aqueous Diels-Alder reactions, show no such effect in formamide and ethylene glycol, which confirms the unique properties of water [37].

At this stage of the discussion, let us examine the activation parameters of Diels-Alder reactions (Scheme 2). In the cycloaddition between cyclopentadiene and methyl vinyl ketone, the Gibbs activation energy in propanol is about 10 kJ mol higher than the Gibbs activation energy in water. Since the standard Gibbs energy of transfer for reactants from 1-propanol to water is slightly smaller $(\Delta G_{tr}(IS) = 9.1$ kJ mol$^{-1})$, it means the Gibbs energy of transfer for the activated complex from 1-propanol to water is slightly negative $(\Delta G_{tr}(AC) = -0.9$ kJ mol$^{-1})$ [38]. In this reaction, the rate of acceleration in water relative to the rate in 1-propanol is mainly caused by destabilization of the initi-

activation parameters (kJ.mol^{-1}) at 25 °C		
$\Delta G^{\neq}$	$\Delta H^{\neq}$	$-T\Delta S^{\neq}$
water[a] 80.3	39.4 ± 0.7	40.9 ± 0.7
water[b] 80.0	38.0 ± 1.7	42.0 ± 1.5
methanol[b] 89.8	38.0 ± 1.0	51.8 ± 1.0
propanol[a] 90.3	45.1 ± 0.7	45.3 ± 0.7

[a] Ref 33   [b] Ref 43

Scheme 2. Activation parameters of the cycloaddition between cyclopentadiene and methyl vinyl ketone

al state. Stabilization of the transition state relative to the initial state was first proposed as a consequence of the reduction of hydrophobic area during the activation process; such an effect was called "enforced hydrophobic interaction", the term "enforced" being used to distinguish the hydrophobic bonding of the reactants during the activation process from hydrophobic interactions not dictated by the activation process [33]. However the abnormally strong stabilization of the transition state due to an optimal accommodation of the active complex in water was finally rationalized as a consequence of an enhancement of the hydrogen bonding of the activating group of the dienophile [39]. Indeed, with methyl vinyl ketone as a dienophile model, computed partial charges displayed greater polarization of the carbonyl group in the transition state and consequently enhanced hydrogen bonding to the transition state [40].

On the basis of Monte Carlo simulations [40] and molecular orbital calculations [26a], hydrogen bonding was proposed as the key factor controlling the variation of the acceleration for Diels-Alder reactions in water. Experimental differences of rate acceleration in water-promoted cycloadditions were recently observed [41]. Cycloadditions of cyclopentadiene with acridizinium bromide, acrylonitrile and methyl vinyl ketone were investigated in water and in ethanol for comparison (Scheme 3). Only a modest rate acceleration of 5.3 was found with acridizinium bromide, which was attributed to the absence of hydrogen-bonding groups in the reactants. The acceleration factor reaches about 14 with acrylonitrile and 60 with methyl vinyl ketone, which is the best hydrogen-bond acceptor [41].

In the retro-Diels-Alder reaction of anthracenedione [42], the volume of activation is small. Acceleration in water cannot come from a change in the hydration shell of the molecule. Hydrophobic interactions are negligible and aqueous acceleration is caused by the hydrogen-bond donating ability of water, which stabilizes the polarized activated complex. The Gibbs energy of activation displays a fair linear correlation with the E_T parameter. Hexafluoroisopropanol with an E_T value of 65.3 is even more efficient as a solvent than water ($E_T = 63.1$) which appears to be less polar [41].

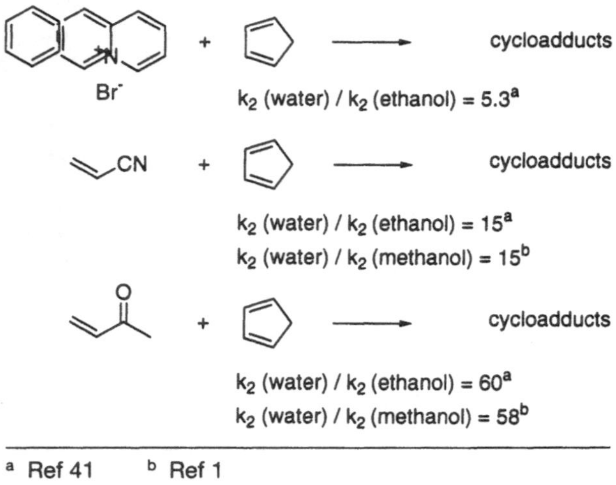

cycloadducts

k_2 (water) / k_2 (ethanol) = 5.3[a]

cycloadducts

k_2 (water) / k_2 (ethanol) = 15[a]

k_2 (water) / k_2 (methanol) = 15[b]

cycloadducts

k_2 (water) / k_2 (ethanol) = 60[a]

k_2 (water) / k_2 (methanol) = 58[b]

[a] Ref 41   [b] Ref 1

Scheme 3. Kinetics of the cycloaddition between cyclopentadiene and acridizinium bromide

Let us return to Scheme 2, which displays activation parameters for cyclo-additions between cyclopentadiene and methyl vinyl ketone. The rate accelera-tion in water versus methanol is entirely entropic in origin [43], whereas both enthalpic and entropic contributions account for the acceleration in water versus 1-propanol [33]. Engberts concluded that this is the result of an entropy-dominated cavity contribution and an enthalpy-dominated polarizability con-tribution. Furthermore Lubineau has investigated rate constants and activated parameters for cycloaddition between a glucose-grafted diene and methyl vinyl ketone in water and in aqueous solvents (Scheme 4) [43]. The rate acceleration in water is enhanced even more in concentrated carbohydrate solutions. The rate variation is fairly correlated with the sugar concentration; the effect of dextran, a polysaccharide containing glucose units, demonstrates the cooperative effect. The acceleration in saccharose is even greater than that observed with saturated β-cyclodextrin solution (Scheme 4). The rate enhancement in glucose solution comes from both enthalpic and entropic contributions, as for lithium chloride solution. By contrast, the rate acceleration in ribose solution results from a decrease in the activation enthalpy. Glucose and saccharose, which are known to stabilize proteins [44], might be acting as prohydrophobic or structure-making additives like lithium chloride [43].

As for surfactants, they have uncertain, sometimes contradictory, consequen-ces on reaction rates [45], but the main advantage of using surfactants as addi-tives lies in their solubilizing effect. Special attention has been paid to the rate-accelerating effect of Lewis acid catalysts. The first study deals with the Diels-Alder reaction between cyclopentadiene and a bidentate dienophile: a large acceleration can be achieved by the combined use of copper(II) nitrate as cata-lyst and water as solvent. The rate enhancement imposed on the catalyzed Diels-Alder reaction is much less pronounced than that for the uncatalyzed reaction

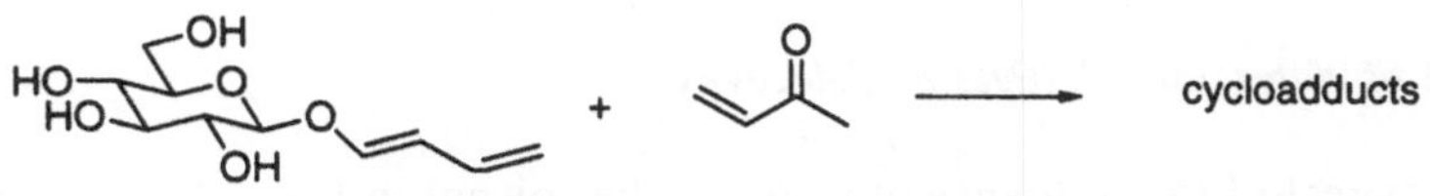

additive	$10^5 k_2$ (M^{-1} s^{-1}) at 25 °C	$\Delta H^{\neq}$ (kJ. mol^{-1})	$-T \Delta S^{\neq}$ (kJ. mol^{-1})
none	28.5	40.0 ± 0.6	53.28 ± 0.62
MeOH (50%)	8.5	33.6 ± 0.8	62.91 ± 0.77
LiCl (2.6 m)	57.8	39.3 ± 1.7	52.18 ± 1.61
glucose (2.6 m)	45.0	39.2 ± 0.3	52.8 7± 0.32
ribose (2.6 m)	35.0	36.7 ± 1.5	56.66 ± 1.46
glucose 1 M	34.0		
glucose 2 M	45.0		
glucose 3 M	61.3		
dextran 9000[a]	34.4		
saccharose 1 M	44.9		
saccharose 2 M	74.9		
sat-β-cyclodextrin	40.2		

[a] 1 M in glucose units

Scheme 4. Kinetics and activation parameters of the cycloaddition between a glycosylated diene and methyl vinyl ketone

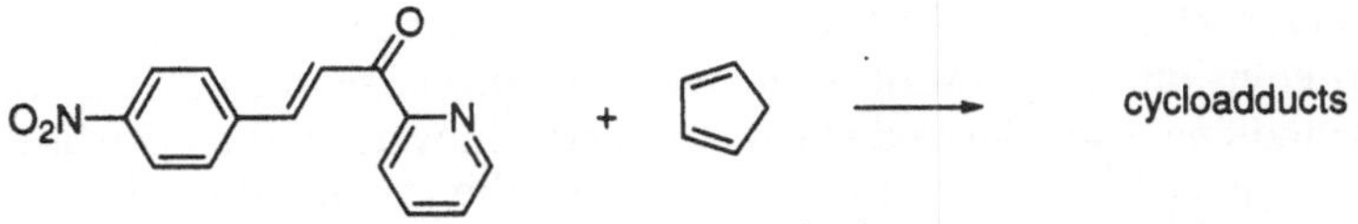

solvent	k_2 (M^{-1} s^{-1})	k_{rel}
ethanol	3.83. 10^{-5}	1
0.01 M Cu(NO$_3$)$_2$ in ethanol	0.769	20078
water	4.02. 10^{-3}	1
0.01 M Cu(NO$_3$)$_2$ in water	3.25	808

Scheme 5. Lewis acid catalyzed cycloaddition in water

(Scheme 5) [46]. Engberts has suggested that the hydrogen-bonding part of the acceleration is largely taken over by the Lewis acid for the catalyzed cycloadditions [27].

Methylrhenium trioxide (CH$_3$ReO$_3$) has proved to be an excellent catalyst in organic solvents, and in water when the dienophile is an α,β-unsaturated ketone (or aldehyde). Nearly exclusively one product isomer was formed, the same one that usually predominates [47]. Likewise, scandium triflate [48] and indium trichloride [49] were found to catalyze the Diels-Alder reaction in a tetrahydrofuran/water mixture and in pure water, respectively.

4.1.2
The Effect of Water and Additives on Selectivity

Another aspect of the influence of water as the solvent in Diels-Alder reactions is the higher endo selectivity observed in aqueous versus organic media (Scheme 1). The increased selectivity in the presence of a prohydrophobic additive such as lithium chloride, and the decreased selectivity in the presence of a antihydrophobic additive such as guanidinium chloride, argue the implication of the hydrophobic effect [50, 51]. The particular effect of surfactants depends both on the nature of substrate and surfactant. They can have a significant effect if used around their critical micellar concentration [52].

Since the Diels-Alder reaction has a negative volume of activation, it is noticeable that the more compact endo transition state should be favored. In the cycloaddition between cyclopentadiene and ethyl maleate, the endo selectivity was directly correlated with solvophobic power (Sp) [53], but in the cycloaddition between cyclopentadiene and methacrylate, the endo selectivity results were accounted for by means of a two parameter Sp/E_T model [36]. Both solvophobicity and hydrogen-bond donating ability are important to account for changes in endo/exo selectivity [35b]. It is worth noting that the difference in compactness between the endo and the exo transition states for the cyclopentadiene/ethyl maleate cycloaddition ($\Delta\Delta V^{\neq} = 0.82$ cm^3 mol^{-1}) is greater than for the cyclopentadiene/methyl acrylate cycloaddition ($\Delta\Delta V^{\neq} = 0.62$ cm^3 mol^{-1}), which could explain the greater hydrophobic influence of the former reaction. Concerning the diastereoselectivity of the Diels-Alder reaction, the influence of water is difficult to envision. Addition of 3-phenylsulfinylprop-2-enoic acids or esters to cyclopentadiene is accelerated when conducted in water but a decrease in diastereoselectivity is observed [54]. By contrast, the cycloaddition between methacrolein and the chiral diene carboxylate depicted in Scheme 6 gave a 65% diastereoselectivity in water whereas no diastereoselectivity was observed in the cycloaddition between neat methacrolein and the parent methyl ester [55].

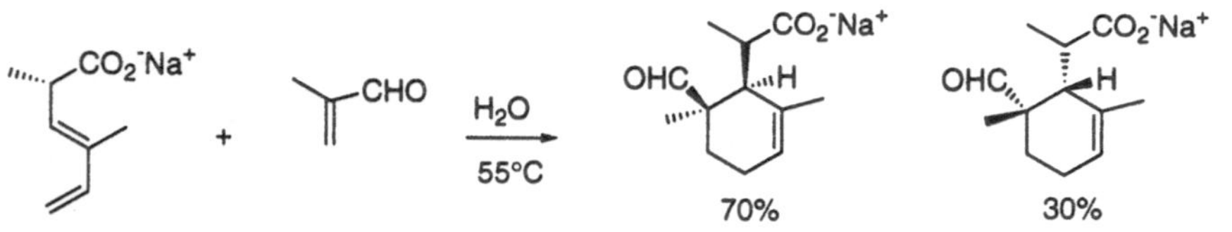

Scheme 6. Cycloaddition between methacrolein and a chiral diene carboxylate in water

The reaction of the simplest diene derived from a carbohydrate (Scheme 7) with acrolein leads to a total endo selectivity; furthermore, water increases the facial discrimination. This might be interpreted by considering the hydrophobicity of the two faces, the attack on the more hydrophobic face (*anti* to hydrophilic functions) being favored in water [56].

In terms of selectivity, water has a beneficial influence essentially on the endo/exo ratio. Accordingly, this has found numerous applications in synthesis,

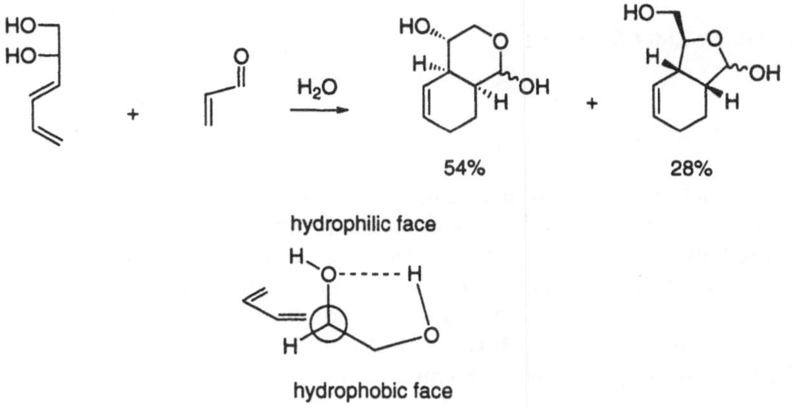

hydrophilic face

hydrophobic face

Scheme 7. Facial selectivity in the cycloaddition between acrolein and a diene derived from a carbohydrate

especially in natural product chemistry and in the pharmaceutical industry, as smooth conditions in water-promoted Diels-Alder reactions lead to a notable increase in yield compared with the conventional thermal reaction [57]. Furthermore, in a seminal contribution, Engberts showed recently that water can enhance enantioselectivity in a copper(II)-catalyzed Diels-Alder reaction. The binding of the coppper(II) salt to the dienophile is enhanced in water by ligands such as tryptophan owing to strengthened arene–arene interactions. In the copper(II) cycloaddition between cyclopentadiene and the bidentate dienophile depicted in Scheme 8, the enantioselectivity of the endo cycloadduct is largely improved in water in comparison to organic solvents [58].

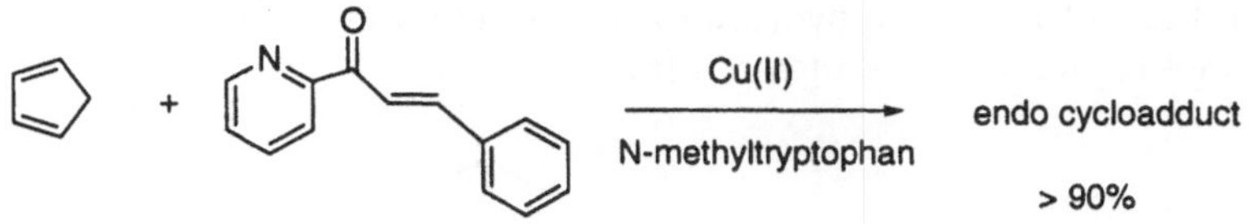

solvent	ee
acetonitrile	17
tetrahydrofuran	24
ethanol	39
chloroform	44
water	74

Scheme 8. Amino acid induced enantioselectivity in a Diels-Alder reaction

4.2
Hetero-Diels-Alder Reactions

4.2.1
Aza-Diels-Alder Reactions

The aza-Diels-Alder reaction combines three components (an aldehyde, an amine salt and a diene) to produce heterocyclic products which are useful synthetic intermediates. Grieco first reported that such a reaction could occur in water. Preliminary studies focused on the reaction of dienes with iminium ions generated by an amine hydrochloride and 37% aqueous formaldehyde solution [59]. When dienylamine hydrochlorides are treated with aqueous formaldehyde at 50 °C, bicyclic ring systems are formed (Scheme 9).

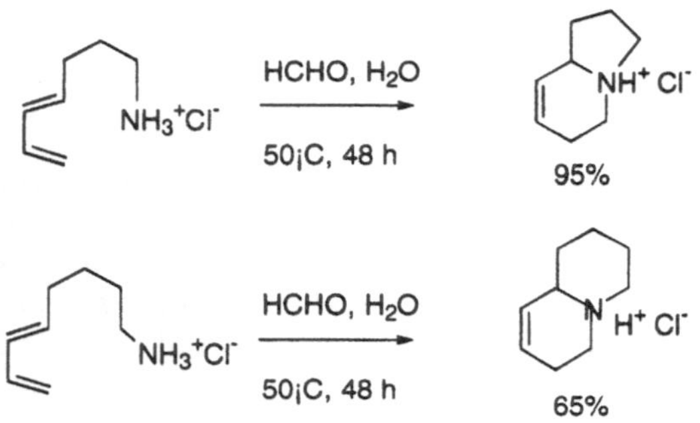

Scheme 9. Use of an aqueous formaldehyde solution in an aza-Diels-Alder reaction

Another intramolecular version of the aza-Diels-Alder reaction was realized with the condensation of a dienyl aldehyde and benzylamine hydrochloride at 70 °C in 50% aqueous ethanol (Scheme 10) [59]. This methodology was used in the synthesis of racemic dihydrocannivonine [60] and substituted octahydroquinolines related to pumiliotoxin C [61].

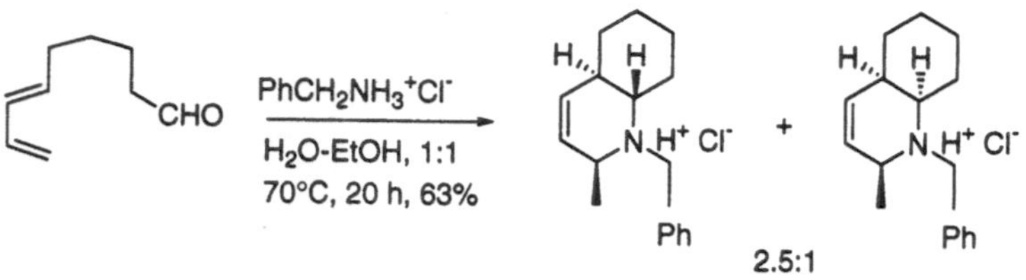

Scheme 10. Intramolecular aza-Diels-Alder reaction with benzylamine hydrochloride

Asymmetric induction occurs when (S)-1-phenylethylamine hydrochloride is used as the ammonium salt [57], which prompted Waldmann to investigate amino acid methyl esters as chiral auxiliaries: the reaction of (S)-isoleucine methyl ester with cyclopentadiene in the presence of 35% aqueous formaldehyde solution/tetrahydrofuran (9:1) afforded the best stereoisomeric ratio (93:7) [62], which is significantly better than that obtained for (S)-1-phenylethylamine hydrochloride (80:20) [59]. The chiral iminium ion generated from this salt and methyl glyoxylate react with cyclopentadiene giving rise to cycloadducts in 52% yield, reaching diastereomeric excesses up to 90:10 for the exo isomers [63]. Synthetic applications of the asymmetric aqueous aza-Diels-Alder reactions with simple protonated iminium ions and with protonated C-acyl iminium ions were recently reviewed [64]. As observed in Diels-Alder reactions, water-tolerant Lewis acid catalysts accelerate the condensation between iminium salts and dienes. The catalytic effect of praseodymium(III) triflate, neodymium(III) triflate and ytterbium(III) triflate on aqueous aza-Diels-Alder reactions was demonstrated with a variety of aldehydes and dienes [65]. In comparison with the uncatalyzed reaction, the lanthanide catalyst does not change the endo/exo ratio of the cycloadducts [65]. An asymmetric version of the neodymium(III) triflate catalyzed aqueous aza-Diels-Alder reaction was tested using unprotected carbohydrates as chiral auxiliaries [66].

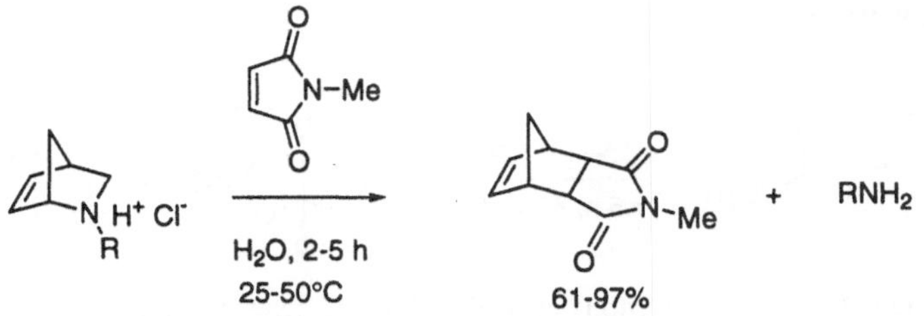

Scheme 11. Cycloreversion of 2-azanorbornene hydrochlorides

The origin of the acceleration of the aqueous aza-Diels-Alder reaction comes from a large decrease in the Gibbs energy of the activated complex, since cycloreversion smoothly occurs when 2-azanorbornene hydrochlorides are kept in water at room temperature in the presence of N-methylmaleimide as a trapping agent for cyclopentadiene formed in the process (Scheme 11) [67].

Grieco took advantage of such a cycloreversion for the N-methylation of amino acid derivatives and dipeptides by trapping the incipient iminium ion with triethylsilane/trifluoroacetic acid without loss of chirality (Scheme 12) [67].

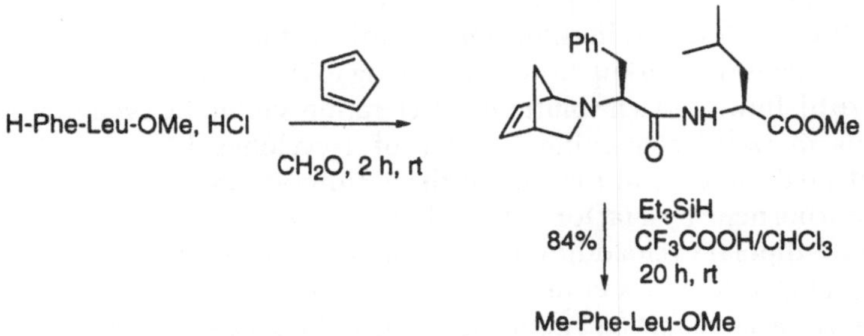

Scheme 12. N-Methylation of amino acid derivatives

Synthetic applications for parent methodologies have been used for the preparation of alkaloids such as pseudotabersonine, physostigmine and tylophorine [64].

4.2.2
Oxa-Diels-Alder Reactions

An aqueous solution of glyoxylic acid reacts with cyclopentadiene to provide α-hydroxy-γ-lactones; the more acidic the solution, the faster the reaction (Scheme 13) [68]. Thus at pH 0.9 (2.25 M glyoxylic acid solution in water) the reaction is complete after 90 min at 40 °C providing a 83 % yield of α-hydroxy-γ-lactones. In the case of cyclohexadiene, the reaction is complete after 2 days at 90 °C in water, compared to 21 h at 120 °C for the reaction with butyl glyoxylate in neat conditions (Scheme 13). These results show that it is possible to exploit the dienophilic character of a carbonyl group in water in spite of its quasi total hydration. Pyruvaldehyde, glyoxal, and even ketones like pyruvic acid, also react with dienes in water [69].

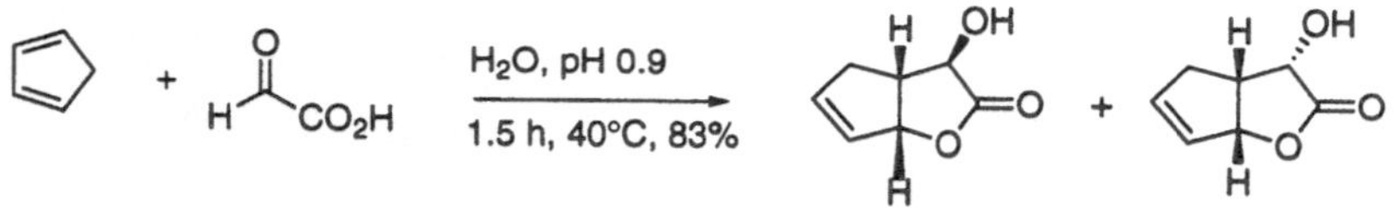

Scheme 13. Oxa-Diels-Alder reaction between cyclopentadiene and glyoxylic acid

The aqueous oxa-Diels-Alder reaction has been successfully exploited in the synthesis of sesbanimides A and B [70], carbovir [71], mevinic acids [72], aristeromycin and carbodine [73], ketodeoxyoctulosonic acid (KDO) and analogs [74], and sialic acids [75].

4.3
Miscellaneous Cycloadditions

The influence of water as a solvent on the rate of dipolar cycloadditions has been reported [76]. Thus the rate of the 1,3-dipolar cycloaddition of 2,6-dichlorobenzonitrile N-oxide with 2,5-dimethyl-p-benzoquinone in an ethanol/water mixture (60:40) is 14-fold that in chloroform [76b]. Furthermore the use of aqueous solvent facilitates the workup procedure owing to the low solubility of the cycloadduct [76b]. In water-rich solutions, acceleration should be even more important. Thus in water containing 1 mol% of 1-cyclohexyl-2-pyrrolidinone an unprecedented increase in the rate of the 1,3-dipolar cycloaddition of phenyl azide to norbornene by a factor of 53 (relative to hexane) is observed [77]. Likewise, the 1,3-dipolar cycloaddition of C,N-diphenylnitrone with methyl acrylate is considerably faster in water than in benzene [78]. Similarly, azomethine ylides generated from sarcosine and aqueous formaldehyde can be trapped by dipolarophiles such as N-ethylmaleimide to provide pyrrolidines in excellent yields

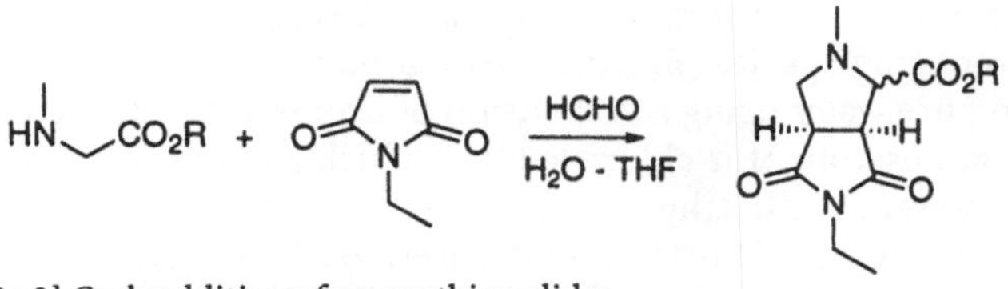

Scheme 14. [2+3] Cycloaddition of azomethine ylides

provided that the reaction is conducted in water/THF mixtures (Scheme 14) [79]. In pure water, the Michael addition of the secondary amine to N-ethyl-maleimide becomes the major process (see Sect. 5.2.2).

Using a Zincke-Bradsher convergent strategy for the synthesis of the ABE tricyclic core of Manzamine A, Magnier and Langlois reported the use of aqueous conditions to achieve, in one pot, the preparation of a naphthylpyridinium salt intermediate which underwent an inverse electron demand heterocyclic Diels-Alder cycloaddition with either ethyl vinyl ether or (Z)-1-ethoxy-1,5-hexadiene to afford the corresponding cycloadduct (Scheme 15) [80].

The most striking effect of water as a solvent in related cycloadditions was observed in the [4+3] cycloaddition of α,α'-dibromo ketones with furan (or pen-

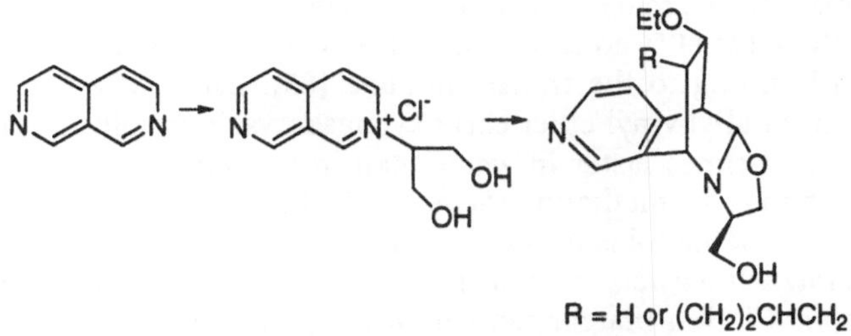

R = H or $(CH_2)_2CHCH_2$

Scheme 15. Zincke-Bradsher one-pot reaction toward Manzamine A

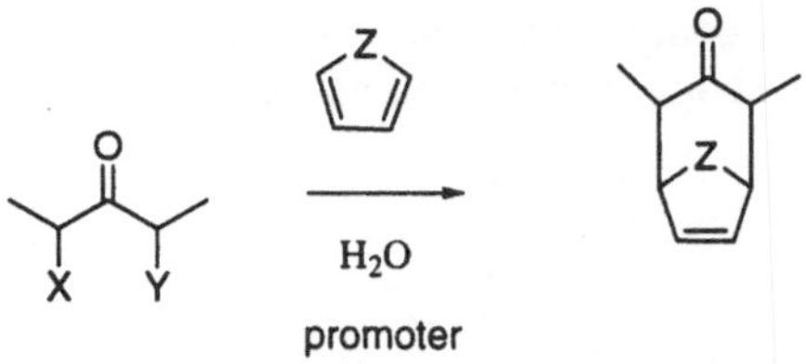

X	Y	Z	promoter	isolated yields
Br	Br	O	Fe	74%
Br	Br	C	Fe	66%
H	Cl	O	Net$_3$	88%
H	Cl	O	Net$_3$	76%

Scheme 16. [4+3] Cycloaddition between α,α'-dibromo ketones and furan (or cyclopentadiene)

tadiene) [81]. Such a cycloaddition, originally described with $Fe_2(CO)_9$ in benzene as the reducing species [82], gave very good yields when the reaction was conducted in pure water using a suspension of iron powder. Related to this reaction is the condensation of α-chloro ketones which add to furan (or cyclopentadiene) in the presence of triethylamine to afford the corresponding cycloadducts in high yields (Scheme 16) [81]. In both cases, the 2-oxyallyl cation, the formation of which is favored in water, was considered as the relative intermediate.

4.4
Claisen Rearrangements

Claisen rearrangements, i. e. the rearrangement of allyl vinyl ethers to γ,δ-unsaturated carbonyl compounds, display a negative activation volume ranging from -13 to -6 cm^3 mol^{-1}. Accordingly, the reaction between the two apolar groups of the molecules should be accelerated in water. In fact, the non-enzymatic rearrangement of chorismate to prephenate occurs 100-times faster in water than in methanol [83]. This value is even greater than that obtained in Diels-Alder reactions (e. g. 60 in the cycloaddition of cyclopentadiene with methyl vinyl ketone). The same factor of 100 in going from the least polar (tetradecane) to the most polar (p-chlorophenol) was observed in the ortho-Claisen rearrangement [84]. The sensibility of the Claisen rearrangement to the substitution pattern was related to polar effects: rate effects due to alkoxy substitution and solvent changes from methanol to 80 % aqueous ethanol were interpreted in terms of increased hydrogen bonding to the transition state [85]. Later, Grieco reported the rearrangement of allyl vinyl ether carboxylates in various solvents at 60 °C: the reaction occurs 7-times faster in water than in trifluoroethanol and 23-times faster in water than in methanol. (Scheme 17) [86]. The comparison between water and trifluoroethanol is particularly interesting since trifluoroethanol is a solvent with low self-association but strong hydrogen-bond donor abilities [87].

The aqueous Claisen rearrangement was exploited in the syntheses of the Inhoffen–Lythgoe diol, aphidicolin and fenestrenes [88]. Another example of a striking rate enhancement in a Claisen rearrangement performed in water is that of α- and β-glucosyl allyl vinyl ether [89]. The water solubility of the reactants was induced by grafting a free carbohydrate onto the allyl vinyl ether moiety; furthermore, the sugar functioned as a chiral template and gave highly

solvent	10^{-5} k (s^{-1})
H_2O	18
CF_3CH_2OH	2.6
CH_3OH	0.79

Scheme 17. Claisen rearrangement of allyl vinyl ether carboxylates in protic solvents

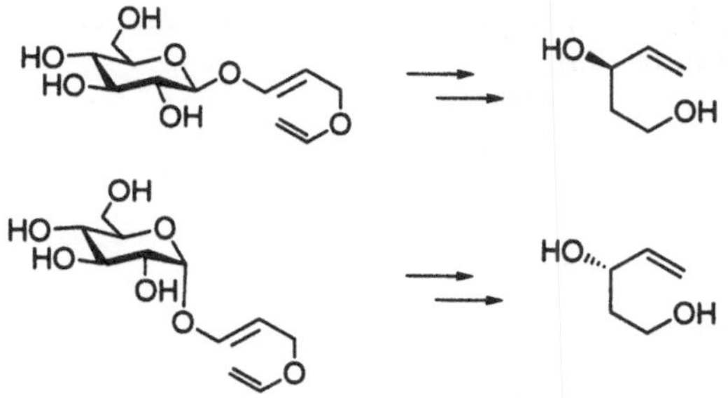

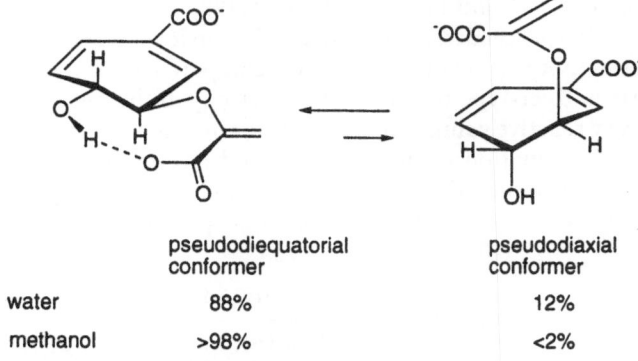

Scheme 18. Claisen rearrangement of a carbohydrate-grafted allyl vinyl ether

crystalline diastereoisomers easily separable to yield pure enantiomers after enzymatic hydrolysis. This method allows the preparation of either enantiomerically pure (R)- and (S)-1,3-diol (Scheme 18) [89].

It is now evident [87] that the origin of the aqueous rate acceleration comes from both the hydrophobic effects and the hydrogen-bond donating ability of water. Applying a self-consistent field solvation model to the aqueous Claisen rearrangement of allyl vinyl ether and all possible methoxy-substituted derivatives, Cramer and Truhlar concluded that the hydrophobic effect is always accelerative even if most of the activation stems from polarization contributions to the activation energy [90]. Monte Carlo simulations showed enhanced hydrogen bonding to the enolic oxygen in the transition state envelope of water molecules [91]. In the rearrangement of allyl vinyl ether, the transition state has the lowest free energy of hydration ΔG_{hyd} along the reaction path, 3.85 kcal mol^{-1} below the reactant. The implied rate increase in water is a factor of 644 over the gas phase [91]. In a related work using combined QM/MM simulation, similar results were interpreted by large, unbalanced polarization effects between the reactant and the transition state [92]. In the uncatalyzed rearrangement of chorismate to prephenate, the intriguing 100-fold rate increase in water over methanol was traced to a shift of the conformational equilibrium towards the reactive pseudodiaxal conformer (Scheme 19); the shift comes from a unique water molecule that

	pseudodiequatorial conformer	pseudodiaxial conformer
water	88%	12%
methanol	>98%	<2%

Scheme 19. Conformational populations for chorismate

bridges between the C4 hydroxyl group and side-chain carboxylate as a double hydrogen-bond donor [93].

5
Carbonyl Additions

5.1
Barbier-Type Additions

The Barbier reaction, described in 1898 [94], consists of a one-pot coupling reaction between an alkyl halide onto a carbonyl compound in the presence of a metal. This reaction which was formerly conducted in an organic solvent with careful exclusion of water is now performed in water in high yields onto a wide variety of substrates. The first aqueous allylation, albeit in low yields, was performed as long as twenty years ago, using zinc in refluxing ethanol containing 5% water [95].

Since then, various metals have been used, but most of the reactions are carried out using zinc, tin or mostly indium which give good results even with masked aldehydes such as in carbohydrates. The stereochemistry of the indium-promoted addition has been extensively studied by Paquette and Isaac who have shown that the reaction in water is generally highly diastereoselective and able to set three contiguous stereogenic centers using α-substituted aldehydes [96]. This diastereoselectivity can in fact be influenced by the choice of the metal and/or allylic halide [97]. The scope, mechanism and synthetic applications of the aqueous Barbier-type reactions have recently been reviewed [98].

5.1.1
Allylation Mediated by Zinc

Reactions using metallic zinc are usually conducted at room temperature without the need for further activation such as ultrasonic activation, strong acidic medium or another metal as activating additive. In 1985, Luche reported the allylation of benzaldehyde with allyl chloride or bromide in a mixture of a saturated solution of NH_4Cl and THF in 95% yields [99]. The reaction under kinetic control, which also proceeded in pure water albeit in lower yields, showed a marked chemoselectivity in favor of aldehyde compared to ketone (Scheme 20) [100]. In these Barbier-type reactions, the reactivity of the metal can be increased using a very reactive submicron crystalline zinc powder which can be generated using a combination of pulsed sono- and electrochemistry [101]. With substituted allyl bromide, the reaction with aldehyde always proceeds at the more substituted carbon atom but with generally poor diastereoselectivity.

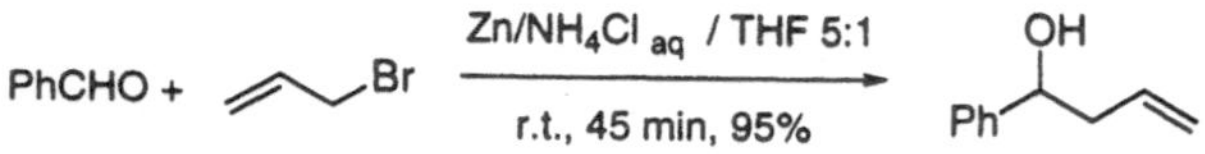

Scheme 20. Allylation reactions under Luche conditions

Various functionalized allylic halides have been used under the Luche conditions. Thus, 2-bromomethyl acrylate reacted with carbonyl compounds to give α-methylene-γ-butyrolactones after acidic treatment of the alcoholic intermediate [102]. The reaction of cinnamyl chloride with aldehyde, unlike cinnamyl bromide which led to phenyl propene leaving the aldehyde unchanged [103], gave the diastereoisomeric γ-products whereas the reaction with ketones gave mixtures of α- and γ-products [104]. In the presence of zinc dust, 1-chloro-3-iodopropene yielded the corresponding chlorohydrin when reacted with aldehydes or ketones under aqueous conditions. In this way, further treatment with base gave vinyl oxiranes whereas zinc in the presence of aqueous HBr led to (E)-buta-1,3-dienes (Scheme 21) [105].

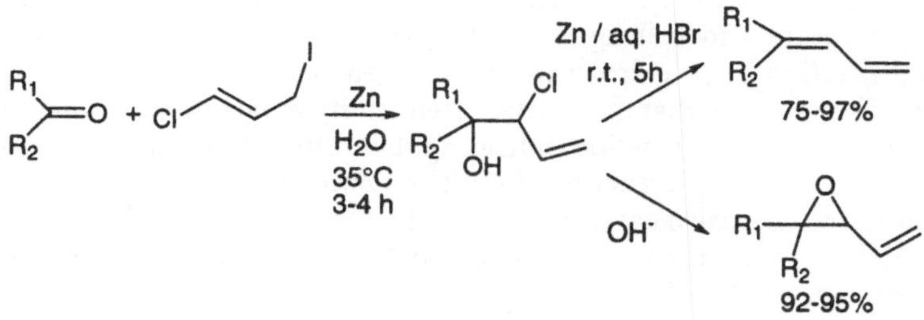

Scheme 21. Formation of vinyl epoxides and 1,3-dienes under Barbier conditions

Alternatively, 2-(chloromethyl)-3-iodo-1-propene reacted with aldehydes or ketones to give after basic treatment methylenetetrahydrofuran in excellent yields. The reaction is as expected chemo- and regioselective: conjugated carbonyl compounds gave 1,2-addition, and aldehydes react more preferably than ketones [106]. Similarly, 3-chloro-homoallylic alcohols were prepared in high yields from the reaction of 2,3-dichloropropene with carbonyl compounds in a two-phase system containing a small amount of acetic acid. In this reaction, the presence of water in the reaction medium was shown to be essential as the reaction did not occur under non-aqueous conditions [107].

Propargyl halides also react with carbonyl compounds in the presence of zinc powder in concentrated aqueous salt solutions to give homopropargylic alcohols with high selectivity. High yields are obtained with aromatic and aliphatic aldehydes whereas ketones react only partially [108].

This zinc-promoted reaction has been used with a variety of carbonyl compounds. Thus, the Luche conditions were applied in a synthesis of (+)-muscarine using an aldehyde derived from ethyl lactate [109]. Allyl halide condensation onto α-ketoamides of proline benzyl ester gave good diastereoselectivity when performed in the presence of zinc dust and pyridinium p-toluenesulfonate in a water/THF mixture. In this way, α-hydroxy ketones were obtained with good enantioselectivity after removal of the chiral auxiliary [110]. Reactions of allyl bromide under the Luche conditions with γ-aldo esters afforded γ-hydroxy esters, which were converted in a one-pot reaction to γ-allyl-γ-butyrolactones (Scheme 22) [111].

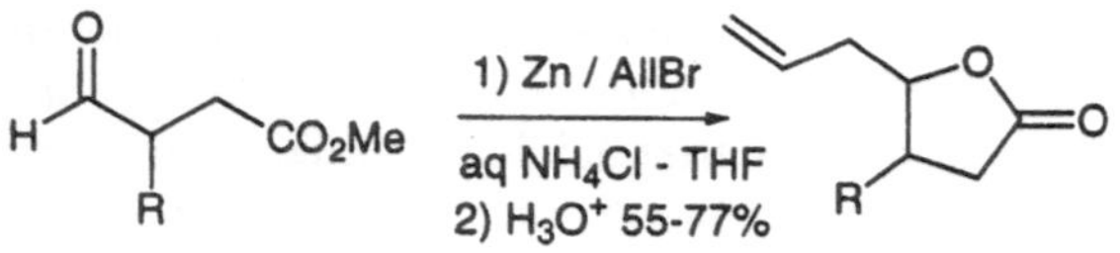

Scheme 22. Formation of γ-allyl-γ-butyrolactones from γ-aldo esters

Hannessian and co-workers recently reported the allylation of N-protected α-amino aldehydes as an efficient route to Phe-Phe hydroxyethylene dipeptide isostere utilized for the design of potential inhibitors of renin and HIV-protease [112]. This reaction was extended to various allyl halides to give *anti-β*-amino alcohols in high yields. Alternatively, enantiopure *syn-β*-amino alcohols were obtained in good yields after Swern oxidation and reduction with NaBH$_4$ [113]. Similarly, Rübsam and colleagues studied the stereoselective outcome of the allylation reaction of β-hydroxy-α-amino aldehydes in the presence of tin or zinc dust. They showed that the diastereoselectivity in fact depends on the nature of the metal and of the halide indicating that different mechanisms can operate in these Barbier-type reactions [97]. The *syn/anti* selectivity and mechanistic implications were also carefully studied by Marton et al. who found that the diastereoselectivity of the zinc-mediated allylation of aldehydes with allyl halides in cosolvent/H$_2$O (NH$_4$Cl), and in cosolvent/H$_2$O (NH$_4$Cl)/haloorganotin media, seems to be determined by the structure of the intermediate anions; it is presumed that these were formed by an electron-transfer process.

Related to these reactions are the zinc-promoted allylation reactions of oximes. Thus, enantiomerically pure or highly enriched allyglycine and its chain-substituted analogs are easily accessible from the reaction of the sultam derivative of benzyl glyoxylic acid oxime with allylic bromides in aqueous ammonium chloride (Scheme 23) [114].

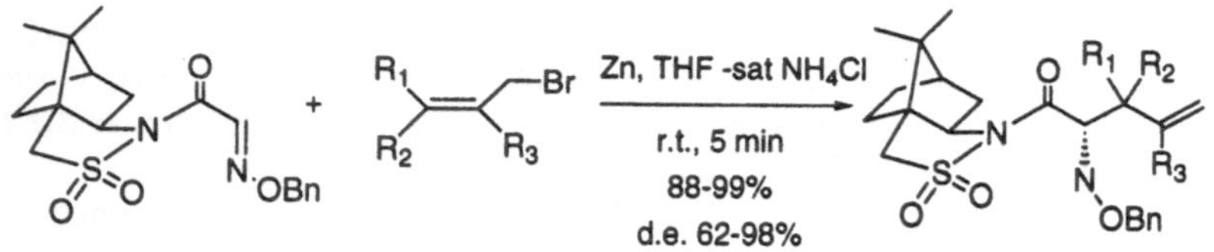

Scheme 23. Allylation of sultam derivatives

5.1.2
Allylation Mediated by Tin

Allyltin compounds were uniformly shown to be less reactive than the zinc or especially indium analogs. Thus, strongly acidic conditions (HBr, AcOH) were required to perform the heterogeneous reaction using metallic tin, allylic bromides and aldehydes, even if the presence of water accelerated the reaction compared to the reaction in pure diethyl ether. Allylation of ketones required metallic aluminum as an additive [115]. In this way, five- or six-membered rings were prepared through intramolecular reactions (Scheme 24) [116].

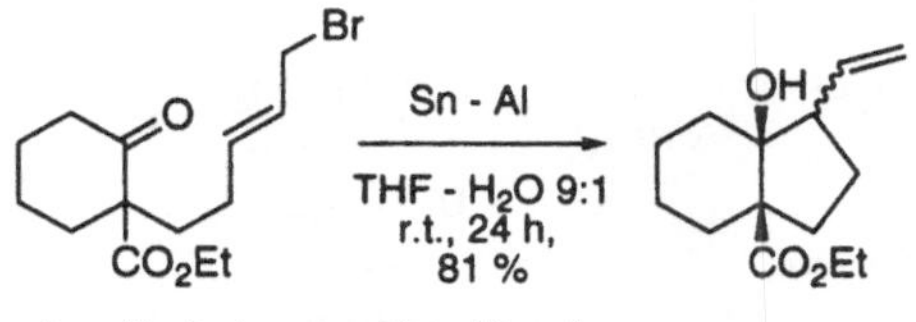

Scheme 24. Intramolecular alkylation mediated by tin

Alternatively, the use of higher temperatures [117] or ultrasonic waves in water/THF (5:1) mixtures were shown to allow the coupling reaction without aluminum powder [100]. When the aldehyde contained a free hydroxyl group, then the reaction under sonication became very rapid and efficient [103]. When applied to carbohydrates, this sono-allylation proceeded with high *threo* selectivity and allowed the preparation of higher-carbon complex sugars directly in aqueous ethanol without protection (Scheme 25) [118]. In this case also sonication can be replaced by heating the reaction mixture for 2 h under reflux [119].

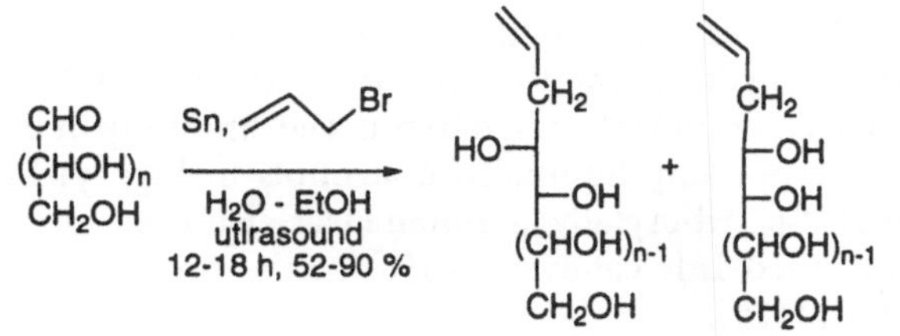

Scheme 25. Highly *threo*-selective tin-mediated allylation of sugars

Active zero-valent tin can also be generated by reduction of tin(II) chloride by aluminum in aqueous THF. In this way, cinnamyl chloride gave high *anti* diastereoselectivity in the reaction with aldehydes [120]. In fact, under acidic conditions, tin(II) halide can be used directly without concomitant reduction. Thus, starting from (*E*)-rich 1-bromo-2-butene, the $SnCl_2$-mediated allylation of aldehydes yielded homoallylic alcohols mainly with *anti* configuration. Curiously, *syn* allylation occured with tin(II) iodide, suggesting in this case an acyclic antiperiplanar transition state even more favored by the addition of tetrabutylammonium bromide which prevents the tin atom from coordinating the oxygen atom of the aldehyde [121]. Homoallylic alcohols have also been prepared in good yields with various allyl chlorides and aldehydes in the presence of $SnCl_2$ and KI. The reaction can be applied to aliphatic aldehydes, conjugated enals and keto aldehydes [122].

The $SnCl_2$-mediated allylation of carbonyl compounds is even possible with allylic alcohol or its corresponding methylcarbonate in the presence of palladium catalyst (Scheme 26) [123]. In the reaction of (*E*)-but-2-enol with benzaldehyde in the presence of $SnCl_2$, the rate and the *anti* selectivity were shown to increase with the amount of water in the reaction mixture [124].

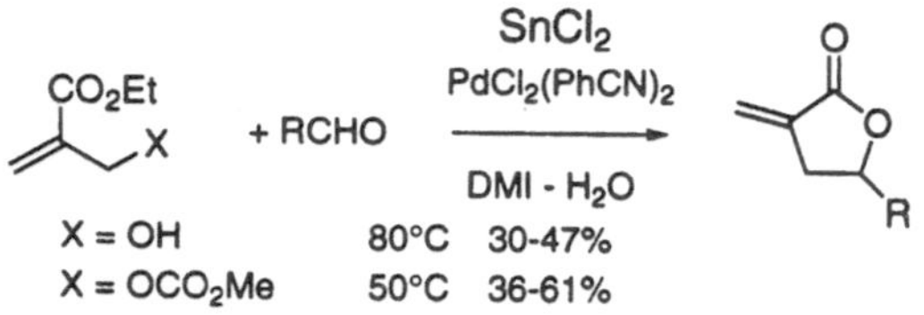

$$SnCl_2$$
$$PdCl_2(PhCN)_2$$

CO₂Et + RCHO $\xrightarrow{\hspace{2cm}}$

DMI - H₂O

X = OH   80°C   30-47%

X = OCO₂Me   50°C   36-61%

Scheme 26. Use of allyl alcohols in Barbier-type reactions

Propargyl bromide reacts with aldehydes in the presence of tin powder in a water/benzene mixture at reflux. Propargylation and allenylation products are formed without marked selectivity [125]. This reaction is also possible using $SnCl_2$ with KI. Whereas the reaction with benzaldehyde gave only the propargylation product, in this case, aliphatic aldehydes gave rise to a mixture of propargylation and allenylation products in a 70:30 ratio [124].

All these reactions can be carried out with preformed organotin(IV) compounds. Thus, allyldibutyltin chloride or allylbutyltin dichloride reacts with aldehydes or ketones in high yields to give the corresponding homoallylic alcohols [126]. However, tetraallyltin requires acidic conditions and reacts exclusively with aldehydes [127]. Recently, the $Sc(OTf)_3$-catalyzed allylation of aldehydes with tetraallyltin was shown to proceed smoothly in micellar systems to afford the corresponding homoallylic alcohols in high yields without any organic solvent [128]. Tributylallyl stannane was also used for the allylation of aldehyde, but required InI_3 catalysis (10%) in the presence of an equimolar amount of Me_3SiCl [129].

In a related reaction, aqueous allylation of imines has been reported by Belluci and co-workers using allyltributylstannane under lanthanide triflate catalysis [130]. Alternatively, the reaction can be performed without catalysis directly with the formaldehyde-generated immonium salts in aqueous media to give, in excellent yields, bis-homoallylamines and tertiary homoallylamines with primary and secondary amines, respectively [131]. These homoallylic amines were also prepared by $Sc(OTf)_3$-catalyzed three-component reactions of aldehydes, amines and allyltributylstannane in micellar systems [132].

5.1.3
Allylation Mediated by Indium

The indium-mediated Barbier reaction has certainly become one of the most popular reactions for creating a carbon–carbon bond under aqueous conditions and has led to spectacular developments in recent years. Compared to other metals, indium is resistant to oxidation, hydrolysis, and has a very low first ionization potential (5.79 eV, in contrast to the second one which is quite normal) which confers on it a remarkable reactivity in Barbier-type reactions. In 1991, Li and co-workers reported the first allylation of aldehydes and ketones mediated by indium in water without any additives or special activation [133]. In particular, the use of indium allowed reactions with acid-sensitive compounds [134] or the preparation of complex carbohydrates such as deaminated

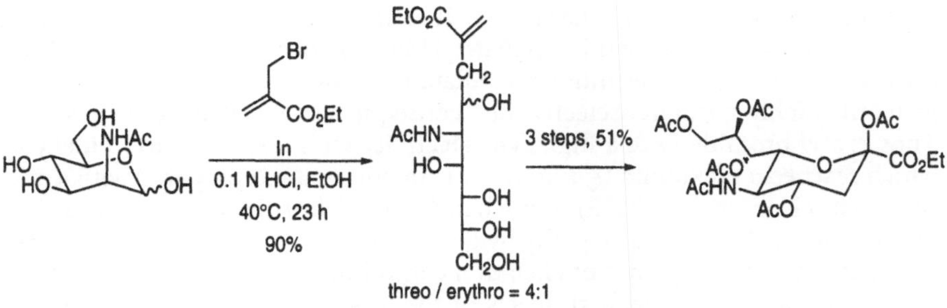

Scheme 27. Synthesis of a *N*-acetyl neuraminic acid derivative

sialic acids (KDN) [135] using α-bromomethylacrylate and mannose or *N*-acetylneuraminic acid with the same halide and *N*-acetyl mannosamine (Scheme 27) [136].

Several analogs of sialic acids have been prepared in this way [137], including a phosphonate analog [138]. Compared with tin, indium-mediated reactions with pentoses and hexoses give uniformly higher yields and higher *threo* selectivity as a result of a chelation-controlled reaction. Interestingly, it has been shown that preformed allyldichloroindium reacts equally with ribose in an ethanol/water mixture [119]. When the reaction was performed with acetonide-protected sugars, the diastereoselectivity was reversed (*erythro/threo* 2:1) [139]. This led, as shown in Scheme 28, to a rapid synthesis of 3-deoxy-D-manno-2-octulosonic acid (KDO) [140].

Zero-valent indium can also be generated in situ by reduction of indium(III) halide with metallic zinc or aluminum [141] or even tin [142]. Thus, β-trifluoromethylated homoallylic alcohols have been prepared in high yields and with excellent stereoselectivity using indium powder [143] or a mixture of InCl$_3$/Sn [144].

The regio- and diastereoselectivity of the reaction using γ-substituted allyl bromides with aldehydes was carefully investigated by Chan and Isaac [145]. The reaction was as expected γ-regioselective except when the γ-substituent was too bulky (i.e. *tert*-butyl group) but gave only moderate *anti* diastereoselectivity.

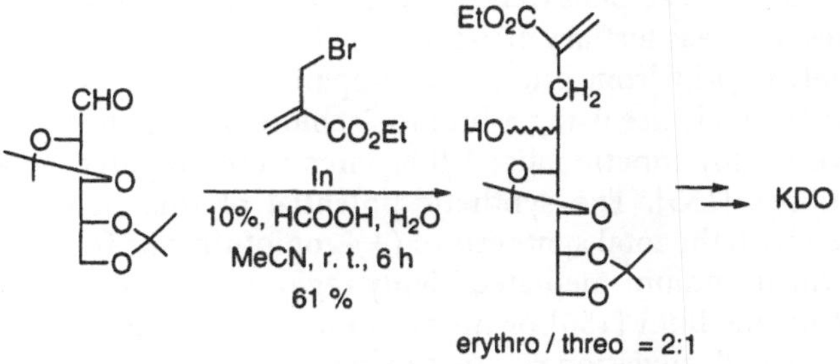

Scheme 28. Synthesis of KDO through *erythro*-selective allylation of a carbohydrate acetonide

This diastereoselectivity, however, as well as the rate of the reaction, could be enhanced by adding lanthanide triflates [146]. In contrast, as shown with the reaction with sugars, the indium-mediated reactions with α-oxyaldehyde is generally highly *syn* stereoselective as a consequence of chelated intermediates. Thus crotyl bromide gave a high *syn* selectivity with a set of α-oxy aldehydes which reached a maximum (*syn/anti* 5.6:1) in pure water (compared with a mixture water/THF or pure THF) and when the neighboring hydroxyl group was unprotected [147]. β-Hydroxy aldehydes also gave good diastereoselectivities (*anti*) also as a consequence of chelation control in water [148]. In a recent study by Isaac and Paquette, high levels of the 3,4-*syn*;4,5-*anti* diastereoisomers were produced in the indium-promoted additions of (Z)-2-bromomethyl-2-butenoate to several α-oxy aldehydes. This stereodifferentiation has been attributed to the strong geometric bias exercised by the allylindium reagent and adherence to a Felkin-Ahn transition state alignment (Scheme 29) [96].

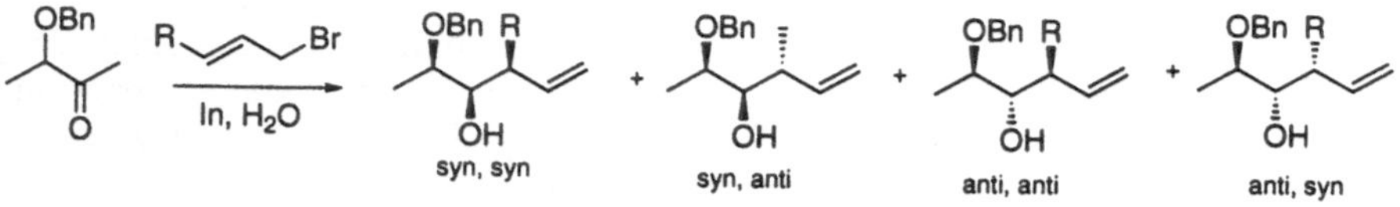

Scheme 29. Diastereoselection in the allylation of α-oxy aldehydes

However, additions of an allylindium reagent to α-thia aldehydes are minimally diastereoselective indicating that the allylindium reagent is not thiophilic. In contrast, as seen in the case of the reaction on N-acetyl mannosamine for the preparation of sialic acid (vide supra), substituted α-amino aldehydes with a not too bulky amino-protecting group, give rise to high levels of diastereoselection resulting from chelation control [149].

The aqueous indium-mediated Barbier reaction can also proceed intramolecularly to give five-, six- and seven-membered rings which can be further cyclized to *cis*-fused α-methylene-γ-lactones [150]. A useful methodology of carbocycle enlargement by one [151] or two carbon atoms [152] has recently been reported using intramolecular indium-mediated allylation of 1,3-dicarbonyl compounds (Scheme 30).

Several other interesting synthetic procedures have recently been reported. Thus, β,γ-unsaturated ketones were synthesized from aqueous allylation of acyloyl pyrazoles, whereas tertiary alcohols were obtained with acyloyl imidazoles [153]. A variety of β-hydroxy esters were prepared efficiently by indium-mediated allylation of aldehydes using 3-bromo-2-chloro-1-propene followed by ozonolysis [154]. Highly functionalized β-lactams were prepared by allylation of 2,3-azetinediones [155]. The synthetic potential of this chemistry was also demonstrated with the total synthesis of (+)-goniofufurone from D-glucurono-6,3-lactone via an indium-mediated highly regio- and stereoselective allenylation in aqueous medium [156] or in the studies towards the total synthesis of Antillatoxin through allylation reaction of carbonyl compounds with β-bromocrotyl bromide in water [157].

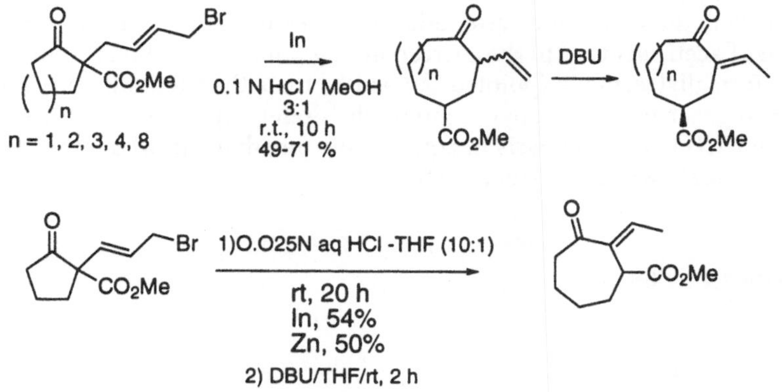

Scheme 30. One and two carbon atom ring enlargements

More unexpected was the formation of cyclopropane during the indium-mediated reaction of dibenzylidene acetone with allyl bromide which gave 1,1-distyryl-2-(but-3-enyl)cyclopropane as a mixture of four isotopomers (Scheme 31) [158]. In a related reaction, a simple and efficient one-pot method was developed to give chiral homoallylic amines and amino acids from the respective aldehydes with high stereoselectivity [159].

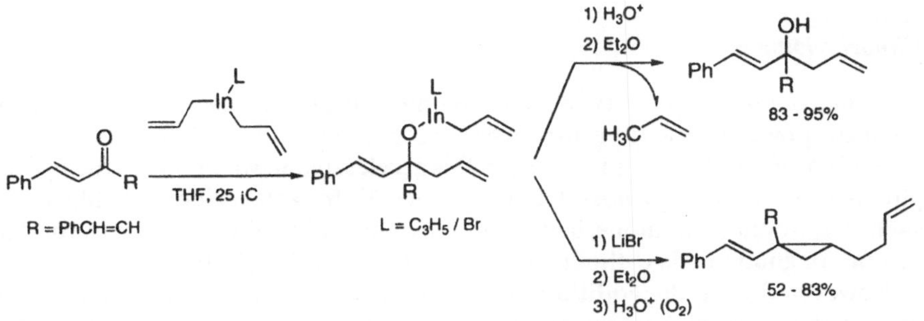

Scheme 31. Formation of cyclopropane during allylation of dibenzylidene acetone

5.1.4
Miscellaneous Allylations Using Other Metals

Several other metals have been used in this Barbier-type reaction. Thus, a mixture of bismuth(III) chloride/aluminum [160] or of bismuth(III)chloride/magnesium [161] was shown to promote allylation of aldehydes under aqueous conditions yielding homoallylic alcohols in good yields. Similarly, the same reaction can be promoted with metallic lead [162], highly reactive antimony prepared by the $NaBH_4$ reduction of $SbCl_3$ [163] or a mixture of manganese and catalytic copper [164]. In the latter case, aromatic aldehydes reacted whereas aliphatic aldehydes were totally inert under the reaction conditions. When copper was omitted, the reaction of aromatic aldehydes in the presence of a catalytic amount of acetic acid led to the corresponding pinacol-coupling product in fair yields. Tetraallylgermane (pointed out as less toxic than tetraallyltin) was also shown to react with carbonyl compounds, preferably with aldehydes, under scandium(III) triflate catalysis in aqueous nitromethane in which the presence of water was shown to be essential [165].

5.2
Conjugate 1,4-Additions

5.2.1
Organometallic Additions

When saturated alkyl halides were used in place of allyl compounds, a zinc/copper couple or zinc dust/copper iodide promoted the 1,4-addition to α-enones or α-enals. Sonication enhanced the efficiency of the process leading to the 1,4-adducts in very good yields [166]. This reaction was later extended to various α,β-unsaturated compounds such as esters, amides and nitriles [167]. The reactivity of the halide followed the order tertiary > secondary ≫ primary and iodide > bromide ≫ chloride making the assumption of a radical process highly probable [168].

5.2.2
Michael Additions

The nucleophilic Michael-type addition onto α,β-unsaturated ketones is one of the most powerful reactions for carbon–carbon bond formation. This reaction which is promoted under pressure [169] is accelerated in water due to the hydrophobic effect as already postulated (see Sect. 3). In fact, various 1,3-diketones, such as 1,3-cyclopentadione, have been shown to add readily to α,β-unsaturated ketones in good yields without any catalyst (Scheme 32) [170].

However, during the addition of β-keto esters onto enals, ytterbium triflate proved to be an efficient catalyst [171] as in the Michael additions of α-nitro esters [172]. Several water-soluble phosphines gave the corresponding phosphonium salts in good yields when added to α,β-unsaturated acids [173] or activated alkynes [174]. With alkynes, vinyl phosphine oxides or alkenes were formed depending on the pH of the aqueous solution. Significantly, the reaction of nitroalkanes with buten-2-one is considerably accelerated when going from

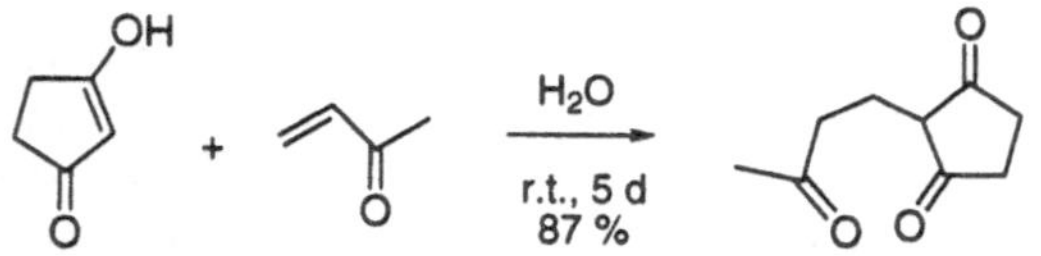

Scheme 32. Michael addition of 1,3-cyclopentadione

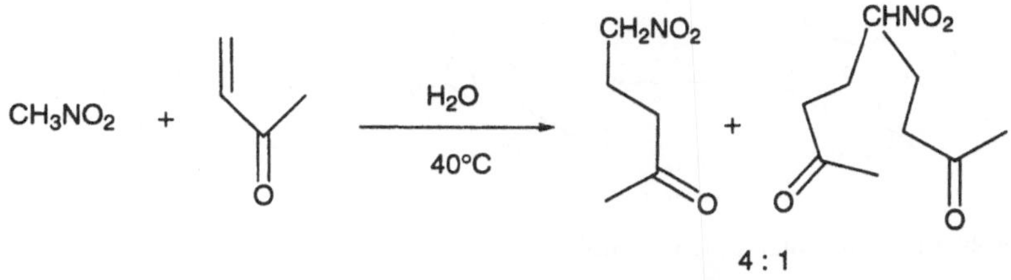

Scheme 33. Michael addition of nitromethane onto methyl vinyl ketone

non-polar solvents or polar solvents such as methanol to water. This uncatalyzed reaction, considered impossible, proceeded smoothly without added base in a completely neutral medium (Scheme 33) [175].

Noteworthy was the increase of the selectivity in water toward the 1:1 adduct when using nitromethane. Under slightly alkaline conditions, cetyltrimethylammonium chloride was shown to catalyze the addition of various nitroalkanes onto conjugated enones [176]. Amines also reacted in aqueous Michael additions, especially with α,β-unsaturated nitriles [177]. The lack of apparent reactivity of α,β-unsaturated esters comes from the reverse reaction which is particularly accelerated in water. In these amine additions, water activation was compared with high pressure giving support to the implication of the hydrophobic effect. A related reaction is the Baylis–Hillman reaction which proceeds readily in water with a good rate enhancement (Scheme 34) [178].

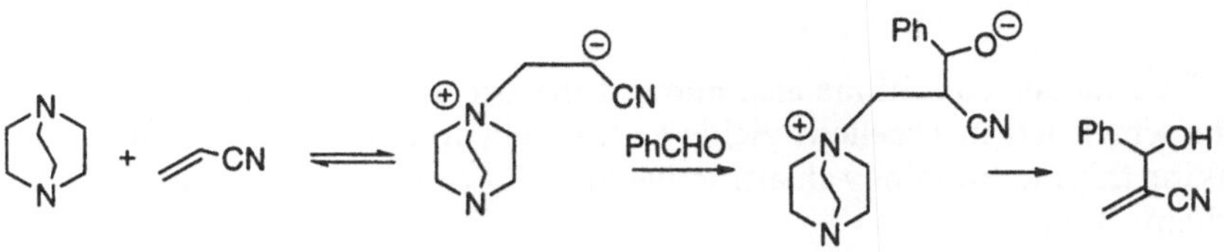

Scheme 34. Baylis–Hillman reaction

5.3
Cross-Aldol and Reformatsky-Type Addition

The aldolization reaction is certainly the most popular reaction for creating carbon–carbon bonds and much effort has been made to achieve the reactions in smooth conditions with a high degree of stereoselectivity. The reaction involves activated carbonyl compounds (enol, enolates, various enol ethers) which add on another carbonyl compound or various electrophiles in related reactions. As most of the activated compounds are compatible with water, it is not surprising that the aldolization in aqueous conditions became an efficient process in view of the negative activation volume of the reaction. The diastereoselectivity depends greatly on the reaction conditions. Thus, in a intramolecular aqueous aldolization, the acid-induced reaction of a keto aldehyde provided a *syn* hydroxy ketone while the base-catalyzed reaction led to the *anti* isomer (Scheme 35) [179].

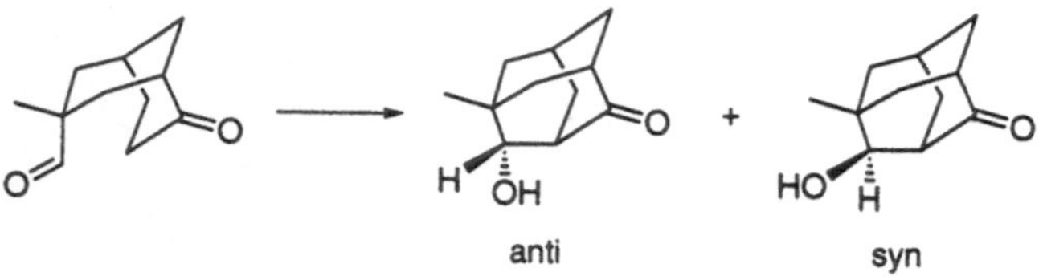

anti   syn

Scheme 35. Intramolecular aqueous aldolization

The aqueous conditions allowed the use of chiral water-soluble catalysts such as Zn^{++} complexes of α-amino acids [180] or β-cyclodextrin [181] with, however, low selectivities. Similarly, the water-soluble unprotected sugars can be extended by one carbon atom by reaction with formaldehyde (Scheme 36) [182].

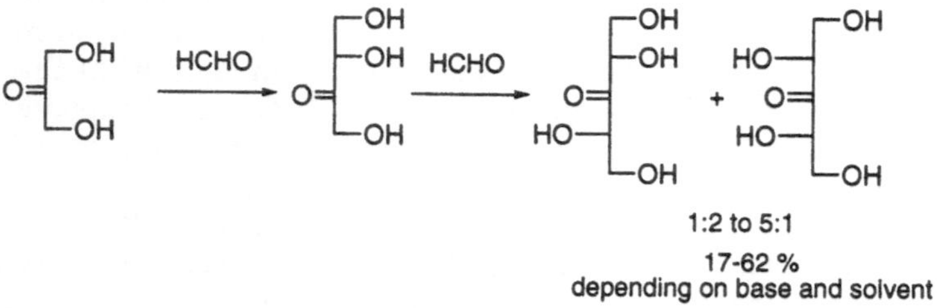

1:2 to 5:1

17-62 %

depending on base and solvent

Scheme 36. One carbon atom extension of ketose

The aqueous conditions also allowed the use of anionic or cationic surfactants which led in excellent yields to the dehydrated condensation products, making this methodology an attractive route for a one-pot synthesis of flavonols (Scheme 37) [183].

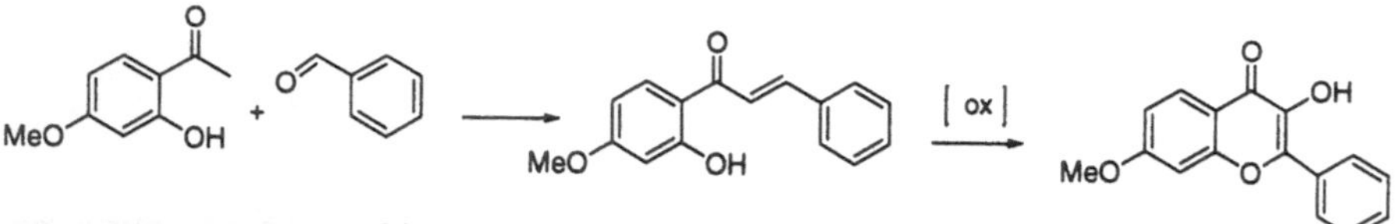

Scheme 37. Preparation of flavonols through aqueous aldolization

The Mukaiyama reaction is the reaction between a silyl enol ether and an aldehyde in CH$_2$Cl$_2$ in the presence of a stoichiometric amount of TiCl$_4$ [184]. The reaction is generally highly diastereoselective and leads to the *anti* diastereoisomer (75:25 in the case of the trimethylsilyl enol ether of cyclohexanone with benzaldehyde). The same reaction in the same solvent proceeded without catalyst but under high pressure (10,000 atm) to give the reverse diastereoselectivity (75:25 now in favor of the *syn* isomer) [185]. As expected, this reaction between two small hydrophobic molecules with a large negative activation volume must be accelerated in water compared to an organic solvent. In fact this

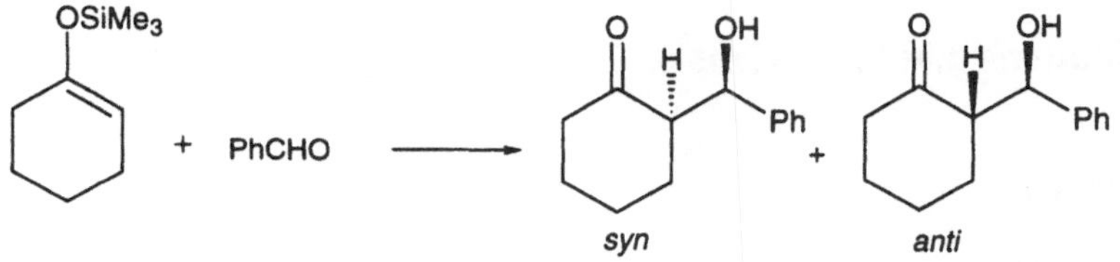

solvent	temp.	time	cond.	yield	syn:anti
CH_2Cl_2	20°C	2 hrs	1 eq $TiCl_4$	82	25:75
CH_2Cl_2	60°C	9 days	10 000 atm	90	75:25
H_2O-THF (1:1)	55°C	2 days	ultrasound	76	74:26

Scheme 38. The Mukaiyama reaction under various conditions

reaction does not work in any organic solvent without an acidic catalyst or without pressure. In water/THF mixtures, the reaction proceeds readily and gives in good yields (76%) the condensation products as a mixture of stereoisomers with the same selectivity as under external pressure (Scheme 38) [186].

Taking into account the competitive hydrolysis of the silyl enol ether, this reaction is remarkable. The method was shown to be general and was extended to a variety of aldehydes and several α,β-unsaturated carbonyl compounds giving uniformly 1,4-addition with aldehydes and a mixture of 1,4- and 1,2-adducts in the case of ketones [187]. Later, this aqueous version of the Mukaiyama reaction was shown to give near quantitative yields in the presence of a water-tolerant Lewis acid such as ytterbium triflate [188]. Keeping with the same concept, copper(II) triflate [189], indium(III) trichloride [190], tris(pentafluorophenyl)boron [191] and scandium(III) triflate in the presence of a surfactant [192] have proved to be active catalysts.

Related to the aldolization reaction is the Mannich reaction, which proceeds readily in water [193]. Allylsilanes [194] and allylstannanes [131] were also shown to add to iminium salts under Mannich-like conditions. As for the Mukaiyama reaction, the reaction between a vinyl ether with iminium salts was catalyzed by ytterbium triflate in a water/THF mixture. The Henry reaction which can be performed in an aqueous medium has been made more efficient in the presence of surfactants. The nitroaldol products were obtained in 65–95% yields [195].

Of interest also is the aqueous Reformatsky cross-coupling reaction between α-halo ketones or aldehydes and carbonyl compounds in the presence of zinc, tin and indium. Like the Barbier-type reactions, indium was found to be the most efficient metal and it strongly reduced undesirable side reactions such as the reduction of the halide [134]. As usual in aqueous reactions implicating carbonyl compounds, lanthanum(III) triflate promoted the reaction [146].

6
Oxidations and Reductions

6.1
Oxidations

Many classical oxidations are performed in water using oxidants such as sodium periodate, potassium permanganate, sodium or calcium hypochlorite or chromic acid under the well-known Jones' conditions. Many efforts have also been made using the inexpensive, but not very reactive, H_2O_2 which must be used most of the time with additives. We will focus here on the reactions that have a special outcome due to the use of water compared to the usual solvents. Peroxybenzoic acid and *meta*-chloroperbenzoic acid (MCPBA) react readily with olefins in aqueous $NaHCO_3$ solutions to give epoxides in high yields [196]. Although the reaction proceeds under heterogeneous conditions, the reaction is often more efficient than in organic solvents and is suitable for acid-sensitive alkenes. In pure water, with water-soluble alkenes, the 1,2-diols are generally obtained with a high degree of *trans* selectivity [197]. Electron-poor olefins, such as α,β-unsaturated acids, react slowly under these conditions, but the corresponding epoxy acids can be obtained in good yields using H_2O_2 in the presence of sodium tungstate [198]. Oxidation of alkenes with H_2O_2 is quite general providing that a co-additive is present. Thus, good results were obtained with diphenylphosphinic anhydride [199]. In the case of α,β-unsaturated ketones, epoxidation was performed in good yields using sodium perborates in slightly alkaline aqueous conditions [200].

The regio- and stereoselectivity of the epoxidations of polyolefinic alcohols are generally low in organic solvents. In contrast, the reaction in water using hydroperoxide in the presence of a transition metal proceeds with good regio- and stereoselectivity [201]. Likewise, the reaction with monoperoxyphthalic acid (MPPA) is also regioselective when the pH of the reaction mixture is controlled [202]. In a similar way, MPPA epoxidizes allylic and homoallylic alcohols in strongly alkaline aqueous medium, with high diastereoselectivity in cycloalkenols in favor of the *syn* adducts except for 2-cycloheptenol which gives the *trans* isomer as the major compound [203]. The selectivity was found to be much higher than in organic solvents using MPPA or MCPBA.

The inexpensive and easy to handle sodium percarbonate in aqueous THF allowed the conversion of various hydroxylated benzaldehydes or acetophenones to the corresponding hydroxy phenols through the so-called Dakin reaction. *ortho*-Hydroxybenzaldehyde is more reactive than the *para* isomer while the *meta* isomer is unreactive [204].

Baeyer-Villiger oxidations of ketones using MCPBA in water were very fast and efficient and afforded the corresponding esters or lactones in very good yields. At room temperature no hydrolysis occurred [205]. Alternatively, lactones have been obtained in good yields by Baeyer-Villiger oxidation with magnesium monoperphthalate hexahydrate in a water/methanol mixture [206]. α-Hydroxy carbonyl compounds are readily oxidized in good yields to carboxylic acids using sodium hypochlorite under ultrasonic activation [207].

Selective oxidation of methyl groups can be achieved by platinum salts in aqueous solution. Thus, p-toluenesulfonic acid is oxidized to the alcohol and then to the aldehyde by the Pt(II)/Pt(IV) system. Likewise the methyl group of ethanol can be oxidized without affecting the hydroxyl group [208]. Potassium or sodium bromate in the presence of cerium ammonium nitrate in a water/dioxane (2:3) mixture can oxidize toluene into a 1:1 mixture of benzaldehyde and benzoic acid. Ethylbenzenes yielded acetophenones [209].

6.2
Reductions

Except sodium borohydride which is commonly used in pure water or water/cosolvent mixtures for the chemoselective reduction of aldehydes and ketones, water is rarely used as the solvent for chemical reduction as most of the usual reducing agents are incompatible and react vigorously with water. However, the use of water could bring not only the usual advantages of low cost, security and easy isolation of the products (e.g. after decantation in homogeneous catalysis), but new reactivity and selectivity compared with organic solvents. For example, the regioselectivity in the reduction of alkadienoic acid which was performed by hydrogen in the presence of $RhCl[P(p\text{-tolyl})_3]_3$, is largely dependent on the solvent [210]. Alkenes and alkynes are also reduced in excellent yields using palladium acetate and triethoxysilane in pure water providing that their solubility in water is as good as unsaturated acids [211], or in water/THF for poorly water-soluble substrates [212]. Chiral water-soluble catalysts have been prepared for asymmetric hydrogenation of 2-acetamidoacrylates. These catalysts, which give moderate to good enantiomeric excesses, were based on Rh(II) and Ru(II) coordinated with sulfonated (R)-BINAP [213]. Alternatively, water-soluble bis(diphenylphosphino)pyrrolidines were shown to give higher enantiomeric excesses [214]. Although water was found to generally decrease the enantioselectivity in the use of chiral sulfonated diphosphines [215], the presence of amphiphiles in the reaction mixture enhanced significantly both the reactivity and the enantioselectivity of the reduction [216].

Pétrier and co-workers have reported an interesting system based on $Zn/NiCl_2$ (9:1) for the chemoselective reduction under aqueous conditions of α,β-unsaturated carbonyl compounds using ultrasonic activation. In this way, good yields were obtained in the selective reduction of the double bond. Water-soluble mono- and trisulfonated triphenylphosphines alone were shown to promote the stereoselective reduction of electron-deficient alkynes. Interestingly, the amount of phosphine controlled the *cis/trans* ratio of the alkenes formed in the reaction [217]. Catalyzed reductions based on these water-soluble sulfonated phenylphosphines in the presence of transition metals are now commonly used. Aromatic and aliphatic aldehydes can be reduced by hydrogen transfer from formate salts under biphasic conditions using water-soluble complexes of ruthenium(II), rhodium(I) and iridium(I) without a phase-transfer agent [218]. The same systems reduced α,β-unsaturated aldehydes to allylic alcohols [219].

Recently, the use of lanthanide derivatives has become increasingly popular in organic synthesis. As a matter of fact, they are now used in aqueous conditions for a variety of reactions. Lanthanides which are strong reducing agents have been used for the aqueous reduction of carboxylic acids, esters nitriles and amides to the corresponding alcohols and amines. The reaction was performed in 10% HCl and gave generally excellent yields [220]. Interestingly, the same system also reduced pyridines, quinolines and isoquinolines in excellent yields [221]. Samarium diiodide has been used in aqueous conditions for the reductions of many functional groups, such as carbonyl compounds, alkyl and aryl halides, carboxylic acids, esters, nitriles, alkenes, nitro compounds and even heteroaromatic rings [222]. Likewise, the α-deoxygenation of unprotected aldonolactones is very efficient when the $SmI_2/THF/water$ system is used (Scheme 39) [223].

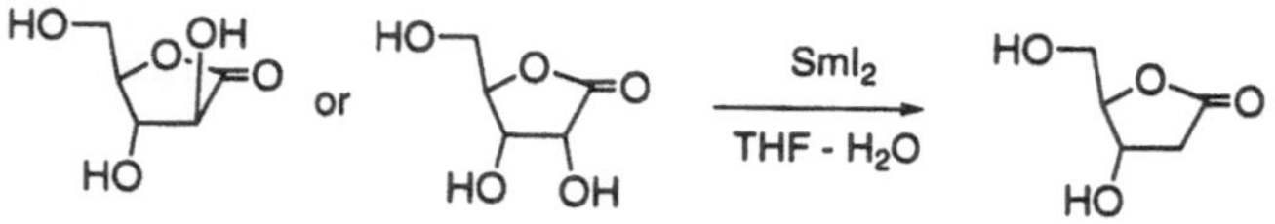

Scheme 39. SmI_2-mediated reductions of aldonolactones

Finally, although the use of tributyltin hydride is possible in water with or without a detergent as a solubilizing agent [224], noteworthy is the preparation of a water-soluble tin hydride which can be used for the reduction of alkyl halides in phosphate buffer in the presence of a radical initiator. In this way, water-soluble compounds such as unprotected sugars can be used directly without tedious protection/deprotection steps [225].

7
Outlook

As seen in this chapter, a great variety of organic reactions of interest can be performed in water. The observed specific reactivity results from the high cohesive energy density and high polarity of water and from its ability to form hydrogen bonds. In the main several of these properties are in operation at the same time, but certainly the unique tridimensional structure of water, still under considerable investigation and known to be essential for life processes, is the key to understanding its role as a solvent. In general, all reactions between two small hydrophobic molecules and which have a negative activation volume are always accelerated in water by destabilization of the initial state. If, in addition, there is an increase of polarity in the transition state, then the rate enhancement can be very important and even facilitate some reactions which would otherwise be impossible. Water can be used with cosolvents in one- or two-phase systems and in the presence of various additives such as surfactants, salts, water-tolerant Lewis acids, etc., which can modulate the reactivity. Then smooth conditions are possible even for highly energy-demanding reactions. At high temperatures and pressures, dramatic changes occur in the properties of water and it becomes a strong acid or base which can then be used for ecological applications in recycling, regeneration, disposal and detoxification of chemicals [21]. The understanding of aqueous chemistry will favor the discovery of new selective transformations which should encourage its use for organic synthesis with benign environmental impact.

8
References

1. Rideout DC, Breslow R (1980) J Am Chem Soc 102:7816
2. Grieco PA (1991) Aldrichimica Acta 24:59; Li CJ (1993) Chem Rev 93:2023; Lubineau A, Augé J, Queneau Y (1994) Synthesis 741
3. Lubineau A, Queneau Y (1989) Tetrahedron 45:6697
4. Grieco PA, Garner P, He ZM (1983) Tetrahedron Lett 24:1897
5. Kuntz EG (1987) Chemtech 17:570; Sinou D (1986) Bull Soc Chim Fr 480
6. Smithrud DB, Sanford EM, Chao I, Ferguson SB, Carcanague DR, Evanseck JD, Houk KN, Diederich F (1990) Pure Appl Chem 62:2227
7. Francks F (1975) In: Water, a comprehensive treatise. Plenum Press, New York, vol 4, p 1
8. Blokzijl W, Engberts JBFN (1993) Angew Chem Int Ed Engl 32:1545
9. Frank HS, Evans MW (1945) J Chem Phys 13:507
10. Nemethy G, Scheraga HA (1962) J Chem Phys 36:3401
11. Ben-Naim A (1965) J Chem Phys 69:3240
12. Bernal JD (1964) Proc R Soc London Soc A 280:299
13. Stanley HE, Teixera J (1980) J Chem Phys 73:3404
14. Davidson DW (1973) In: Water, a comprehensive treatise, Plenum Press, New York, vol 2, p 115
15. (a) Matubayasi N (1994) J Am Chem Soc 116:1450; (b) Hecht D, Tadesse L, Walters L (1992) J Am Chem Soc 114:4336
16. (a) Privalov PL, Gill SJ (1989) Pure Appl Chem 61:1097; (b) Dill KA (1990) Biochemistry 29:7133; Muller N (1990) Acc Chem Res 23:23
17. Graziano G, Barone G (1996) J Am Chem Soc 118:1831
18. Sharp KA, Madan B (1997) J Phys Chem B 101:4343
19. Tanford CH (1979) Proc Natl Acad Sci USA 76:4175
20. Kauzmann W (1987) Nature 325:763
21. (a) Siskin M, Katritzky AR (1991) Science 254:231; (b) Katritzky AR, Allin SM, Siskin M (1996) Acc Chem Res 29:399
22. Dack MRJ (1975) Chem Soc Rev 4:211
23. Reichardt C (1988) Solvents and solvent effects in organic chemistry, VCH, Weinheim
24. Dack MRJ (1974) J Chem Ed 51:231
25. Lubineau A (1986) J Org Chem 51:2142
26. (a) Blake JF, Lim D, Jorgensen WL (1994) J Org Chem 59:803; (b) Furlani TR, Gao J (1996) J Org Chem 61:5492
27. Otto S, Bertoncin F, Engberts JBFN (1996) J Am Chem Soc 118:7702
28. Isaacs NS, Maksimovic L, Laila A (1994) J Chem Soc Perkin Trans 2 495
29. Diels O, Alder K (1931) Ann Chem 490:243
30. Hopff H, Rantenstrauch CW (1942) Chem Abstr 36:1046
31. Breslow R (1991) Acc Chem Res 24:159 and Ref 1
32. Breslow R (1994) Hydrophobic and antihydrophobic effects on organic reactions in aqueous solution. In: Cramer CJ, Truhlar DG (eds) Structure and reactivity in aqueous solution. ACS Symposium Series 568:291
33. Blokzijl W, Blandamer MJ, Engberts JBFN (1991) J Am Chem Soc 113:4241
34. Schneider H-J, Sangwan NK (1986) J Chem Soc Chem Commun 1787
35. (a) Sangwan NK, Schneider H-J (1989) J Chem Soc Perkin Trans 2 1223; (b) Cativiela C, Garcia JI, Gil J, Martinez RM, Mayoral JA, Salvatella L, Urieta JS, Mainar AM, Abraham MH (1997) J Chem Soc Perkin Trans 2 653
36. Cativiela C, Garcia JI, Mayoral JA, Salvatella L (1996) Chem Soc Rev 209
37. Breslow R, Guo T (1988) J Am Chem Soc 110:5613

38. Blokzijl W, Engberts JBFN (1992) J Am Chem Soc 114:5440
39. Otto S, Blokzijl W, Engberts JBFN (1994) J Org Chem 59:5372
40. Blake JF, Jorgersen WL (1991) J Am Chem Soc 113:7430
41. Van der Wel GK, Wijnen JW, Engberts JBFN (1996) J Org Chem 61:9001
42. Wijnen JW, Engberts JBFN (1997) J Org Chem 62:2039
43. Lubineau A, Bienaymé H, Queneau Y, Scherrmann MC (1994) New J Chem 18:279
44. (a) Blake JF, Oakenfull D, Smith MB (1979) Biochemistry 18:5191; (b) Lee JC, Timasheff SN (1981) J Biol Chem 256:718
45. (a) Hunt I, Johnson CD (1991) J Chem Soc Perkin Trans 2 1051; (b) Braun R, Schuster F, Sauer J (1986) Tetrahedron Lett 27:1285
46. Otto S, Engberts JBFN (1995) Tetrahedron Lett 36:2645
47. Zhu Z, Espenson JH (1997) J Am Chem Soc 119:3507
48. Kobayashi S, Hachiya I, Araki M, Ishitani H (1993) Tetrahedron Lett 34:3755
49. Loh T-P, Pei J, Lin M (1996) J Chem Soc Chem Commun 2315
50. Breslow R, Maitra U, Rideout D (1983) Tetrahedron Lett 24:1901
51. Breslow R, Maitra U (1984) Tetrahedron Lett 25:1239
52. Diego-Castro MJ, Hailes HC (1998) Tetrahedron Lett 39:2211
53. Schneider H-J, Sangwan NK (1987) Angew Chem Int Ed Engl 26:896
54. Proust SM, Ridley DD (1984) Aust J Chem 37:1677
55. Brandes E, Grieco PA, Garner P (1988) J Chem Soc Chem Commun 500
56. Lubineau A, Augé J, Lubin N (1990) J Chem Soc Perkin Trans 1 3011
57. Garner PP (1998) Diels-Alder reactions in aqueous media. In: Grieco PA (ed) Organic synthesis in water. Chapman & Hall, London, chap 1
58. Otto S, Boccaletti G, Engberts JBFN (1998) J Am Chem Soc 120:4238
59. Larsen SD, Grieco PA (1985) J Am Chem Soc 107:1768
60. Grieco PA, Larsen SD (1986) J Org Chem 51:3553
61. Grieco PA, Parker DT (1988) J Org Chem 53:3658
62. Waldmann H (1988) Angew Chem Int Ed Engl 27:274
63. Waldmann H, Braun M (1991) Liebigs Ann Chem 1045
64. Parker DT (1998) Hetero-Diels-Alders reactions. In:Grieco PA (ed) Organic synthesis in water. Chapman & Hall, London, chap 2
65. Yu L, Chen D, Wang PG (1996) Tetrahedron Lett 37:2169
66. Yu L, Li J, Ramirez J, Chen D, Wang PG (1997) J Org Chem 62:903
67. Grieco PA, Bahsas A (1987) J Org Chem 52:5746
68. Lubineau A, Augé J, Lubin N (1991) Tetrahedron Lett 32:7529
69. Lubineau A, Augé J, Grand E, Lubin N (1994) Tetrahedron Lett 50:10265
70. Grieco PA, Henry KJ, Nunes JJ, Matt Jr JE (1992) J Chem Soc Chem Commun 368
71. Mackeith RA, McCague R, Olivo HF, Palmer CF, Roberts SM (1993) J Chem Soc Perkin Trans 1 313
72. McCague R, Olivo HF, Roberts SM (1993) Tetrahedron Lett 34:3785
73. Burlina F, Favre A, Fourrey J-L, Thomas M (1997) Bio & Med Chem Lett 7:247
74. Lubineau A, Augé J, Lubin N (1993) Tetrahedron 49:4639
75. Lubineau A, Queneau Y (1995) J Carbohydrate Chem 14:1295
76. (a) Dignam KJ, Hegarty AF, Quain PL (1978) J Org Chem 43:388; (b) Inoue Y, Araki K, Shiraishi S (1991) Bull Chem Soc Jpn 64:3079
77. Wijnen JW, Steiner RA, Engberts JBFN (1995) Tetrahedron Lett 36:5389
78. Pandey PS, Pandey IK (1997) Tetrahedron Lett 38:7237
79. Lubineau A, Bouchain G, Queneau Y (1995) J Chem Soc Perkin Trans 1 2433
80. Magnier E, Langlois Y (1998) Tetrahedron Lett 39:837
81. Lubineau A, Bouchain G (1997) Tetrahedron Lett 38:8031
82. Takaya H, Makino S, Hayakawa Y, Noyori R (1978) J Am Chem Soc 100:1765
83. Copley SD, Knowles JR (1987) J Am Chem Soc 109:5008
84. White WN, Wolfarth EF (1970) J Org Chem 35:2196
85. Coates RM, Rogers BD, Hobbs SJ, Peck DR, Curran DP (1987) J Am Chem Soc 109:1160
86. Brandes E, Grieco PA, Gajewski JJ (1989) J Org Chem 54:515

87. Gajewski JJ (1997) Acc Chem Res 30:219
88. Grieco PA, Brandes EB, McCann S, Clark JD (1989) J Org Chem 54:5849
89. Lubineau A, Augé J, Bellanger N, Caillebourdin S (1992) J Chem Soc Perkin Trans 1 1631
90. Cramer CJ, Truhlar DG (1992) J Am Chem Soc 114:8794
91. Severance DL, Jorgensen WL (1992) J Am Chem Soc 114:10966
92. Gao J (1994) J Am Chem Soc 116:1563
93. Carlson HA, Jorgensen WL (1996) J Am Chem Soc 118:8475
94. Barbier P (1898) Comptes Rendus Acad Sci Paris 128:110
95. Killinger TA, Boughton NA, Runge TA, Wolinsky J (1977) J Organomet Chem 124:131
96. Isaac MB and Paquette LA (1997) J Org Chem 62:5333
97. Rübsam F, Seck S and Giannis A (1997) Tetrahedron 53:2823
98. Li CJ (1996) Tetrahedron 52:5643
99. Pétrier C, Luche J-L (1985) J Org Chem 50:910
100. Pétrier C, Einhorn J, Luche J-L (1985) Tetrahedron Lett 26:1449
101. Durant A, Delplanche JL, Winand R, Reisse J (1995) Tetrahedron Lett 36:4257
102. Mattes H, Benezra C (1985) Tetrahedron Lett 26:5697
103. Einhorn C, Luche J-L (1987) J Organomet Chem 322:177
104. Sjöholm R, Rairama R, Ahonen M (1994) J Chem Soc Chem Commun 1217
105. Chan TH, Li CJ (1990) Organometallics 9:2649
106. Li CJ, Chan TH (1991) Organometallics 10:2548
107. Oda Y, Matsuo S, Saiko K (1992) Tetrahedron Lett 33:97
108. Bieber LW, Da Silva MF, Da Costa RC, Silva LOS (1998) Tetrahedron Lett 39:3655
109. Chan TH, Li CJ (1992) Can J Chem 70:2726
110. Waldmann H (1990) Synlett 627
111. Kunz T, Reissig H-U (1989) Liebiegs Ann Chem 891
112. Hanessian S, Park H, Yang RY (1997) Synlett 351
113. Hanessian S, Park H, Yang RY (1997) Synlett 353
114. Hanessian S, Yang RY (1996) Tetrahedron Lett 37:5273
115. Nokami J, Otera J, Sudo T, Okawara R (1983) Organometallics 2:191
116. Nokami J, Wakabayashi S, Okawara R (1984) Chem Lett 869
117. Wu SH, Huang BZ, Zhu TM, Yiao DZ, Chu YL (1990) Act Chim Sinica 48:372
118. Schmid W, Whitesides GM (1991) J Am Chem Soc 113:6674
119. Kim E, Gordon DM, Schmid W, Whitesides GM (1993) J Org Chem 58:5500
120. Uneyama K, Nanbu H, Torii S (1986) Tetrahedron Lett 27:2395
121. Masuyama Y, Kishida M, Kurusu Y (1996) Tetrahedron Lett 37:7103
122. Houllemare H, Outurquin F, Paulmier C (1997) J Chem Soc Perkin Trans 1 1629
123. Masuyama Y, Nimura Y, Kurusu Y (1991) Tetrahedron Lett 32:225
124. Masuyama Y, Takahara JP, Kurusu Y (1989) Tetrahedron Lett 30:3437
125. Wu SH, Huang BZ, Gao X (1990) Synth Commun 20:1279
126. Boaretto A, Marton D, Tagliavini G (1985) J Organomet Chem 286:9
127. Yanagisawa A, Inoue H, Morodome M, Yamamoto H (1993) J Am Chem Soc 115:10356
128. Kobayashi S, Wakabayashi T, Oyamada H (1997) Chem Lett 831
129. Miyai T, Inoue K, Yasuda M, Baba A (1997) Synlett 699
130. Belluci C, Cozzi PG, Umani-Ronchi A (1995) Tetrahedron Lett 36:7289
131. Grieco PA, Bahsas A (1987) J Org Chem 52:1378
132. Kobayashi S, Busujima T, Nagayama S (1998) J Chem Soc Chem Commun 19
133. Li CJ, Chan TH (1991) Tetrahedron Lett 32:7017
134. Chan TH, Li CJ, Lee MC, Wei ZY (1994) Can J Chem 72:1181
135. Chan TH, Li CJ (1992) J Chem Soc Chem Commun 747
136. Gordon DM, Whitesides GM (1993) J Org Chem 58:7937
137. Choi SK, Lee S, Whitesides GM (1996) J Org Chem 61:8739
138. Gao J, Martichonok V, Whitesides GM (1996) J Org Chem 61:9538
139. Binder WH, Prenner RH, Schmid W (1994) Tetrahedron 50:749
140. Gao J, Härter R, Gordon DM, Whitesides GM (1994) J Org Chem 59:3714

141. Araki S, Jin SJ, Idou Y (1992) Bull Chem Soc Jpn 65:1736
142. Loh TP, Li XR (1996) J Chem Soc Chem Commun 16:1929
143. Loh TP, Li XR (1997) Tetrahedron Lett 38:869
144. Loh TP, Li XR (1997) Angew Chem Int Ed Engl 36:980
145. Isaac MB, Chan TH (1995) Tetrahedron Lett 36:8957
146. Diana SCH, Sim KY, Loh TP (1996) Synlett 3:263
147. Paquette LA, Mitzel TM (1996) J Org Chem 61:8799
148. Paquette LA, Mitzel TM (1996) J Am Chem Soc 118:1931
149. Paquette LA, Mitzel TM, Isaac MB, Crasto CF, Schomer WW (1997) J Org Chem 62:4293
150. Bryan VJ, Chan TH (1996) Tetrahedron Lett 37:5341
151. Haberman JX, Li CJ (1997) Tetrahedron Lett 38:4735
152. Li CJ, Chen DL, Lu YQ, Haberman JX, Mague JT (1998) Tetrahedron 54:2347
153. Bryan VJ, Chan TH (1997) Tetrahedron Lett 38:6493
154. Yi XH, Meng Y, Li CJ (1997) Tetrahedron Lett 38:4731
155. (a) Jayaraman M, Manhas MS, Bose AK (1997) Tetrahedron Lett 38:709; (b) Paquette LA,
     Isaac MB (1998) Heterocycles 47:107
156. Yi XH, Meng Y, Li CJ (1998) J Chem Soc Chem Commun 449
157. Loh TP, Cao GQ, Pei J (1998) Tetrahedron Lett 39:1453
158. Capps SM, Lloyd-Jones GC, Murray M, Peakman TM, Walsh KE (1998) Tetrahedron Lett
     39:2853
159. Loh TP, Ho DSC, Xu KC, Sim KY (1997) Tetrahedron Lett 38:865
160. Wada M, Ohki H, Akiba K (1987) J Chem Soc Chem Commun 708
161. Wada M, Fukuma T, Morioka M, Takahashi T, Miyoshi N (1997) Tetrahedron Lett 38:8045
162. Zhou JY, Jia Y, Sun GF, Wu SH (1997) Synth Commun 27:1899
163. Ren PD, Jin QH, Yao ZP (1997) Synth Commun 27:2761
164. Li CH, Meng Y, Yi XH, Ma J, Chan TH (1997) J Org Chem 62:8632
165. Akiyama T, Iwai J (1997) Tetrahedron Lett 38:853
166. Pétrier C, Dupuy C, Luche J-L (1986) Tetrahedron Lett 27:3149
167. Dupuy C, Pétrier C, Sarandeses LA, Luche J-L (1991) Synth Commun 21:643
168. Luche J-L, Allavena C (1988) Tetrahedron Lett 29:5373
169. Matsumoto K (1981) Angew Chem Int Ed Engl 20:770
170. Lavallée JF, Deslongchamps P (1988) Tetrahedron Lett 29:6033
171. Keller E, Feringa BL (1996) Tetrahedron Lett 37:1879
172. Keller E, Feringa BL (1997) Synlett 842
173. Larpent C, Patin H (1988) Tetrahedron 44:6107
174. Larpent C, Meignan G, Patin H (1990) Tetrahedron 46:6381
175. Lubineau A, Augé J (1992) Tetrahedron Lett 33:8073
176. Ballini R, Bosica G (1996) Tetrahedron Lett 37:8027
177. Jenner G (1996) Tetrahedron 52:13557
178. Augé J, Lubin N, Lubineau A (1994) Tetrahedron Lett 35:7947
179. Denmark SE, Lee W (1992) Tetrahedron Lett 33:7729
180. (a) Nakagawa M, Nakao H, Watanabe KI (1985) Chem. Lett 3:391; (b) Buonora PT, Rosauer
     KG, Dai L (1995) Tetrahedron Lett 36:4009
181. (a) Watanabe KI, Yamada Y, Goto K (1985) Bull Chem Soc Jpn 58:1401; (b) Zhang Y, Xu W
     (1989) Synth Commun 19:1291
182. Shigemasa Y, Yokohama K, Sashira H, Saimoto H (1994) Tetrahedron Lett 35:1263
183. Fringuelli F, Pani G, Piermatti O, Pizzo F (1994) Tetrahedron 50:11499
184. Mukaiyama T, Banno K, Narasaka T (1974) J Am Chem Soc 96:7503
185. Yamamoto H, Maruyama K, Matsumoto K (1983) J Am Chem Soc 105:6963
186. Lubineau A (1986) J Org Chem 51:2142
187. Lubineau A, Meyer E (1988) Tetrahedron 44:6065
188. Kobayashi S., Hachiya I (1992) Tetrahedron Lett 33:1625
189. Kobayashi S, Nagayama S, Busujima T (1997) Chem Lett 959
190. Loh TP, Pei J, Cao GQ (1996) J Chem Soc Chem Commun 1819
191. Ishihara K, Hananki N, Yamamoto M (1993) Synlett 577

192. Kobayashi S, Wakabayashi T, Nagayama S, Oyamada H (1997) Tetrahedron Lett 38:4559
193. Tychopoulos V, Tyman JHP (1986) Synth Commun 16:1401
194. Larsen SD, Grieco PA, Fobare WF (1986) J Am Chem Soc 108:3512
195. Ballini R, Bosica G (1997) J Org Chem 62:425
196. Fringuelli F, Germani R, Pizzo F, Savelli G (1989) Tetrahedron Lett 30:1427
197. Fringuelli F, Germani R, Pizzo F, Savelli G (1989) Synth Commun 19:1939
198. Kirshenbaum KS, Sharpless KB (1985) J Org Chem 50:1979
199. Kende AS, Delair P, Blass BE (1994) Tetrahedron Lett 35:8123
200. Reed KL, Gupton JT, Solarz TL (1989) Synth Commun 19:3579
201. Gao Y, Hanson RM, Klunder JM, Ko SY, Masamune H, Sharpless KB (1987) J Am Chem Soc 109:5765
202. Fringuelli F, Germani R, Pizzo F, Santinelli F, Savelli G (1992) J Org Chem 57:1198
203. Ye D, Fringuelli F, Piermetti O, Pizzo F (1997) J Org Chem 62:3748
204. Kabalka GW, Reddy NK, Narayana C (1992) Tetrahedron Lett 33:865
205. Fringuelli F, Germani R, Pizzo F, Savelli G (1989) Gazz Chim Ital 119:249
206. Mino T, Masuda S, Nishio M, Yamashita M (1997) J Org Chem 62 2633
207. Yang DTC, Zhang CJ, Fu PP, Kabalka GW (1997) Synth Commun 27:1601
208. Labinger JA, Herring AM, Bercaw JE (1990) J Am Chem Soc 112:5628
209. Ganin E, Amer I (1995) Synth Commun 25:3149
210. Okano T, Kaji M, Isotani S, Kiji J (1992) Tetrahedron Lett 33:5547
211. Tour JM, Pendalwar SL (1990) Tetrahedron Lett 31:4719
212. Tour JM, Cooper JP, Pendalwar SL (1990) J Org Chem 55:3452
213. Wan K, Davis ME (1993) Tetrahedron: Asymmetry 4:2461
214. Nagel U, Kinzel E (1986) Chem Ber 119:1731
215. Sinou D, Safi M, Claver C, Masdeu A (1991) J Mol Catal 68 L9
216. Grassert I, Paetzold E, Oehme G (1993) Tetrahedron 49:6605
217. Larpent C, Meignan G (1993) Tetrahedron Lett 34:4331
218. Benyei A, Joo F (1990) J Mol Catal 58:151
219. Grosselin JM, Mercier C (1990) J Mol Catal 63:L25
220. Kamochi Y, Kudo T (1994) Chem Pharm Bull 42:402
221. Kamochi Y, Kudo T (1995) Chem Pharm Bull 43:1422
222. (a) Hasegawa E, Curran DP (1993) J Org Chem 58:5008; (b) Kamochi Y, Kudo T (1993) Chem Lett 1495
223. Hannessian S, Girard C (1994) Synlett 861
224. Maitra U, Sarma KD (1994) Tetrahedron Lett 35:7861
225. Light J, Breslow R (1990) Tetrahedron Lett 31:2957

Metal Catalysis in Water

Denis Sinou

Laboratoire de Synthèse Asymétrique, associé au CNRS, Université Claude Bernard Lyon 1, CPE Lyon, 43, Boulevard du 11 novembre 1918, F-69622 Villeurbanne cedex, France.
E-mail: sinou@univ-lyon1.fr

Organometallic catalysis in aqueous systems is now a very active field or research, both from an academic and an industrial point of view. The use of transition-metal catalysts in water or in a two-phase system offers the same advantages as in a usual organic medium. However they simplify the separation of the catalyst from the products, eventually for its recycling, and this is very important for large-scale chemical processes. The use of water as the solvent can also exhibit different selectivities to those shown in an organic medium. A comprehensive review concerning organometallic catalysis in aqueous solution is presented here, with particular emphasis on the application in organic synthesis.

Keywords: Catalysis, Aqueous, Transition metals, Hydrogenation, Hydroformylation, Alkylation, Coupling, Oxidation, Polymerization.

Topics in Current Chemistry, Vol. 206
© Springer-Verlag Berlin Heidelberg 1999

List of Abbreviations

BDPP	2,4-bis(diphenylphosphino)pentane
Binap	2,2'-bis(diphenylphosphino)-1,1'-binaphthalene
Binas-Na	octasodium salt of 2,2'-bis[(m-sulfonatodiphenylphosphino)-methyl]-4,4',8,8'-tetrasulfonato-1,1'-binaphthalene
Bisbis-Na	hexasodium salt of 2,2'-bis[(m-sulfonatodiphenylphosphino)-methyl]-disulfonato-1,1'-biphenyl
CBD	1,2-bis[(diphenylphosphino)methyl]cyclobutane
Chiraphos	2,3-bis(diphenylphosphino)butane
dppp	1,3-bis(diphenylphosphino)propane
Norbos-Na	trisodium salt of 3,4-dimethyl-2,5,6-tris(p-sulfonatophenyl)-1-phosphanorborna-2,5-diene
PPM	4-diphenylphosphino-2-(diphenylphinomethyl)pyrrolidine
PTA	1,3,5-triaza-7-phosphaadamantane
tppms	sodium salt of m-sulfonatophenyldiphenylphosphine
tppts	trisodium salt of tris(m-sulfonatophenyl)phosphine

1
Introduction

Homogeneous organometallic catalysis is now a well-used tool in organic synthesis, probably due to the high activities and selectivities generally achieved under mild reaction conditions. However, due to the use of a generally costly and toxic transition metal, one of the most important developments in homogeneous catalysis in the last 15 years is the introduction of aqueous two-phase catalysis. Effectively the homogeneous water-soluble catalyst, dissolved in water, is easily and quantitatively separated in its active form from the reactants/reaction products by simple decantation and can eventually be recycled. This methodology has been extensively studied since its discovery in 1975 [1, 2] and is now used in industrial processes as well as in laboratory organic synthesis. However other advantages, such as different selectivities to those observed in usual organic-media, can be expected using water as the solvent. Finally the use of water is environmentally attractive. Since some reviews have appeared in the literature on the synthesis and applications of water-soluble phosphines in organometallic catalysis [3 – 12], this chapter will focus on current applications in organic synthesis using aqueous-organic two-phase as well as aqueous organometallic catalysis, with emphasis on the actual developments in this field, mainly the literature of the 1990s.

2
Hydrogenation

Hydrogenation was one of the first organometallic catalytic reactions studied in aqueous solution and continues to attract interest.

2.1
Hydrogenation with Molecular Hydrogen

2.1.1
General Hydrogenation

Hydrogenation of various unsaturated substrates has been carried out in water or in a two-phase system using preformed or in situ rhodium and ruthenium catalysts associated with water-soluble ligands. Complexes such as $RhCl(PTA)_3$ [13], $[Rh(nbd)(n\text{-phophos})_2](NO_3)_3$ [14], and $[Ru(\eta^6\text{-}C_6H_6)(CH_3CN)_3](BF_4)_2$ [15] (Scheme 1) are very active catalysts for the hydrogenation of simple alkenes, as well as keto acids, unsaturated mono- and diacids and esters, under very mild conditions.

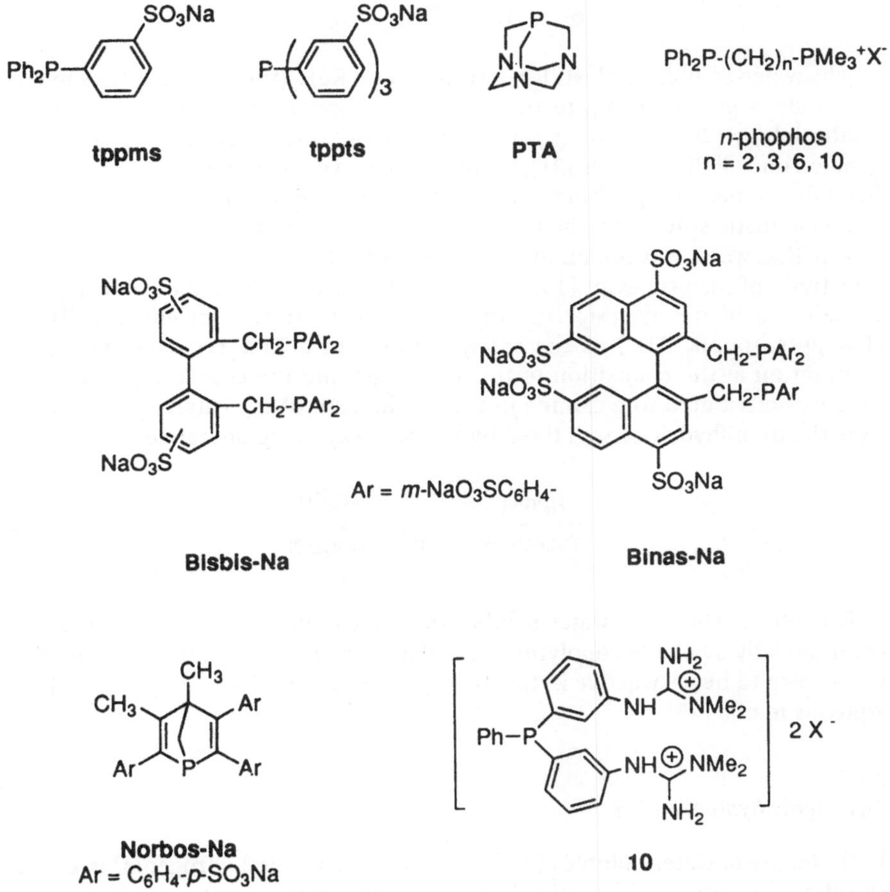
SO3Na
Ph2P
tppms
SO3Na
P
3
tppts
P
N
N
N
PTA
Ph2P-(CH2)n-PMe3+X-
n-phophos
n = 2, 3, 6, 10
NaO3S
CH2-PAr2
CH2-PAr2
NaO3S
Bisbis-Na
SO3Na
NaO3S
NaO3S
CH2-PAr2
CH2-PAr
SO3Na
Ar = m-NaO3SC6H4-
Binas-Na
CH3
CH3
Ar
Ar
P
Ar
Norbos-Na
Ar = C6H4-p-SO3Na
NH2
NH
NMe2
Ph–P
NH
NMe2
NH2
2 X-
10

Scheme 1. Achiral water-soluble phosphines

Water-soluble ruthenium complexes $RuHCl(tppts)_3$, $RuCl_2(tppts)_3$, RuH_2 (tppts)$_3$, or the rhodium complex $RhCl(PTA)_3$, are also effective catalysts for the hydrogenation of the carbonyl function of aldehydes [16], carbohydrates [17], and keto acids [13], provided that the iodide salt NaI is added for ruthenium complexes.

One of the most interesting applications of these catalytic systems is the regioselective reduction of α,β-unsaturated aldehydes to unsaturated alcohols or saturated aldehydes [18, 19]. For example, 3-methyl-2-buten-1-al or prenal was selectively reduced to prenol with a selectivity up to 97% using ruthenium complexes associated with tppts in a mixture of water/toluene at 35 °C and 20 bars hydrogen [Eq. (1)]; conversely, the saturated aldehyde was obtained with a selectivity up to 90% using $RhCl(tppts)_3$ as the catalyst at 80 °C and 20 bars hydrogen. The same selectivities were observed for (E)-cinnamaldehyde, 2-buta-nal and citral.

$$R \diagup\!\!\!\diagdown\!\!\!\diagup CH_2OH \xleftarrow[H_2]{Ru^{III}/tppts} R \diagup\!\!\!\diagdown\!\!\!\diagup O \xrightarrow[H_2]{Rh^{I}/tppts} R' \diagup\!\!\!\diagup\!\!\!\diagdown O \qquad (1)$$

The water-soluble catalysts $RuCl_2(tppts)_3$ and $RuH_2(tppts)_4$ were also support-ed on silica gel according to the concept of supported aqueous phase (SAP) methodology [20]. Various α,β-unsaturated aldehydes were selectively hydro-genated into allylic alcohols [21]; however, the recycling of these catalysts was dif-ficult due to poisoning adsorption of organic compounds at the catalyst surface.

Mechanistic studies on the hydrogenation reaction in these systems showed clearly that water was not an inert solvent, but influenced both the rate and the selectivity of the processes [13, 15, 22–24]. The hydrogenation of α-acetamido-cinnamic acid methyl ester in ethyl acetate/D_2O as the solvent and $RhClL_3$ (L = tppts, tppms, PTA) as the catalyst indicated a 75% regiospecific mono-deuteration at the α-position to the acetamido and the ester groups [Eq. (2)]. This was attributed to a change in the mechanism of the catalytic hydrogena-tion, the monohydric as well the dihydric pathway being operative.

$$\underset{Ph \quad\quad NHCOCH_3}{\overset{CO_2CH_3}{\diagup\!\!\!\diagdown}} \xrightarrow[D_2O/AcOEt]{H_2/Rh/L} \underset{Ph \quad\quad NHCOCH_3}{\overset{CO_2CH_3}{\diagup\!\!\!\diagup D}} \qquad (2)$$

Recently a rhodium water-soluble polymer-bound catalyst, based on the commercially available copolymer of maleic anhydride and methyl vinyl ether, was shown to be very active in the hydrogenation of various substrates in basic aqueous media [25].

2.1.2
Asymmetric Hydrogenation

With the use of water-soluble chiral diphosphine ligands, hydrogenation of pro-chiral olefins can provide optically active compounds. Asymmetric hydrogena-tion in aqueous media or in an aqueous-organic two-phase system with rhodium and ruthenium complexes associated with several water-soluble chiral ligands (Scheme 2) has been investigated, particularly with α-amino acid precursors [Eq. (3)]. Chiral sulfonated phosphines derived from Chiraphos **1**, BDPP **2**, and Cyclobutanediop **3** [26, 27], as well as ligands **4**–**6** bearing quaternized amino groups [28–31] were used as ligands. While rhodium complexes of Chiraphos derivatives **1** and **4** gave high enantioselectivities, up to 96%, under two-phase catalysis or in water alone, rhodium complexes of BDPP **2** and **5** or Cyclobutanediop **3** or Diop **6** gave lower enantioselectivities (up to 71% for BDPP **2** and **5** and 34% for Cyclobutanediop **3** or Diop **6**). More recently, enantioselectivities up to 70% have been obtained in the reduction of these α-amino acid precursors using rhodium complexes associated with sulfonated Binap **7** [32, 33]; the enantioselectivities increased to 88% by changing from a rhodium to a ruthenium complex [34].

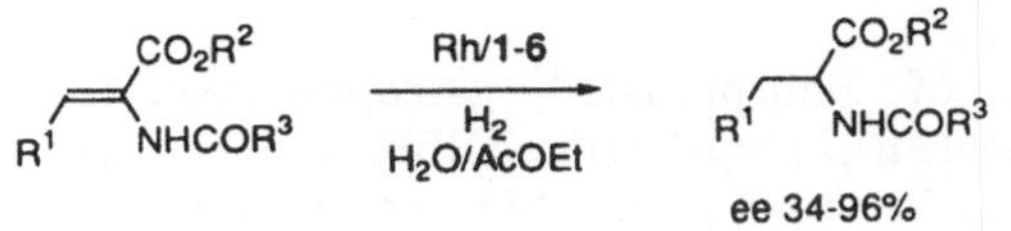
CO2R2
R1
NHCOR3
Rh/1-6
H2
H2O/AcOEt
CO2R2
R1
NHCOR3
ee 34-96%

(3)

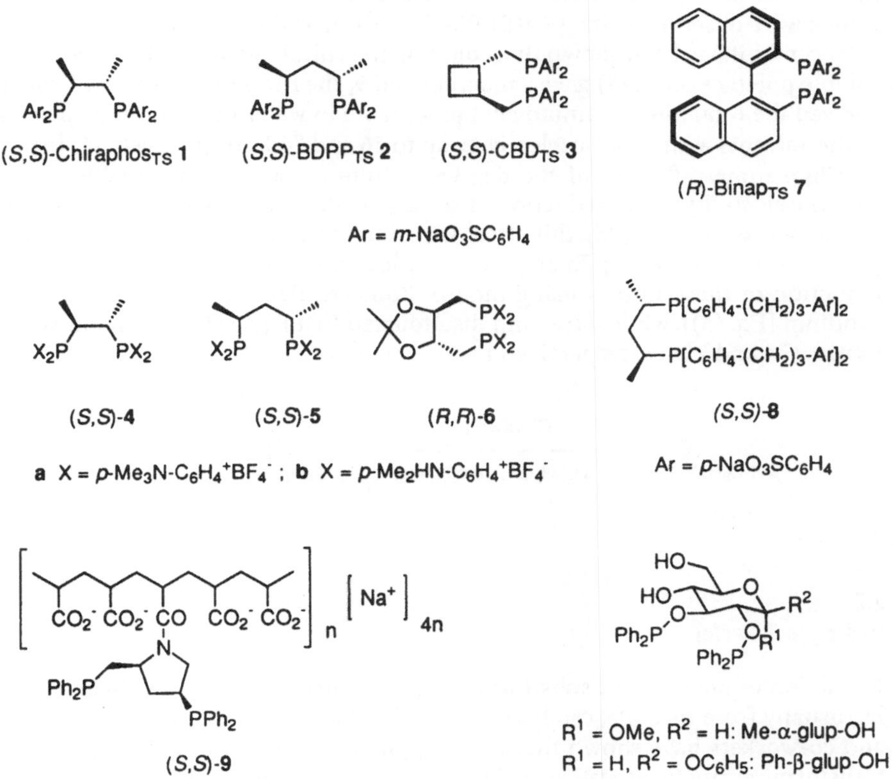
Ar2P
PAr2
(S,S)-ChiraphosTS 1
(S,S)-BDPPTS 2
PAr2
PAr2
(S,S)-CBDTS 3
PAr2
PAr2
(R)-BinapTS 7
Ar = m-NaO3SC6H4
X2P
PX2
(S,S)-4
(S,S)-5
O
O
PX2
PX2
(R,R)-6
P[C6H4-(CH2)3-Ar]2
P[C6H4-(CH2)3-Ar]2
(S,S)-8
a X = p-Me3N-C6H4+BF4- ; b X = p-Me2HN-C6H4+BF4-
Ar = p-NaO3SC6H4
CO2- CO2- CO CO2- CO2-
N
n
Na+
4n
Ph2P
PPh2
(S,S)-9
HO
HO
O
O
R2
Ph2P
O
R1
Ph2P
R1 = OMe, R2 = H: Me-α-glup-OH
R1 = H, R2 = OC6H5: Ph-β-glup-OH

Scheme 2. Chiral water-soluble phosphines

Davis and co-workers [35, 36] prepared the supported phase catalyst SAP-Ru-Binap(4-SO$_3$Na) which allowed the reduction of 2-(6'-methoxy-2'-naphthyl)acrylic acid to the commercially important antiinflammatory agent naproxen with 96% enantioselectivity [Eq. (4)]. The recycling of the catalyst was easily achieved without any leaching of ruthenium in the organic phase.

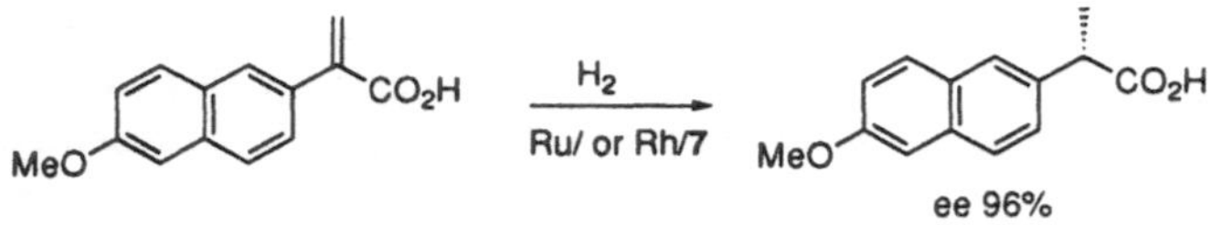

A surface-active tetrasulfonated chiral diphosphine 8 derived from BDPP showed improved reactivity and similar selectivity to the non-modified BDPP (with ee up to 69%) in the reduction of α-acetamidocinnamic acids in an ethyl acetate/water system [37].

Rhodium complexes of phosphinated glucopyranosides, [Rh(Me-α-glupOH)(cod)]BF$_4$ and [Rh(Ph-β-glup-OH)(cod)]BF$_4$, reduced prochiral dehydroamino acid derivatives in water in the presence of surfactants [38, 39]; addition of dodecyl sulfate increased both the rate and enantioselectivity of the hydrogenation, enantiomeric excesses up to 83% being obtained. The same trends were observed using [Rh(BPPM)(cod)]BF$_4$ as the catalyst [40].

Recently, it has been shown that coupling the chiral ligand PPM with a water-soluble poly(acrylic acid) gave a macroligand 9, the rhodium complex of which allowed the reduction of amino acid precursors in water or water/ethyl acetate as the solvents with enantioselectivity up to 56 and 74%, respectively [41].

While some influence of the degree of sulfonation of chiral BDPP on the enantioselectivity in the reduction of some dehydroamino acids was noticed by Sinou and co-workers [26], this effect was more pronounced in the reduction of prochiral imines [42, 43]. Enantioselectivities up to 96% were observed in the reduction of these imines using monosulfonated BDPP$_{MS}$ in association with rhodium [Eq. (5)], while tetra- and disulfonated BDPP gave lower enantioselectivities (34 and 2% ee, respectively).

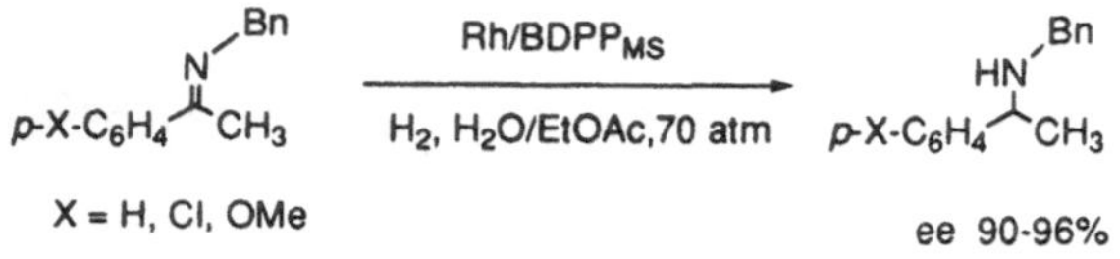

2.2
Hydrogen Transfer

Reduction of unsaturated substrates can also be performed by hydrogen transfer, usually from formate, catalyzed by rhodium and ruthenium complexes. Joò and co-workers have shown that RuCl$_2$(tppms)$_2$ [44] and RuCl$_2$(PTA)$_4$ [45, 46] transforms aromatic as well as α,β-unsaturated aldehydes to the corresponding aromatic or unsaturated alcohols, with a selectivity up to 98% in the latter case, in a biphasic aqueous-organic medium using sodium formate as the hydrogen source. By way of contrast, the rhodium complex [Rh(PTA)$_3$ HCl]Cl is an effective catalyst for the regioselective reduction of these α, β-unsaturated aldehydes to saturated aldehydes [47].

Aldoses (D-mannose, D-glucose) were also reduced to the corresponding alditols in water in the presence of RuCl$_2$(tppts)$_3$ as the catalyst with sodium formate, or an azeotropic mixture of formic acid and triethylamine, as the hydrogen donor [17].

Enantiomeric excesses up to 43% were obtained in the catalytic transfer hydrogenation of some α, β-unsaturated carboxylic acids in water using sodium formate in the presence of rhodium complexes associated with chiral sulfonated ligands such as Cyclobutanediop 3 [48].

3
Carbonylation Reactions

The carbonylation of organic compounds catalyzed by organometallic complexes is a useful tool in organic synthesis for the preparation of carbonyl compounds starting from olefins or halogeno compounds.

3.1
Hydroformylation Reactions

Hydroformylation is a major industrial process producing aldehydes from olefins, carbon monoxide, and hydrogen in the presence of a cobalt or better a rhodium catalyst combined with phosphorus ligands [Eq. (6)].

$$\diagdown\!\!\diagup\diagdown + \text{CO/H}_2 \xrightarrow{\text{Rh/tppts}} \diagdown\!\!\diagup\diagdown\text{CHO} + \underset{iso}{\overset{\text{CHO}}{\diagup\!\!\diagdown}} \qquad (6)$$
$$\hspace{4cm} n$$

The highly water-soluble RhH(CO)(tppts)$_3$ was used in a two-phase system in the hydroformylation of propene [1] and the industrial process was developed by Ruhr-Chemie AG [9, 12, 49, 50]. However, due to a need to develop new water-soluble ligands with higher efficiency in hydroformylation of propene, Herrmann and co-workers prepared sulfonated ligands Bisbis-Na, Norbos-Na, and Binas-Na (Scheme 1) [51–53]. In the biphasic hydroformylation of propene, these sulfonated ligands associated with rhodium yielded exceptionally high n/iso ratios of the resulting butyraldehyde with very high activities and productivities at low phosphine/rhodium ratios. The relatives activities are tppts/Bisbis-Na/Norbos-Na/Binas-Na = 1/5.6/7.4/11.1, with n/iso ratios up to 97:3 and 98:2 using Bisbis-Na and Binas-Na, respectively. Two-phase hydroformylation of styrene with an in situ catalyst prepared from Rh(CO)$_2$(acac) and chiral Binas-Na proceeded with good regioselectivity to 2-phenylpropionaldehyde (95%) but low enantioselectivity (18%) [54].

Unfortunately, the rhodium/sulfonated phosphine catalyst showed very low catalytic activities in the hydroformylation of higher olefins such as 1-octene in a two-phase system, due to mass transfer limitations resulting from the lower solubilities of such longer-chain derivatives in water [55]. In order to improve reaction rates, Russell [56] used rhodium catalysts containing $Ph_2PC_6H_4CO_2Na$ or tppms in the presence of transfer agents or surfactants such as $(PhCH_2)$ $Bu_3N^+Cl^-$ or $Me_3(C_{12}H_{25})N^+Br^-$; higher olefins, such as 1-dodecene or 1-hexadecene, were converted to aldehydes with high conversion and n/iso ratios up to 95:5. Fell and Papadogianakis [57] used a water-soluble catalyst consisting of $Rh_4(CO)_{12}$ and surface-active sulfobetaine derivatives of tris(2-pyridyl)phosphine. More recently, Hanson and co-workers have prepared surface-active phosphines of general structures $P[(CH_2)_n\text{-}C_6H_4\text{-}p\text{-}SO_3Na]_3$ and $P[C_6H_4\text{-}(CH_2)_n\text{-}C_6H_4\text{-}p\text{-}SO_3Na]_3$ [58–60]. In the two-phase hydroformylation of 1-octene, the rhodium catalysts derived from these new phosphines exhibited faster initial rates and higher selectivities; for example, at a ligand/rhodium ratio of 10, the n/iso values are 8.0–9.5 compared to 3.6 for tppts. However salt concentration had a considerable influence on the activity and selectivity of these catalysts; addition of Na_2SO_4 or Na_2HPO_4 enhanced both the rate and selectivity in the hydroformylation of 1-octene [61]. Although styrene was efficiently hydroformylated with a rhodium complex of the surfactant phosphine P(menthyl) $[(CH_2)_8\text{-}C_6H_4\text{-}p\text{-}SO_3Na]_2$, no optical induction was observed [62].

Recently the water-soluble tripodal phosphane ligand cis,cis-1,3,5-$(PPh_2)_3$-1,3,5-$[CH_2(OCH_2CH_2)_nOCH_3]C_6H_6$ (n = 30–160) has been prepared [63] and exhibited comparable catalytic activity in the hydroformylation of 1-hexene in a single-phase or in a biphasic system.

Chaudhari and co-workers [64] showed that the rate of hydroformylation of 1-octene using $[Rh(cod)Cl]_2$/tppts as the catalyst in a two-phase toluene/water system was enhanced by a factor of 10–50 by introducing the promoter ligand PPh_3 in the organic phase. Conversely, addition of tppts as a promoter to the catalyst $[Rh(cod)Cl]_2$/PPh_3 in the hydroformylation of allyl alcohol in toluene/water increased the rate by a factor of 5.

Monflier et al. reported very high conversion (up to 100%) and regioselectivity (≤95%) in the hydroformylation of various water-insoluble terminal olefins such as 1-decene with Rh/tppts catalyst system in water in the presence of per(2,6-di-O-methyl)-β-cyclodextrin (or Me-β-CD) [Eq. 7] [65, 66]. These high activities and selectivities were attributed to the formation of an alkene/cyclodextrin inclusion complex and to the solubility of the cyclodextrin in both the aqueous and organic layers; the cyclodextrin probably plays the role of an inverse phase transfer catalyst.

$$R\diagup\!\!\!\diagdown\; + \;CO/H_2 \quad\xrightarrow[\text{Me-}\beta\text{-CD, }H_2O]{\text{Rh(acac)(CO)}_2/\text{tppts}}\quad R\diagup\diagdown\diagup CHO \;+\; \underset{R}{\diagup\diagdown} CHO \tag{7}$$

R = $n\text{-}C_8H_{17}$, $n\text{-}C_{10}H_{21}$, $n\text{-}C_{12}H_{25}$ n/iso 95/5

Supported aqueous phase catalysts were developed for hydroformylation by Davis and co-workers [67]. Very hydrophobic alkenes, such as 1-octene, 1-tetradecene, or 1-heptadecene and also oleic alcohol, were hydroformylated using supported HRh(CO)(tppts)$_3$ with quite high activities and selectivities, and a n/iso ratio ranging from 1.8 to 2.9, depending on the water content of the catalyst and on the ligand/rhodium ratio [68–70]. The use of this supported catalyst for the hydroformylation of methyl acrylate and other α,β-unsaturated esters greatly improved the formation of 2-formylpropanoate esters [Eq. (8)] with selectivity up to 95% and a α/β ratio of 150 [Eq. (8)] [71, 72]. Other supported metal complexes containing Co [73] or Pt [74] were also used in some cases.

$$\text{\textbackslash\textbackslash}CO_2Me \xrightarrow[\text{Rh/tppts/SiO}_2]{H_2 + CO} \quad \begin{matrix} CO_2Me \\ | \\ CHO \end{matrix} \quad + \quad CHO\text{\textasciitilde}CO_2Me \tag{8}$$

$$\alpha\text{-aldehyde} \qquad \beta\text{-aldehyde}$$

3.2
Other Carbonylation Reactions

Although hydroxycarbonylation of alkenes is a very attractive route to carboxylic acids, it is only recently that this reaction has been successfully performed in a two-phase water/toluene system with yields up to 98% in acid using PdCl$_2$/tppts as the catalyst [Eq. (9)], provided that a Brönsted acid was added as promoter [75].

$$R\text{\textasciitilde} + CO + H_2O \xrightarrow[\text{HX}]{\text{Pd/tppts}} R\text{\textasciitilde}CO_2H \ + \ \begin{matrix} CO_2H \\ | \\ R\text{\textasciitilde} \end{matrix} \tag{9}$$

$$R = \text{alkyl, aryl} \qquad X = Cl, Br, CF_3CO_2, PF_6 \qquad \text{Yield up to 98\%}$$

Hydrocarbonylation of higher α-olefins also occurred under the same conditions with high conversions (up to 99%) and high selectivities (up to 90%) in the presence of chemically modified cyclodextrins [76].

Carbonylation of allylic as well as benzylic halides occurred readily in a two-phase aqueous sodium hydroxide/heptane system at atmospheric pressure and room temperature in the presence of PdCl$_2$(tppms)$_2$ as the catalyst giving β,γ-unsaturated acids in moderate yields [77, 78]. The addition of surfactants such as n-C$_7$H$_{15}$SO$_3$Na or n-C$_7$H$_{15}$CO$_2$Na accelerated the carbonylation reaction.

Bromobenzene was carbonylated in a biphasic toluene/water system in the presence of Pd(tppts)$_3$ giving the expected benzoic acid exclusively with high conversion [79].

By using PdCl$_2$/tppts as the catalyst, 5-hydroxymethylfurfural was selectively carbonylated to 5-formylfuran-2-acetic acid in an acidic aqueous medium at 70°C under 5 bar CO pressure, while performing the reaction in the presence of HI gave the reduced compound 5-methylfurfural [Eq. (10)] [80].

$$\text{HOCH}_2\text{-}\!\!\left\langle\!\!\!\begin{array}{c}\\O\end{array}\!\!\!\right\rangle\!\!\text{-CHO} \quad \xrightarrow{\text{CO}} \quad \begin{cases} \xrightarrow[\substack{\text{Pd/tppts}\\ \text{H}_2\text{SO}_4}]{} \quad \text{HO}_2\text{CCH}_2\text{-}\!\!\left\langle\!\!\!\begin{array}{c}\\O\end{array}\!\!\!\right\rangle\!\!\text{-CHO} \\[2em] \xrightarrow[\substack{\text{Pd/tppts}\\ \text{HI}}]{} \quad \text{CH}_3\text{-}\!\!\left\langle\!\!\!\begin{array}{c}\\O\end{array}\!\!\!\right\rangle\!\!\text{-CHO} \end{cases} \qquad (10)$$

4
Alkylation and Coupling Reactions

The palladium-catalyzed reaction is now a common tool in organic reactions and, in particular, in carbon–carbon bond formation; this is mainly due to the high chemio-, regio- and stereoselectivity and the very mild experimental conditions needed for this procedure. However it is only recently that coupling reactions involving π-allylpalladium as well as σ-alcenyl or σ-aryl intermediates have been investigated in aqueous media.

4.1
Alkylation Reactions

Substitution of allylic acetates or carbonates has been performed by Sinou et al. [81–83] in a water/nitrile medium using a Pd(OAc)$_2$/tppts or a Pd$_2$(dba)$_3$/tppts catalyst [Eq. (11)].

$$R\!\!\diagup\!\!\diagdown\!\!\diagup\!\!^{\text{OCOX}} \quad \xrightarrow[\text{RCN/H}_2\text{O}]{\text{Nu}^-,\ \text{Pd(OAc)}_2/\text{tppts}} \quad R\!\!\diagup\!\!\diagdown\!\!\diagup\!\!^{\text{Nu}}$$

X = CH$_3$, OCH$_3$

Yield 50-95%   (11)

Nu = CH(CO$_2$Me)$_2$, CH(COMe)$_2$, CH(COCH$_3$)CO$_2$CH$_3$, CH(NO$_2$)CO$_2$Et, NHR', NR'$_2$, N$_3$, SO$_2$C$_6$H$_4$-p-CH$_3$

Allylic carbonates coupled with various carbon nucleophiles, such as ethyl acetoacetate, in quite good yield without added base, although allylic acetates reacted with nitro ethyl acetate in the presence of NEt$_3$ or better DBU. It is noteworthy that the regio- and stereoselectivities are the same in an organic phase or in water. Heteroatomic nucleophiles can also be used for this reaction. For example, reaction of primary amines with cinnamyl acetate selectively generated the product of mono N-allylation, in contrast to the mono- and diallylated products observed in an organic phase. Heteronucleophiles such as azide and toluene sulfinate also reacted giving the corresponding allylic azide and toluenesulfone in high yields.

More recently, Bergbreiter et al. have used palladium complexes associated with phosphines derived from poly(N-isopropyl)acrylamide-c-(N-acryloxysuccinimide) in the allylation reaction in water [84]. Hayashi and co-workers have also shown that palladium–phosphine complexes bound to an amphiphilic polymer resin exhibited high activities [85]. The catalysts were easily separated and recycled by precipitation and filtration.

Heterogenization of the catalyst Pd/tppts or Pd/tppms was also performed by deposition on silica (SAP catalyst) [86–88]. The activity of the catalyst was found to be dependent on the water content of the support; however, the SAP catalyst was found to be drastically more active using PhCN as the co-solvent than its homologous biphasic system.

Although the palladium-catalyzed reaction of uracils with cinnamyl acetate in DMSO gave a complex mixture of all *N*-allylation products, performing the reaction in an aqueous acetonitrile (9:1) medium allowed the highly regio-selective formation of 1-cinnamyl uracils in reasonable yields [Eq. 12)] [89,90]. For the 2-thiouracils, only the formation of the product of monoallylation at sulfur was observed, although a complex mixture was also obtained in dioxane.

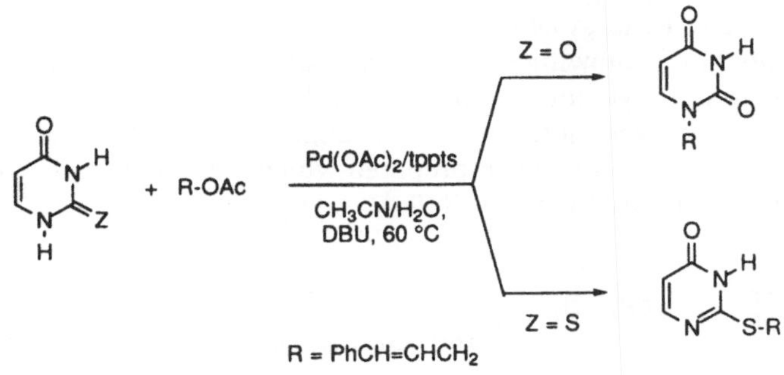

$$ \tag{12} $$

The Pd(OAc)$_2$/tppts system was very efficient for the removal of the allyloxy-carbonyl protecting group under very mild conditions in water/nitrile. For example, the allyloxycarbonyl moiety of allylic esters, carbamates and carbonates were selectively cleaved in a few minutes in quite good yields (70–99%), the π-allyl scavenger being diethylamine [Eq. (13)] [91]. Moreover, selective removal of allyloxycarboxyl and allyloxycarbonyl groups in the presence of dimethylallyl and cinnamyl carboxy groups was observed, as well as allyloxycarbamates in the presence of substituted allyloxycarbamates and allyloxycarbonates in the presence of dimethylallylcarbonates by careful choice of the amount of catalyst and the solvent system used [92–94]. Removal of allyloxycarbonyl groups also occurred under essentially neutral conditions using NaN$_3$ as the π-allyl scavenger [95].

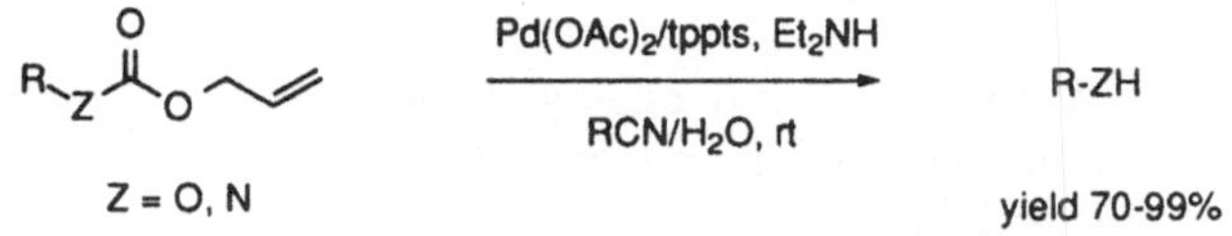

$$ \tag{13} $$

4.2
Coupling Reactions

The reaction between aryl or alkenyl halides or arenediazonium salts and al-kenes catalyzed by palladium complexes, the so-called Heck reaction, has been performed in aqueous media. Arylation of styrene or acrylic acid derivatives occurred in high yields in the presence of a free-ligand palladium complex as catalyst and a base (Na$_2$CO$_3$ or K$_2$CO$_3$) [96–98] and, eventually, a quaternary ammonium salt [Eq. (14)] [99, 100].

$$\text{ArX} + \underset{}{=\!\!\!/}^{E} \xrightarrow[\text{K}_2\text{CO}_3, \text{R}_4\text{NX}]{\text{Pd(OAc)}_2, \text{H}_2\text{O}} \underset{\text{Ar}}{\nearrow\!\!=\!\!/}^{E} \tag{14}$$

yield 87-97%

X = Br, I, N$_2^+$BF$_4^-$; E = Ph, CO$_2$Et; R$_4$NX = n-Bu$_4$NCl, n-Bu$_4$NBr, n-Bu$_4$NHSO$_4$

Recently, these coupling reactions have been carried out in water or in a two-phase water/acetonitrile system in the presence of PdCl$_2$(tppms)$_2$ [101] or Pd(OAc)$_2$/tppts [102], allowing easy recycling of the catalyst, and the reaction was extended to the arylation of ethylene [103]; in the latter case, the yields of the product styrenes were in the 60–100% range [Eq. (15)]. New carbohydrate-phosphane ligands have also been prepared which exhibited better yields and higher catalytic activities than tppts [104].

$$\underset{R}{\overset{X}{\bigcirc}} + \text{CH}_2\text{=CH}_2 \xrightarrow[\text{H}_2\text{O or H}_2\text{O/CH}_3\text{CN, 100 °C}]{\text{PdCl}_2(\text{tppms})_2} \underset{R}{\overset{}{\bigcirc}\!\!\!/} \tag{15}$$

yield 60-100%

X = I, Br, Cl; R = CO$_2$H, CO$_2$Me, CN, NO$_2$, NH$_2$, Br, Cl, OH

Drastic changes in regioselectivities were observed when intramolecular Heck-type cyclization was performed in an aqueous medium using Pd (OAc)$_2$/tppts as the catalyst; the selective formation of the *endo*-cyclized compound was observed instead of the *exo* compound obtained in a usual organic medium [Eq. (16)] [105].

$$\underset{Z}{\overset{}{\diagdown}}\!\!\!X \xrightarrow[\text{CH}_3\text{CN/H}_2\text{O, (\textit{i}-Pr)}_2\text{NEt}]{\text{PdCl}_2/\text{tppts}} \underset{}{\overset{Z}{\bigcirc}}\!\! + \underset{Z}{\overset{}{\diagdown}}\!\! \tag{16}$$

Z = N-R, C(CO$_2$Me)$_2$; X = Br, I   65-100%/35-0%

The reaction of various aryl and alkenyl halides with organostannanes (RSnCl$_3$) (Stille reaction) occurred in aqueous alkaline solution using a catalyst generated from PdCl$_2$ or PdCl$_2$(tppms)$_2$ to give the cross-coupling products in high yields [Eq. 17)] [106, 107].

$$\text{(aryl-I, R)} + \text{PhSnCl}_3 \xrightarrow[\text{H}_2\text{O/KOH, 100 °C}]{\text{PdCl}_2 \text{ or PdCl}_2(\text{tppms})_2} \text{(aryl-Ph, R)} \qquad (17)$$

R = CH$_3$, CO$_2$H, COCH$_3$, OH, NH$_2$ yield 45-98%

The cross-coupling reactions of various aryl halides and triflates with vinyl- or arylboronic acids and esters (Suzuki cross-coupling reaction) was also carried out in water in the presence of tetrabutylammonium bromide and a base such as Na$_2$CO$_3$, using a phosphine-free palladium catalyst to give biaryl derivatives [Eq. 18)] [108, 109]. More recently, Casalnuovo [101] and Genêt [102, 110] have performed this reaction using water-soluble palladium catalysts PdCl$_2$ (tppms)$_2$ and Pd(OAc)$_2$/tppts in water/acetonitrile.

$$\text{(aryl-Br, R)} + (\text{HO})_2\text{B-(aryl, R')} \xrightarrow[\text{Na}_2\text{CO}_3, 70\ °\text{C}]{\text{Pd(OAc)}_2,\ n\text{-Bu}_4\text{NX}} \text{(biaryl, R, R')} \qquad (18)$$

yield 80-100%

R = NO$_2$, CO$_2$CH$_3$, OCH$_3$, COCH$_3$, NHAc; R' = CH$_3$, CF$_3$, F, OH

This methodology was used to synthesize a high-molecular-weight, rigid-rod polymer from a dibromo compound and a bis-boronic ester in water [Eq. (19)] [111].

$$\text{Br-(aryl, CO}_2\text{H, HO}_2\text{C)-Br} + (\text{MeO})_2\text{B-(aryl)-B(OMe)}_2 \xrightarrow[\text{H}_2\text{O/DMF, NaHCO}_3]{\text{Pd(tppms)}_3}$$

$$\left[\text{(aryl, CO}_2\text{H, HO}_2\text{C)-(aryl)}\right]_n \qquad (19)$$

A variety of aryl iodides, aryl iodonium salts, vinyl iodides and acetylenic iodides bearing a broad range of functional groups, were coupled with terminal alkynes (Sonogashira reaction) in aqueous media in the presence of a water-soluble catalyst PdCl$_2$(tppms)$_2$ and Pd(OAc)$_2$/tppts [101, 102, 112], or non-water-soluble catalysts Pd(OAc)$_2$ and PdCl$_2$(PPh$_3$)$_2$ [113, 114] to give the substituted alkynes in high yields under very mild conditions. For example, unprotected nucleosides and nucleotides such as 5-iodo-2'-deoxycytidine 5'-monophosphate and 5-iodo-2'-deoxyuridine were coupled to propargylamine and propargyltri-fluoroacetamide in 73 and 95% yield, respectively.

More recently, complexes obtained by association of Pd(OAc)$_2$ and the cation-ic guanidino phosphine **10** were also shown to be very active catalysts for this coupling reaction [115–118]. They were mainly used in the synthesis of modi-fied proteins by coupling amino acids containing alkyne or iodoaryl groups with the corresponding reaction partners in aqueous acetonitrile; for example, iodobenzoate reacted with propargylglycine to give the expected cross-coupled product in 75 % yield [Eq. (20)].

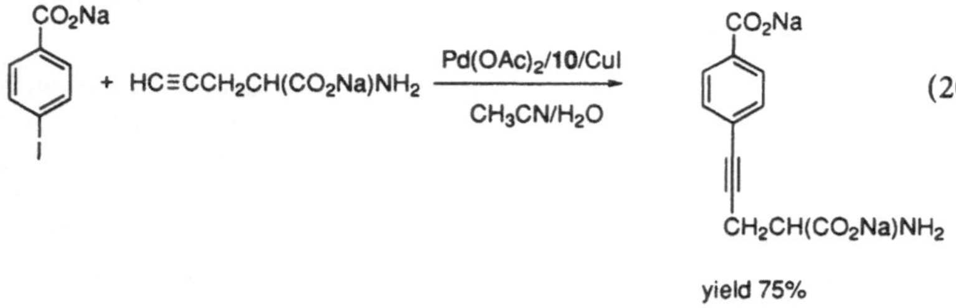

$$
\begin{array}{c}
\text{CO}_2\text{Na} \\
\text{(ring, para-I)}
\end{array}
+ \text{HC} \equiv \text{CCH}_2\text{CH(CO}_2\text{Na)NH}_2
\xrightarrow[\text{CH}_3\text{CN/H}_2\text{O}]{\text{Pd(OAc)}_2/10/\text{CuI}}
\begin{array}{c}
\text{CO}_2\text{Na} \\
\text{(ring)} \\
||| \\
\text{CH}_2\text{CH(CO}_2\text{Na)NH}_2
\end{array}
\qquad (20)
$$

yield 75%

Coupling 2-iodophenols and 2-iodoanilines and terminal alkynes in water in the presence of Pd(OAc)$_2$/tppts gave the corresponding indoles and benzofurans in good yields [Eq. (21)] [112].

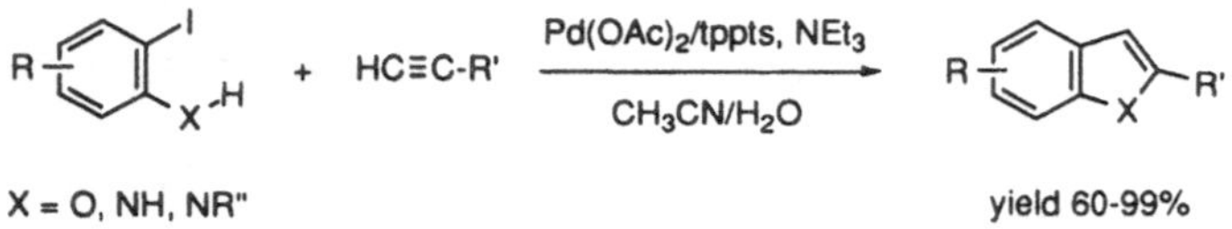

$$
R\underset{X}{\overset{I}{\text{--}}}{}^{\text{-H}}
+ \text{HC} \equiv \text{C-R'}
\xrightarrow[\text{CH}_3\text{CN/H}_2\text{O}]{\text{Pd(OAc)}_2/\text{tppts, NEt}_3}
R\text{--}\underset{X}{\boxed{}}\text{-R'}
\qquad (21)
$$

X = O, NH, NR"   yield 60-99%

A new cyclization and hydrofunctionalization of 1,6-enynes has also been reported recently [119]

5
Other Reactions

5.1
Oxidation

Wacker oxidation of olefins to ketones catalyzed by palladium complexes is a well-known process which has been applied to numerous olefins [120]. However, selective oxidation of C_8–C_{16} α-olefins remains a challenge. Recently, Mortreux et al. have developed a new catalytic system for the quantitative and selective oxidation of higher α-olefins in an aqueous medium [121–123]. For example, 1-decene was oxidized to 2-decanone in 98% yield using PdSO$_4$/H$_9$PV$_6$Mo$_6$O$_{40}$/CuSO$_4$ as the catalyst in the presence of *per*(2,6-di-O-methyl)-β-cyclodextrin, which probably played the role of a reverse phase transfer reagent [Eq. (22)].

$$
\text{C}_8\text{H}_{17}\text{-CH=CH}_2 + \text{O}_2
\xrightarrow[\text{H}_2\text{O, Me-}\beta\text{-CD, 80 °C}]{\text{PdSO}_4/\text{CuSO}_4/\text{H}_9\text{PV}_6\text{Mo}_6\text{O}_{40}}
\text{C}_8\text{H}_{17}\text{-CO-CH}_3
\qquad (22)
$$

yield 98%

A supported aqueous phase Wacker oxidation catalyst was also used by Davis et al. [124]; however, conversion of up to only 24 % was observed in the oxidation of 1-heptene.

5.2
Telomerization

Although the telomerization of dienes in a two-phase system has been intensively investigated with compounds containing active hydrogen such as alcohols, amines, phenols, acids, etc., the selective and productive telomerization of butadiene continues to be a challenge. It is only recently that primary octadienylamines have been obtained with selectivity up to 88 % in the telomerization of butadiene with ammonia using a two-phase toluene/water system and Pd(OAc)$_2$/tppts as the catalyst [Eq. (23)] [125].

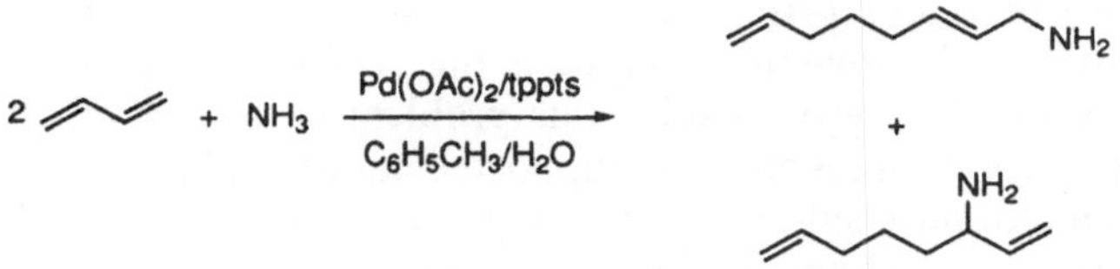

$$2 \diagup\!\!\!\diagdown + NH_3 \xrightarrow[C_6H_5CH_3/H_2O]{Pd(OAc)_2/tppts} \quad \diagdown\!\!\diagup\!\!\diagdown\!\!\diagup\!\!\diagdown\!\!\diagup NH_2 \quad + \quad NH_2 \diagup\!\!\diagdown\!\!\diagup\!\!\diagdown\!\!\diagdown \tag{23}$$

Telomerization of butadiene into 2,7-octadien-1-ol was also performed in neat water in the presence of carbon dioxide and certain trialkylamines in the presence of Pd(OAc)$_2$/tppts or Pd(OAc)$_2$/tppms, the structure of these amines having an important influence on the rate and the selectivity of the reaction [126].

5.3
Polymerization

Grubbs and co-workers reported the ring-opening metathesis polymerization (ROMP) of norbornene derivatives in water using Ru(H$_2$O)$_6$(ts)$_2$ as the catalyst [127, 128]. More recently, these authors have described the first example of a homogeneous living polymerization in water using a water-soluble ruthenium carbene [Eq. (24)] [129].

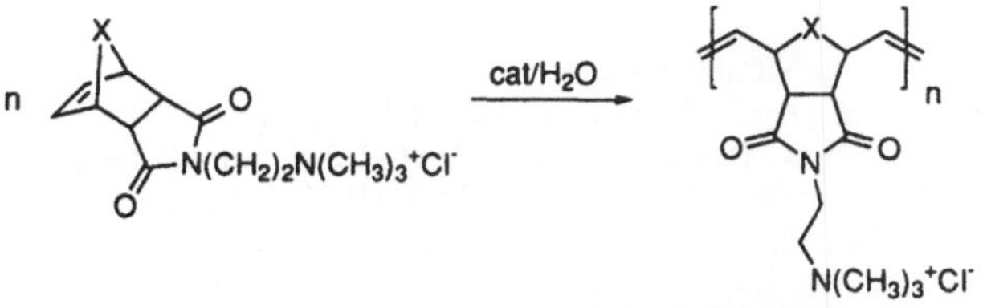

$$n \quad \xrightarrow{cat/H_2O} \quad \tag{24}$$

cat = [(C$_6$H$_{11}$)$_2$(CH$_2$CH$_2$NMe$_3$$^+Cl^-$)P]$_2$RuCl$_2$=CHPh

Neoglycopolymers for intercellular recognition studies were also prepared from carbohydrates bearing 7-oxo-norbornene derivatives by Kiessling et al. [130, 131] using the same methodology.

The formation of alternating copolymers of ethylene and carbon monoxide proceeded rapidly in water in the presence of $Pd(OTs)_2(CH_3CN)_2/dppp_{TS}$ and a Brönsted acid [132].

6
Perspectives

During the last decade there have been many new applications of water-soluble organometallic catalysts in the field of organic synthesis. Although the first aim of this new concept was the easy separation of the catalyst from the reaction products for its eventual recycling, new selectivities and sometimes higher activities were found using water as the reaction medium.

Although very high activities have been found in hydroformylation, there is a need for the synthesis of new tailored soluble phosphines exhibiting high activities for their use in other reactions such as hydrogenation or coupling reactions.

Another problem in aqueous catalysis is the very low solubility of organic substrates in water. One way to solve this problem is the use of a "promoter ligand"; another solution seems to be supported aqueous phase catalysis.

Although enantiopure sulfonated diphosphines have been successfully used in the hydrogenation of some prochiral substrates, very low enantioselectivities were obtained in other reactions. Thus elaboration of new chiral water-soluble catalysts is of utmost importance for the future. The use of micelles should also be extended to reactions other than asymmetric hydrogenation.

7
References

1. Kuntz EG (1975) FR Patent 2.314.910 Rhône-Poulenc Recherche
2. Joò F, Beck MT (1975) React Kin Catal Letters 2:357
3. Joò F, Tòth Z (1980) J Mol Catal 8:369
4. Kuntz EG (1987) CHEMTECH 17:570
5. Sinou D (1987) Bull Soc Chim Fr 480
6. Southern TG (1989) Polyhedron 8:407
7. Barton M, Atwood JD (1991) J Coord Chem 24:43-67
8. Kalck P, Monteil F (1992) Adv Organomet Chem 34:219
9. Herrmann WA, Kohlpaintner CW (1993) Angew Chem Int Ed Engl 32:1524
10. Li C-J, Chan TH (1997) Organic Reactions in aqueous media. Wiley-Interscience, New York, p 115
11. Joò F, Kathò A (1997) J Mol Catal A 116:3
12. Cornil B, Herrmann WA, Eckl RW (1997) J Mol Catal A 116:27
13. Joò F, Nàdasdi L, Bènyei AC, Darengsbourg DJ (1996) J Organomet Chem 512:45
14. Renaud E, Russell RB, Fortier S, Brown SJ, Baird MC (1991) J Organomet Chem 419:403
15. Chan WC, Lau CP, Cheng L, Leung YS (1994) J Organomet Chem 464:103
16. Fache E, Santini C, Senocq F, Basset JM (1992) J Mol Catal 72:337
17. Kolaric S, Sunjic V (1996) J Mol Catal A 110:189
18. Grosselin JM, Mercier C, Allmang G, Grass F (1991) Organometallics 10:2126
19. Hernandez M, Kalck P (1997) J Mol Catal A 116:131
20. Davis ME (1992) CHEMTECH 498
21. Fache E, Mercier C, Pagnier N, Despeyroux B, Panster P (1993) J Mol Catal 79:117
22. Laghmari M, Sinou D (1991) J Mol Catal 66:L15

23. Bakos J, Karaivanov R, Laghmari M, Sinou D (1994) Organometallics 13:2951
24. Joò F, Csiba P, Bènyei A (1993) J Chem Soc Chem Commun 1602
25. Bergbreiter DE, Liu Y-S (1997) Tetrahedron Lett 38:3703
26. Amrani Y, Lecomte L, Sinou D, Bakos J, Toth I, Heil B (1989) Organometallics 8:542
27. Laghmari M, Sinou D, Masdeu A, Claver C (1992) J Organomet Chem 438:213
28. Toth I, Hanson BE, Davis ME (1990) Cat Lett 5:183
29. Tòth I, Hanson BE (1990) Tetrahedron: Asymmetry 1:895
30. Toth I, Hanson BE (1990) Tetrahedron: Asymmetry 1:913
31. Tòth I, Hanson BE, Davies ME (1990) J Organomet Chem 396:363
32. Wan KT, Davis ME (1994) J Catal 148:1
33. Wan KT, Davis ME (1993) J Chem Soc Chem Commun 1262
34. Wan KT, Davis ME (1993) Tetrahedron: Asymmetry 4:2461
35. Wan KT, Davis ME (1994) Nature 370:449
36. Wan KT, Davis ME (1995) J Catal 152:25
37. Ding H, Hanson BE, Bakos J (1995) Angew Chem Int Ed Engl 34:1645
38. Kumar A, Oehme G, Roque JP, Schwarze M, Selke E (1994) Angew Chem Int Ed Engl 33:2197
39. Flach HN, Grassert Y, Oehme G (1994) Macromol Chem Phys 195:3289
40. Grassert I, Paetzolt E, Oehme G (1993) Tetrahedron 49:6605
41. Malmström T, Anderson C (1996) J Chem Soc Chem Commun 1135
42. Bakos J, Orosz A, Heil B, Laghmari M, Lhoste P, Sinou D (1991) J Chem Soc Chem Commun 1684
43. Lensink C, De Vries JG (1992) Tetrahedron: Asymmetry 3:235
44. Bènyei A, Joò F (1990) J Mol Catal 58:151
45. Darensbourg DJ, Joò F, Kannisto M, Kathò A, Reibenspies JH (1992) Organometallics 11:1990
46. Darensbourg DJ, Joò F, Kannisto M, Kathò A, Reibenspies JH, Daigle DJ (1994) Inorg Chem 33:200
47. Darensbourg DJ, Stafford NW, Joò F, Reibenspies JH (1995) J Organomet Chem 488:99
48. Sinou D, Safi M, Claver C, Masdeu A (1991) J Mol Catal 68:L9
49. Cornils B, Kuntz EG (1995) J Organomet Chem 502:177
50. Cornils B, Wiebus E (1995) CHEMTECH 33
51. Herrmann WA, Kohlpaintner CW, Bahrman H, Konkol W (1992) J Mol Catal 73:191
52. Herrmann WA, Kohlpaintner CW, Manetsberger RB, Bahrman H, Kottmann H (1995) J Mol Catal A 97:65
53. Cornils B, Wiebus E (1996) Recl Trav Chim Pays-Bas 115:211
54. Eckl RW, Priermeier T, Herrmann WA (1997) J Organomet Chem 532:243
55. Bartik T, Bunn BB, Bartik B, Hanson BE (1994) Inorg Chem 33:164
56. Russell MJH (1988) Platinum Met Rev 32:179
57. Fell B, Papadogianakis G (1991) J Mol Catal 66:143
58. Ding H, Hanson BE, Bartik T, Bartik B (1994) Organometallics 13:3761
59. Bartik T, Bartik B, Hanson BE (1994) J Mol Catal 88:43
60. Bartik T, Bartik B, Guo I, Hanson BE (1994) J Organomet Chem 480:15
61. Ding H, Hanson BE (1994) J Chem Soc Chem Commun 2747
62. Bartik T, Ding H, Bartik B, Hanson BE (1995) J Mol Catal A 98:117
63. Stössel P, Mayer HA, Auer F (1998) Eur J Inorg Chem 37
64. Chaudhari RV, Bhanage BM, Desplande RM, Delmas H (1995) Nature 373:501
65. Monflier E, Tilloy S, Fremy G, Castanet Y, Mortreux A (1995) Tetrahedron Lett 36:9481
66. Monflier E, Fremy G, Castanet Y, Mortreux A (1995) Angew Chem Int Ed Engl 34:2269
67. Arhancet JP, Davis ME, Merola JS, Hanson BE (1989) Nature 339:454
68. Arhancet JP, Davis ME, Merola JS, Hanson BE (1990) J Catal 121:327
69. Horvath IT (1990) Catal Lett 6:43
70. Arhancet JP, Davis ME, Hanson BE (1991) J Catal 129:100
71. Fremy E, Monflier E, Carpentier JF, Castanet Y, Mortreux A (1995) Angew Chem Int Ed Engl 34:1474

72. Fremy G, Monflier E, Carpentier JF, Castanet Y, Mortreux A (1996) J Catal 162:339
73. Guo I, Hanson BE, Tòth I, Davis ME (1991) J Organomet Chem 403:221
74. Guo I, Hanson BE, Toth I, Davis ME (1991) J Mol Catal 70:363
75. Tilloy S, Monflier E, Bertout F, Castanet Y, Mortreux A (1997) New J Chem 21:529
76. Monflier E, Tilloy S, Bertout F, Castanet Y, Mortreux A (1997) New J Chem 21:857
77. Okano T, Okabe N, Kiji J (1992) Bull Chem Soc Jpn 65:2589
78. Okano T, Hayashi T, Kiji J (1994) Bull Chem Soc Jpn 67:2339
79. Monteil F, Kalck P (1994) J Organomet Chem 482:45
80. Papadogianakis BE, Maat L, Sheldon RA (1994) J Chem Soc Chem Commun 2659
81. Safi M, Sinou D (1991) Tetrahedron Lett 32:2025
82. Blart E, Genêt JP, Safi M, Savignac M, Sinou D (1994) Tetrahedron 50:505
83. Sigismondi S, Sinou D (1997) J Mol Catal 116:289
84. Bergbreiter DE, Liu Y-S (1997) Tetrahedron Lett 38:7843
85. Uozumi Y, Danjo H, Hayashi T (1997) Tetrahedron Lett 38:3557
86. Schneider P, Quignard F, Choplin A, Sinou D (1996) New J Chem 20:545
87. Dos Santos S, Tong Y, Quignard F, Choplin A, Sinou D, Dutasta JP (1998) Organometallics 17:78
88. Tonks L, Anson MS, Hellgardt K, Mirza AR, Thompson DF, Williams JMJ (1997) Tetrahedron Lett 38:4319
89. Sigismondi S, Sinou D, Moreno-Mañas M, Pleixats R, Villaroya M (1994) Tetrahedron Lett 35:7085
90. Goux C, Sigismondi S, Sinou D, Moreno-Mañas M, Pleixats R, Villaroya M (1996) Tetrahedron 52:9521
91. Genêt JP, Blart E, Savignac M, Lemeune S, Paris JM (1993) Tetrahedron Lett 34:4189
92. Lemaire-Audoire S, Savignac M, Blart E, Pourcelot G, Genêt JP (1994) Tetrahedron Lett 35:8783
93. Genêt JP, Blart E, Savignac M, Lemeune S, Lemaire-Audoire S, Paris JM, Bernard JM (1994) Tetrahedron 50:497
94. Lemaire-Audoire S, Savignac M, Pourcelot G, Genêt JP, Bernard JM (1997) J Mol Cat 116:247
95. Sigismondi S, Sinou D (1996) J Chem Res S 46
96. Bumagin NA, More PG, Beletskaya IP (1989) J Organomet Chem 371:397
97. Sengupta S, Bhattacharya S (1993) J Chem Soc Perkin Trans 1 1943
98. Bumagin NA, Bykov VV, Sukhomlinova LI, Tolstaya TP, Beletskaya IP (1995) J Organomet Chem 486:259
99. Jeffery T (1994) Tetrahedron Lett 35:3051
100. Jeffery T (1996) Tetrahedron 52:10113
101. Casalnuovo AL, Calabrese JC (1990) J Am Chem Soc 112:4324
102. Genêt JP, Blart E, Savignac M (1992) Synlett 715
103. Kiji J, Okano T, Hasegawa T (1995) J Mol Catal A 97:73
104. Beller M, Krauter JGE, Zapf A (1997) Angew Chem Int Ed Engl 36:772
105. Lemaire-Audoire S, Savignac M, Dupuis C, Genêt JP (1996) Tetrahedron Lett 37:2003
106. Roshchin AI, Bumagin NA, Beletskaya IP (1995) Tetrahedron Lett 36:125
107. Rai R, Aubrecht MKB, Collum DB (1995) Tetrahedron Lett 36:3111
108. Bumagin NA, Bykov VV (1997) Tetrahedron 53:14437
109. Badone D, Baroni M, Cardamone R, Ielmini A, Guzzi U (1997) J Org Chem 62:7170
110. Genêt JP, Linquist A, Blart E, Mouries V, Savignac M, Vaultier M (1995) Tetrahedron Lett 36:1443
111. Wallow TI, Novak BM (1991) J Am Chem Soc 113:7411
112. Amatore C, Blart E, Genêt JP, Jutand A, Lemaire-Audoire S, Savignac M (1995) J Org Chem 60:6829
113. Kang S-K, Lee H-W, Jang S-B, Ho P-S (1996) J Chem Soc Chem Commun 835
114. Bumagin NA, Sukhomlinova LI, Luzikova EV, Tolstaya TP, Beletskaya I (1996) Tetrahedron Lett 37:897
115. Dibowski H, Schmidtchen FP (1995) Tetrahedron 51:2325

116. Hessler A, Stelzer O, Dibowski H, Worm K, Schmidtchen FP (1997) J Org Chem 62:2362
117. Dibowski H, Schmidtchen FP (1998) Tetrahedron Lett 39:525
118. Dibowski H, Schmidtchen FP (1998) Angew Chem Int Ed Engl 37:476
119. Galland JC, Savignac M, Genêt JP (1997) Tetrahedron Lett 38:8695
120. Parshall GW, Ittel SD (1992) Homogeneous catalysis. The applications and chemistry of catalysis by soluble transition metal complexes, 2nd edn. Wiley-Interscience, New York
121. Monflier E, Blouet E, Barbaux Y, Mortreux A (1994) Angew Chem Int Ed Engl 33:2100
122. Monflier E, Tilloy S, Fremy G, Barbaux Y, Mortreux A (1995) Tetrahedron Lett 36:387
123. Monflier E, Tilloy S, Blouet E, Barbaux Y, Mortreux A (1996) J Mol Catal A 109:27
124. Arhancet JP, Davis ME, Hanson BE (1991) Catal Lett 11:129
125. Prinz T, Keim W, Driessen-Hölscher D (1996) Angew Chem Int Ed Engl 35:1708
126. Monflier E, Bourdauducq P, Couturier JL, Kervennal J, Mortreux A (1995) J Mol Catal 97:29
127. Nguyen ST, Johnson LK, Grubbs RH, Ziller JN (1992) J Am Chem Soc 114:3974
128. Nguyen ST, Grubbs RH (1993) J Am Chem Soc 115:9858
129. Lynn DM, Mohr B, Grubbs RH (1998) J Am Chem Soc 120:1627
130. Mortell KH, Gingras M, Kiessling LL (1994) J Am Chem Soc 116:12053
131. Mortell KH, Weatherman RV, Kiessling LL (1996) J Am Chem Soc 118:2297
132. Verspui G, Papadogianakis G, Sheldon RA (1998) Chem Commun 401

Perfluorinated Solvents – a Novel Reaction Medium in Organic Chemistry

Bodo Betzemeier · Paul Knochel

Ludwig-Maximilians-Universität München, Institut für Organische Chemie,
Butenandtstraße 5–13, D-81377 München, Germany

Although perfluorinated hydrocarbons are well known, they have only recently found application in organic chemistry as a useful class of solvents. Their physical properties make them unique for their use as reaction media. Since perfluorocarbons are immiscible with many common organic solvents, they are suitable for the formation of biphasic systems. This overview describes the utilization of perfluorocarbons as reaction media for various kinds of reaction such as oxidation, bromination, etc. In addition, a novel biphase reaction system based on perfluorinated solvents as well as its application in organic synthesis is presented.

Keywords. Perfluorocarbons, Catalysis, Biphase System, Oxidation, C-C-Coupling, Phosphanes.

Topics in Current Chemistry, Vol. 206
© Springer-Verlag Berlin Heidelberg 1999

Abbreviations

COD   cyclooctadiene
conv.   conversion
dba   dibenzylidene acetone
FC-75   fluorocarbon: mainly perfluorobutyltetrahydrofuran
FC-77   fluorocarbon: mainly perfluoro-2-butyltetrahydrofuran
FG   functional group
solub.   solubility
TEMPO   2,2,6,6-tetramethylpiperidine-N-oxide

1
Introduction

Most organic reactions are carried out in a solvent which has several important roles. At the molecular level, it breaks the crystal lattice of solid reagents, interacts with gaseous reagents and often lowers considerably the transition state of many reactions. Because of the intermolecular interactions between a solvent and organic reagents, it may not only enhance the reaction rate but also change the product distribution [1]. From the macroscopic point of view, the solvent also removes the excess heat produced during a reaction or allows an uniform supply of calories to the reagents. All these advantages of using a solvent for carrying out a reaction are also valuable for large-scale reactions. However, in this case the separation of the product from the solvent at the completion of the reaction may be costly and tedious, especially if other by-products have been formed or if an expensive catalyst has to be separated and recovered as well. Since the solubility of reagents in a solvent S is strongly dependent on the temperature, the question arises if a solvent cannot be chosen in such a way that this solvent S solubilizes the reagents $R^1, R^2, \ldots, R^n$ at the reaction temperature but not these reagents or more importantly the products $P^1, P^2, \ldots, P^n$ at room temperature. This would allow a facile separation of the products from the reaction mixture simply by decantation. If the reaction is selective, i.e. only one product is formed, no special work-up conditions are required. Especially important are the variations of this concept when one regenerable reagent R^a or a catalyst C is selectively soluble in the solvent S at room temperature and reaction temperature whereas the starting materials $R^1, \ldots, R^n$ and products $P^1, \ldots, P^n$ are insoluble. This leads to a biphase system using two different solvents. S^1 solubilizes the organic reagents and products, whereas the second solvent S^2 solubilizes the regenerable reagent R^a or the catalyst C. Such a process is of industrial interest since, on the one hand, it allows a facile separation of the often costly catalyst and, on the other hand, a run of cyclic operations with the same catalyst solution are possible. Furthermore the product phase S^1 should not be contaminated with the catalyst C or the regenerable reagent R^a, which is often a complex problem in industry, Scheme (1).

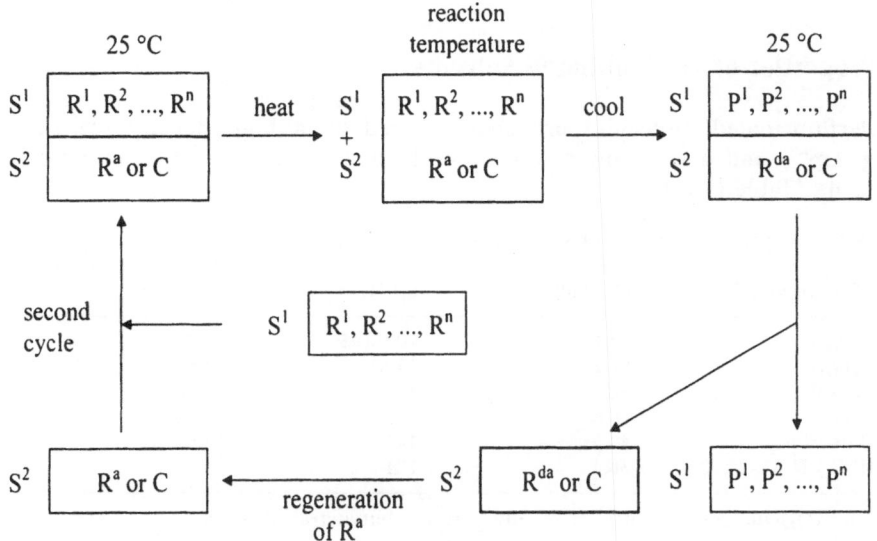

$R^1, R^2, ..., R^n$: Reagents

$P^1, P^2, ..., P^n$:   Products

R^a: Recoverable Reagent in activated form

R^{da}:   Recoverable Reagent in desactivated form

C: Catalyst

S^1, S^2: Solvents

Scheme 1. Representation of a reaction performed in a biphase system

To be applicable to a broad range of substrates, the natures of the two solvents S^1 and S^2 have to be very different in physical properties and chemical behaviour so that low miscibility is observed. The use of water as solvent S^2 (S^1 being an organic solvent) has found many applications including industrial processes [2]. However, the incompatibility of many organometallic reagents with water and the difficulty of removing metal catalysts from water for disposal led to the search for an alternative solvent. Recently, due to the initial studies of Vogt and Kaim [3] as well as the work of Horváth and Rábai [4], the concept of biphase systems has been extended to perfluorinated solvents [5]. These solvents display low chemical reactivity, low toxicity and low miscibility with many organic solvents. Furthermore, these solvents are inflammable and show a relatively low volatility. In this chapter, the basic physical and chemical properties will be indicated and the applications of perfluorinated solvents for the performance of organic reactions will be presented in detail as well as the reactions performed under conditions of a fluorous biphase system.

2
Properties of Perfluorinated Solvents

Perfluorinated molecules are characterized by a high density (ca. 1.7–1.9 $g \cdot cm^{-3}$) and a very low solubility both in water and in most organic solvents (Table 1) [6].

Table 1. Physical properties of some typical perfluorinated solvents

PF-Compound	O_2-solub.[a]	bp/°C	$\varrho/g \cdot cm^{-3}$
C_8F_{18}	52.1	100–105	1.74
$C_8F_{17}Br$	52.7	140.5	1.89
FC-75[b]	52.2	102	1.78
C_6F_6	48.8	81.5	1.60
F-decalin	40.3	142	1.95
$(F_9C_4)_3N$	38.4	174	1.90

[a] ml (O_2)/100 ml (solvent);   [b] mainly perfluorobutyltetrahydrofuran.

Fluorocarbons undergo very weak van-der-Waals interactions due to the low polarizability of the electrons of a C-F bond and low availability of the lone-pair of fluorine [7]. This implies that, in contrast to most organic solvents, the replacement of a molecule of perfluoroalkane by another molecule which has little interaction with his neighbours costs little energy. It is therefore expected that gases should have an exceptionally high solubility in perfluorinated compounds. This is in fact the case and especially molecular oxygen has an excellent solubility [6, 8] in most perfluorinated solvents (up to 57 ml of gaseous oxygen in 100 ml of C_7F_{14}) leading to numerous synthetic applications in reactions which involves oxygen as well as to their use as artificial blood substitutes [9]. Their chemical inertness, their low toxicity and easy recovery after the reaction makes these compounds an interesting class of solvents with potential uses for both industrial and academic chemists.

3
Organic Reactions in Perfluorinated Solvents

Although perfluorocarbons have been known for a very long time, only recently have applications such as a reaction medium in organic synthesis been found [10]. Due to their high density, their low miscibility with water and common organic solvents, the perfluorocarbons can be separated on completion of the reaction simply by filtration (when solid products are formed) or decantation. The latter is very useful when low-boiling compounds are involved since they can easily be distilled from the reaction mixture. In addition to the advantages in work-up of reactions perfluorocarbons show also other interesting solvent properties. They are extremely nonpolar, inert and are available in a wide range of boiling points (from 56 °C for C_6F_{14} to 220 °C for $(C_5F_{11})_3N$) which makes them also useful for carrying out reactions under vigorous conditions.

3.1
Transesterification Reactions

One of the first reactions carried out in perfluorinated solvents was the transesterification described by Zhu in 1993 [10]. He used the perfluorinated solvent FC-77 (mainly perfluoro-2-butyltetrahydrofuran) for an azeotropic distillation of the formed methanol or propanol in the transesterification of methyl- or propylesters 1 with different alcohols 2 using a Dean-Stark apparatus Eq. (1).

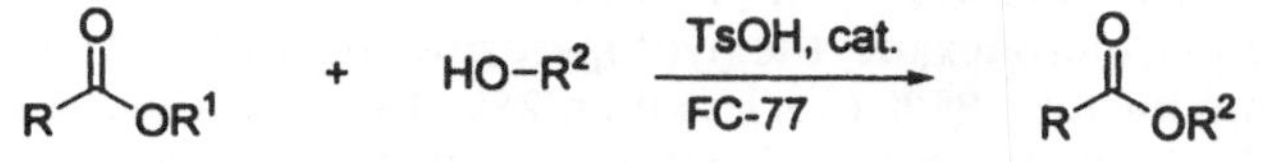

$$\text{1} \qquad \text{2} \qquad\qquad \text{3: 67- 92 \%} \tag{1}$$

$R^1 = Me, Pr \qquad R^2 = alkyl, benzyl$
FC-77: mainly perfluoro-2-butyltetrahydrofuran

After the completion of the reaction the product 3 was separated from the solvent simply by decantation. Purification by distillation resulted in various esters in 67–92% yield. Under these conditions also enamine formation as well as acetalization of ketones has been studied Eq. (2).

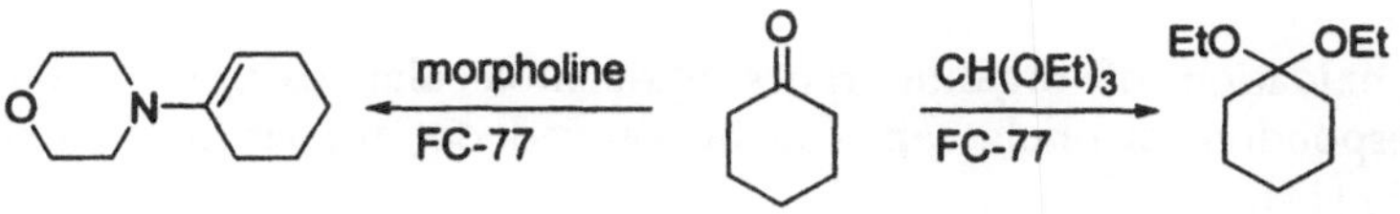

$$\tag{2}$$

FC-77: mainly perfluoro-2-butyltetrahydrofuran

3.2
Bromination of Olefins

Normally, the solvent of choice for bromination of alkenes is carbon tetrachloride which has the disadvantage of high toxicity and destruction of the ozone layer. Despite that, an international agreement for the production of reduced amounts of carbon tetrachloride exists and there is still a need for an alternative reaction medium. Savage et al. have performed the bromination reaction of functionalized olefins in perfluorohexanes [11]. With only one equivalent of bromine, the dibromide **4** was observed after 1 h in nearly quantitative yield, Eq. (3).

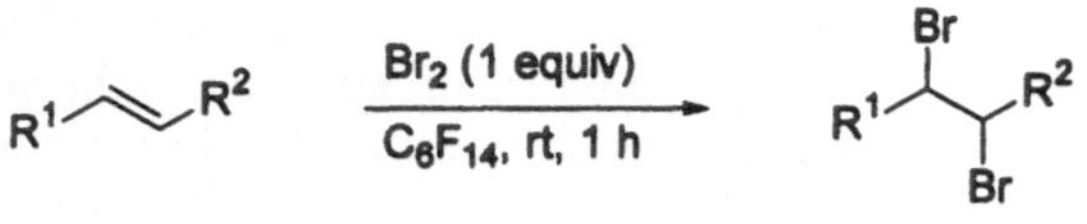

$$R^1 \diagdown R^2 \xrightarrow[\text{C}_6\text{F}_{14},\text{ rt, 1 h}]{\text{Br}_2\text{ (1 equiv)}} R^1 \diagdown R^2 \tag{3}$$

4

R^1: alkyl, alkenyl, aryl
R^2: ester, aryl, H

3.3
Oxidation of Organozinc Bromides to Hydroperoxides

In common organic solvents, the oxidation of organozinc halides with molecular oxygen produces only mixtures of hydroperoxides and the corresponding alcohols. The best results (80:20 ratio) were obtained using a very high dilution of the zinc organometallic in ether (3 mmol · L^{-1}) [12]. An improvement in the reaction conditions was found by using perfluorohexane as the reaction medium. After addition of a solution of organozinc bromide **5** in THF to oxygensaturated perfluorohexane, the hydroperoxides **6** can be obtained in good yields and in a purity > 98% (i.e. less than 2% of the corresponding alcohol). This method also tolerates functional groups such as esters, silyl ethers and halides Eq. (4) [13].

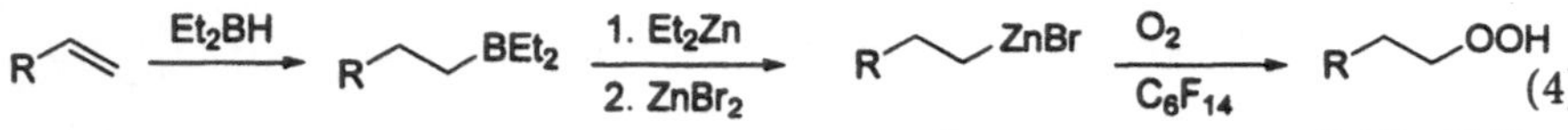

$$R \diagup\diagdown \xrightarrow{\text{Et}_2\text{BH}} R \diagdown\diagup \text{BEt}_2 \xrightarrow[\text{2. ZnBr}_2]{\text{1. Et}_2\text{Zn}} R \diagdown\diagup \text{ZnBr} \xrightarrow[\text{C}_6\text{F}_{14}]{\text{O}_2} R \diagdown\diagup \text{OOH} \tag{4}$$

R = alkyl, FG-alkyl   **5**   **6**: 58 - 85 %

3.4
Direct Oxidation of Organoboranes with Molecular Oxygen

The oxidation of organoboranes with molecular oxygen leading to the corresponding alcohols can also be carried out in perfluorinated solvents Eq. (5) [14].

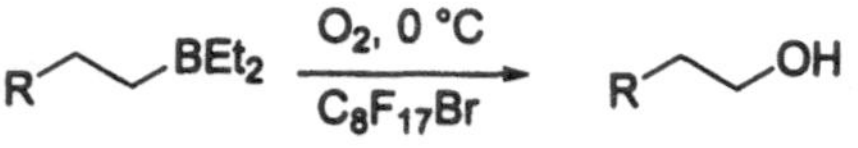

$$R \diagdown\diagup \text{BEt}_2 \xrightarrow[\text{C}_8\text{F}_{17}\text{Br}]{\text{O}_2, \text{ 0 °C}} R \diagdown\diagup \text{OH}$$

7   **8**: 71 - 91 %

R = alkyl, FG-alkyl   (5)

PivO(CH$_2$)$_6$OH   91 %   HO$_{\prime\prime\prime}$ ⟨Me⟩   75 %

Br(CH$_2$)$_6$OH   85 %

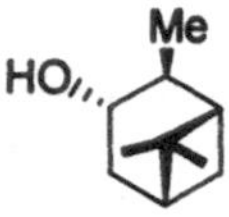

Normally, triorganoboranes do not react very readily with oxygen. Harsh reaction conditions are often required and only a partial transfer of an organic group attached to the boron is observed with often only moderate yields [15]. In bromoperfluorooctane, diethylorganoboranes of type 7 which can easily be obtained by hydroboration of alkenes with diethylborane are oxidized very cleanly to the corresponding alcohols 8 in good to excellent yields. Despite the diradical character of oxygen, the oxidation of boranes proceeds with retention of the stereochemistry of the secondary carbon center under this conditions. This can be explained by the high reactivity of the ethyl-boron bond towards oxygen [16]. Dioxygen inserts into the ethyl-boron bond of the borane 9 leading to the peroxide 10 via a radical mechanism, but the migration of the cyclohexyl group takes place with retention of configuration affording trans-2-phenyl-cyclohexanol 11 Eq. (6).

$$\underset{\textbf{9}}{\text{[cyclohexyl-Ph, BEt}_2]} \xrightarrow[\text{C}_8\text{F}_{17}\text{Br}]{\text{O}_2} \underset{\textbf{10}}{\text{[cyclohexyl-Ph, BEt-O-OEt]}} \xrightarrow{\text{H}_2\text{O}} \underset{\textbf{11}}{\text{[cyclohexyl-Ph, OH]}} \qquad (6)$$

4
Fluorous Biphase Catalysis

In the reactions described in the previous section the strategy using perfluorinated solvents was focused on the facile separation of the reaction products and the high solubility of gases in those media. During the last few years, another useful application has been developed: the specific solubilization of an organometallic catalyst in perfluorinated solvents. In this case, the separation of the catalyst from the reaction mixture is the main problem since most of transition-metal-catalyzed organic reactions require relatively large quantities of often costly transition metal catalysts [2]. Furthermore, the removal of residual traces of these catalysts from the reaction product is often time consuming and expensive. This has prevented the application of this methodology to large scale synthesis. To overcome this problem, a specific solubilization of the catalyst is necessary which allows its facile separation from the reaction mixture. This could be achieved either by immobilization of the catalyst on a solid phase (e.g. resin) or, as described in the introduction, by liquid-liquid biphase catalysis [2–5].

4.1
Hydroformylation of Terminal Olefins

The initial studies of Horváth and Rábai were concentrated on the synthesis of the fluorinated trialkylphosphane 12 which is a suitable ligand for many transition metals [4, 17]. This phosphane was prepared by a hydrophosphination of the corresponding fluorinated alkene 13 Eq. (7).

$$F_{13}C_6 \diagup\!\!\!\diagup \quad \xrightarrow[\text{AIBN, 100 °C}]{\text{PH}_3\ (0.25\ \text{equiv})} \quad \left(F_{13}C_6 \diagup\!\!\!\!\diagdown \right)_3\!\!P \qquad (7)$$

13   **12**: 37 %

The ethylene spacer is necessary as a shield from the strong electron-withdrawing effect of the perfluoroalkyl chain, which would decrease the donor properties of the phosphane. This phosphorous-ligand is extremely soluble in perfluorinated solvents such as perfluoromethylcyclohexane ($CF_3C_6F_{11}$) and only trace amounts of it can be extracted with organic solvents. The in situ prepared rhodium(I) complex **14** is a useful catalyst for the hydroformylation of terminal alkenes under FBS-conditions. The aldehydes **16** and **17** were formed in 85 % yield by hydroformylation of 1-octene (**15**) with an linear to iso ratio of nearly 3:1 Eq. (8).

$$Rh(CO)_2(acac) + \mathbf{12} \quad \xrightarrow{\text{CO / H}_2} \quad HRh(CO)[P(C_2H_4\text{-}C_6F_{13})_3]_3$$

14

(8)

$$C_6H_{13} \diagup\!\!\!\diagup \quad \xrightarrow[\substack{\text{CO / H}_2,\ \text{10 bar, 24 h} \\ CF_3C_6F_{11}\ /\ \text{toluene, 100 °C}}]{\mathbf{14}\ (5\ \text{mol \%})} \quad C_6H_{13}\diagup\!\!\!\!\diagdown\!\!\diagup CHO \ + \ C_6H_{13}\diagdown\!\!\!\diagup\!\!\!\underset{Me}{\overset{CHO}{|}}$$

15   **16**   **17**

85 %; 3:1

The reaction was carried out in a solvent system of $CF_3C_6F_{11}$ and toluene under an atmosphere of CO/H_2 (10 bar) at 100 °C. Although relatively harsh reaction conditions were used, no leaching of the catalyst was observed.

4.2
Hydrogenation of Olefins

The perfluorinated phosphane **12** has also found application in the hydrogenation of various alkenes [18]. Therefore, a Wilkinson-type rhodium complex **18** has been prepared by treatment of RhCl(COD)$_2$ with the phosphane **12** in $CF_3C_6F_{11}$ Eq. (9).

$$RhCl(COD)_2 + \left(F_{13}C_6 \diagup\!\!\!\!\diagdown \right)_3\!\!P \quad \xrightarrow{\quad\quad} \quad RhCl[P(C_2H_4\text{-}C_6F_{13})_3]_3 \quad (9)$$

12   **18**

This complex is selectively soluble in perfluorinated solvents and catalyzes the hydrogenation of e.g. cyclododecene under FBS-conditions at 45 °C (H_2, 1 atm) affording cyclododecane (**19**) in 94 % yield Eq. (10).

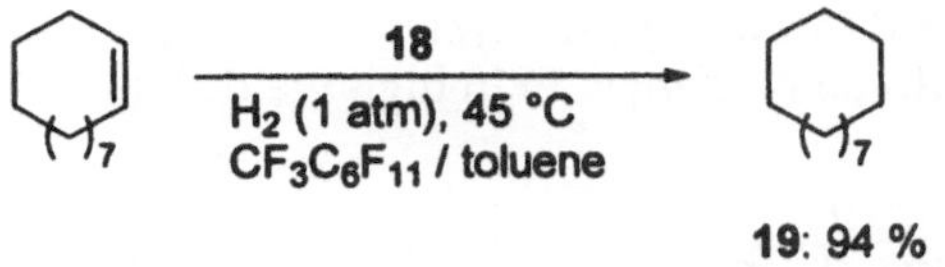

$$(10)$$

19: 94 %

4.3
Hydroboration of Olefins

Another application of the Wilkinson-type catalyst **18** is the rhodium catalyzed hydroboration of olefins [19]. Various alkenes **20** (internal, terminal, styrenes, etc.) have been successfully hydroborated with catecholborane (**21**) providing the corresponding boronic esters **22** in nearly quantitative yield. Oxidative work-up ($NaOH/H_2O_2$) led to the corresponding alcohols **23** in 76–90 % yield Eq. (11).

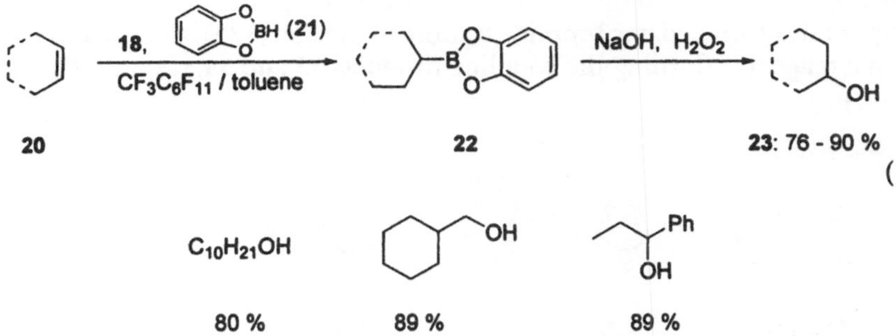

20 **22** **23**: 76 - 90 %

$$(11)$$

$C_{10}H_{21}OH$

80 % 89 % 89 %

The hydroboration of non-aromatic alkenes is regio- and diastereoselective. Only the reaction of styrene derivatives such as **24** results in mixtures of regio-isomers (**25** and **26**) Eq. (12).

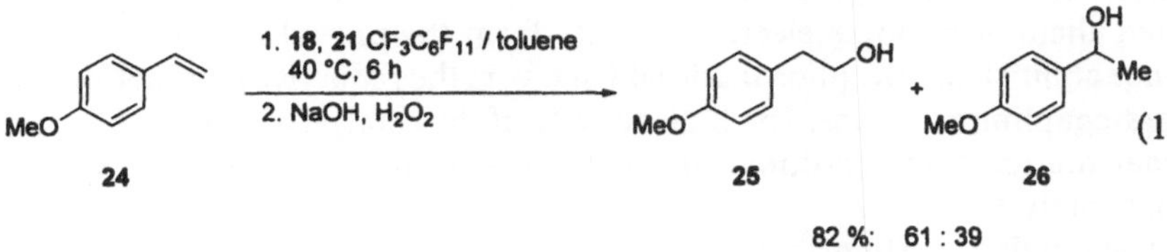

24 **25** **26**

$$(12)$$

82 %: 61 : 39

4.4
Palladium-Catalyzed Cross-Coupling Reactions

Carbon-carbon bond formation is one of the most important reaction in organic synthesis [20]. However, most of the reactions require relatively large amounts of a costly transition metal catalyst and its removal from the reaction mixture is difficult. With the perfluoroalkyl substituted triarylphosphane **27** palladium catalyzed cross-coupling reactions of aryl iodides **28** with arylzinc bromides **29** are possible in a fluorous biphase system (bromoperfluorooctane/toluene) affording polyfunctional biphenyls **30** in high yields Eq. (13) [21].

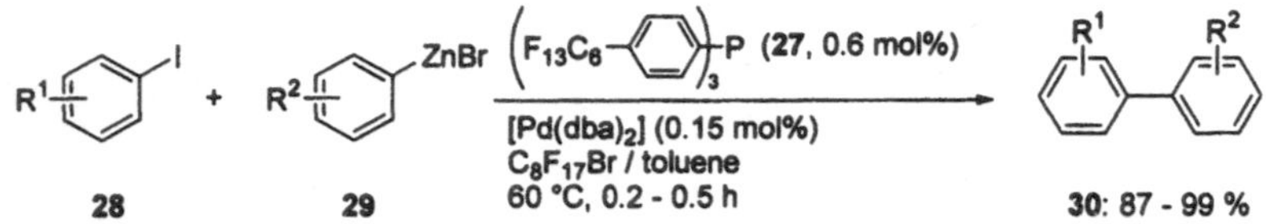

$$
R^1 \text{—} \langle \text{—} \rangle \text{—} I \;+\; R^2 \text{—} \langle \text{—} \rangle \text{—} ZnBr \quad
\xrightarrow[\substack{[Pd(dba)_2]\,(0.15\,mol\%)\\ C_8F_{17}Br\,/\,toluene\\ 60\,°C,\,0.2\text{-}0.5\,h}]{\left(F_{13}C_6\text{—}\langle\text{—}\rangle\right)_3 P\ (27,\,0.6\,mol\%)}\quad
R^1 \text{—}\langle\text{—}\rangle\text{—}\langle\text{—}\rangle\text{—} R^2 \tag{13}
$$

28   **29**   **30**: 87 - 99 %

The use of the triarylphosphane **27** is essential for the success of this reaction since with the trialkylphosphane **12** introduced by Horváth only a low reactivity was observed. Interestingly, this reaction shows a high selectivity for aryl iodides. The reaction between $(3\text{-}CF_3)C_6H_4ZnBr$ and 4-bromo-1-iodobenzene provides the corresponding biphenyl derivative **31** in 92% yield. No substitution of the bromine atom was observed. Functional groups such as esters, silyl ethers, chlorides, nitro- and methoxy groups are tolerated as well as heterocyclic zinc compounds (2-thienylzinc bromide leading to the biaryl **32**), alkenyl and benzylic zinc reagents affording the coupling products **33** and **34**, respectively [Scheme (2)].

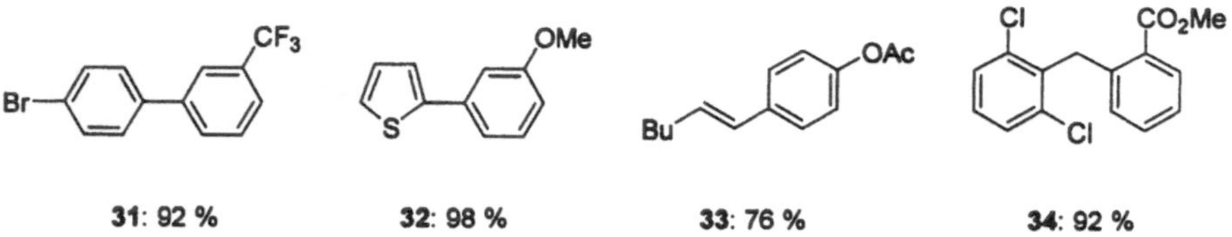

31: 92 %   **32**: 98 %   **33**: 76 %   **34**: 92 %

Scheme 2

With this phosphane, the cross-coupling reaction can be repeated several times without significant decrease of yield. Interestingly, the palladium catalyst formed with perfluoroalkyl substituted triarylphosphane **27** shows a higher activity than $Pd(PPh_3)_4$ which may due to the electron-withdrawing perfluorinated chain. It removes electron density from the phenyl ring and lowers the donor ability from the phosphane which favors the reductive elimination step in cross-coupling reactions. The high activity of the catalyst and his good stability under the reaction conditions allows the use of only 0.15 mol% of the palladium catalyst.

The required perfluorinated phospane ligand can easily be prepared in three steps starting from 4-iodoaniline. The fluorinated chain was introduced via an Ullmann-type reaction with $F_{13}C_6I$ in presence of copper-bronze in DMSO at 120°C for 1 h leading to the perfluoroalkyl substituted aniline **35** in 86% yield [22]. The functional group transformation from NH_2 to Br was performed with a Sandmeyer-reaction. Bromine-lithium exchange with n-BuLi (–78°C, 12 h) and reaction with $PhOPCl_2$ led to the phosphane **27** in 78% yield Eq. (14) [23].

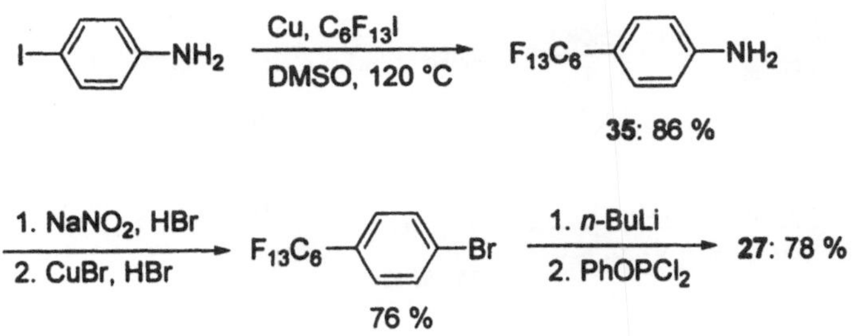

$$35: 86 \%$$

(14)

4.5
Palladium-Catalyzed Allylic Substitution

A similar triarylphosphane was introduced by Leitner for the performance of transition metal catalyzed reactions in supercritical CO_2 [24]. Recently, this phosphane was used for palladium catalyzed substitutions of allylic substrates in perfluorinated solvents under fluorous biphase conditions [25]. Therefore, the reaction of cinnamyl methylcarbonate (36) with several nucleophiles (Nu-H) in presence of a catalyst prepared in situ from $Pd_2(dba)_3$ and the fluorinated phosphane 37 (1:3 ratio) has been investigated. With 5 mol% of the catalyst in THF/$CF_3C_6F_{13}$ as solvent system, the substitution product 38 was formed at 50 °C after only 15 min with quantitative conversion and in moderate to good yield Eq. (15).

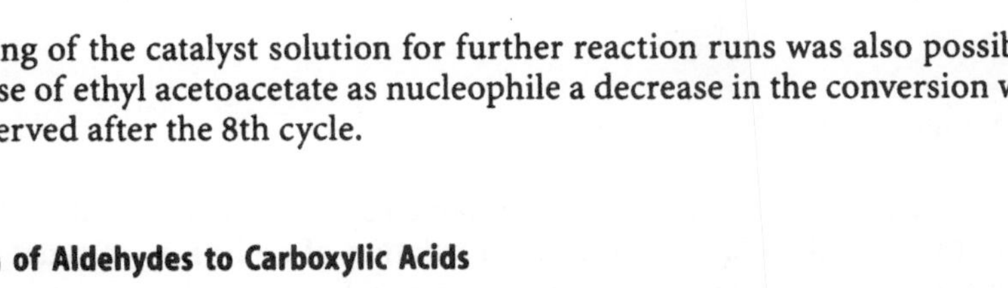

(15)

A recycling of the catalyst solution for further reaction runs was also possible. In the case of ethyl acetoacetate as nucleophile a decrease in the conversion was only observed after the 8th cycle.

4.6
Oxidation of Aldehydes to Carboxylic Acids

It has been shown that perfluorinated β-diketonates are excellent catalysts for oxidation reactions in perfluorinated solvents [26]. The nickel catalyst $Ni(39)_2$, prepared by reaction of the β-diketone 39 [27] with $NiCl_2$, catalyzes the oxidation of various (functionalized) aromatic and aliphatic aldehydes to the corresponding carboxylic acids in 76 to 87% yield in a solvent system of perfluorodecaline and toluene at 64 °C under O_2-atmosphere Eq. (16).

Without the catalyst, no or, in some cases, very slow oxidation was observed. The catalyst solution can be reused several times for further reaction runs. In the

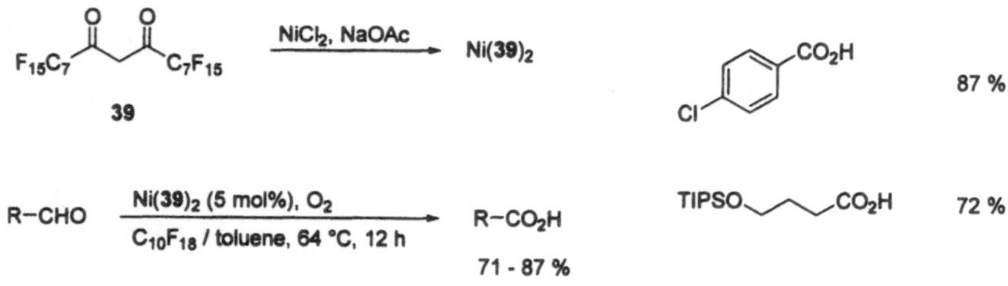

$$\text{(16)}$$

case of 4-chlorobenzaldehyde, the yield decreases only slightly from 87% (first run) to 70% after six runs.

4.7
Oxidation of Sulfides to Sulfoxides and Sulfones

With the same catalyst, the oxidation of sulfides to the corresponding sulfoxides and sulfones with molecular oxygen is possible as well [26]. However, an addition of 2-methylpropanal is necessary for the success of the reaction in order to generate a peracid as oxidation agent in situ [28]. With 1.6 equiv of this aldehyde the sulfoxides **40** are formed in 60–91% yield whereas a large excess (5 equiv) solely led to the corresponding sulfones **41** in 83–87% yield Eq. (17).

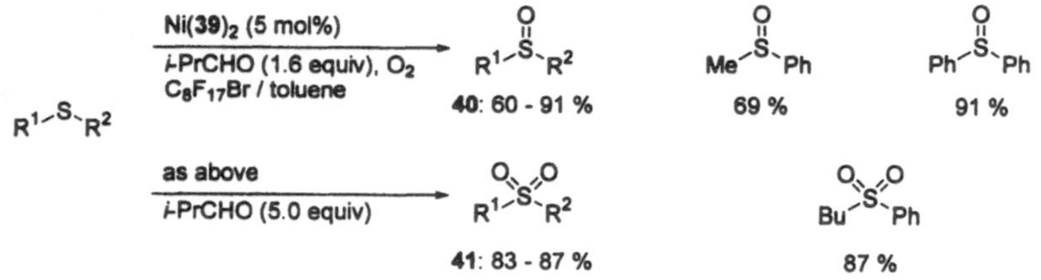

$$\text{(17)}$$

4.8
Ruthenium-Catalyzed Epoxidation of Olefins

The ruthenium catalyst $K[Ru(39)_3]$, obtained by the reaction of diketone **39** with in situ prepared $RuCl_2$, catalyzes the oxidation of olefins in presence of 2-methylpropanal and oxygen gas to the corresponding epoxides in good yields [26, 29]. Interestingly, this oxidation is selective for di- and trisubstituted double bonds whereas terminal olefins remain unreacted under this condition Eq. (18).

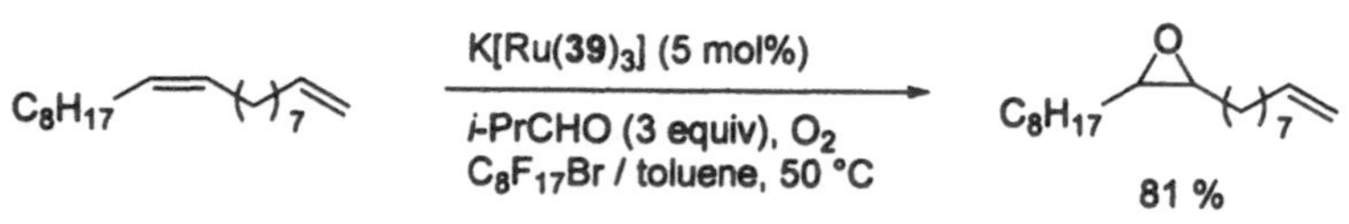

$$\text{(18)}$$

4.9
Oxidation of Terminal Olefins to Methyl Ketones

In presence of the palladium catalyst Pd(39)$_2$ the oxidation of various terminal alkenes 42 to the corresponding methyl ketones 43 can be carried out in a biphasic solvent system containing benzene and bromoperfluorooctane as well [30]. In this case *tert*-butylhydroperoxide is used as the oxidizing agent [31]. Styrene derivatives can be oxidized in good to excellent yields whereas alkene derivatives require a longer reaction time and an increased amount of *t*-BuOOH affording methylketones in moderate to good yields Eq. (19).

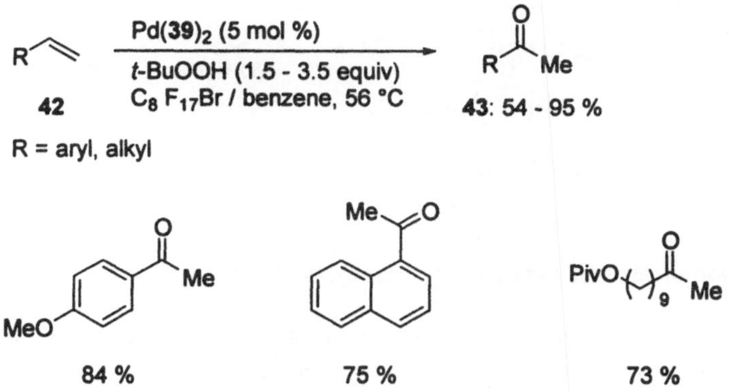

(19)

Not only terminal olefins have been oxidized but also stilbene and ethyl cinnamate have been converted to benzyl phenylketone 44 and the corresponding β-ketoester 45, respectively Eq. (20).

With the catalyst solution, the reaction of 4-methoxystyrene to 4-methoxyacetophenone has been repeated eight times without a significant decrease in yield (78 to 72%).

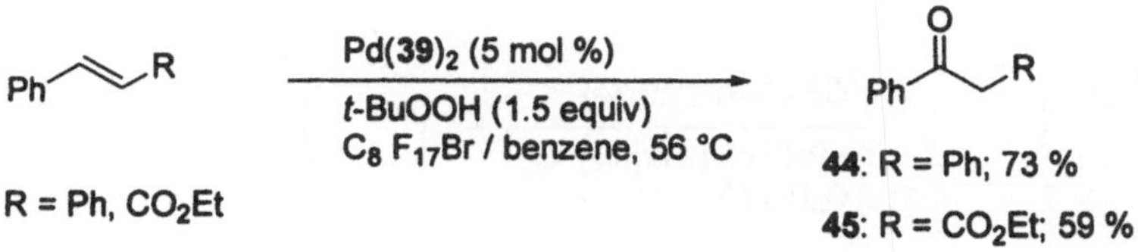

(20)

4.10
Functionalization of Alkanes and Alkenes

The insertion of oxygen into a C-H-bond via allylic oxidation or alkane oxidation is a difficult reaction and not many efficient methods are known in the literature [32]. Recently, a fluorinated manganese catalyst solubilized in a perfluorinated solvent has found application in the oxidation of cyclohexene to cyclohexenol and cyclohexenone [33, 34]. As catalyst precursor the manganese salt of perfluorooctylpropionic acid (**46**) complexed by the fluorinated cyclic triamine **47** for a better solubilization in perfluorinated solvents has been used Eq. (21).

Cyclohexene, which was used as solvent as well, was only oxidized after addition of a catalytic amount of *t*-BuOOH in an oxygen atmosphere. After 3 h, a conversion of 650 % of cyclohexene (relatively to *t*-BuOOH) to cyclohexenol and cyclohexenone (1:3 ratio) has been taken place. However, no other substrate has given similar results. Neither with toluene nor with cyclohexane yields above 100% (relatively to *t*-BuOOH) have been observed.

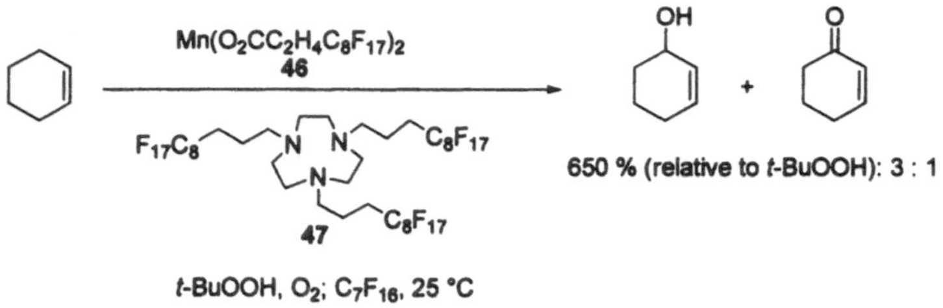

$$650 \text{ \% (relative to } t\text{-BuOOH): 3 : 1}$$

t-BuOOH, O_2; C_7F_{16}, 25 °C

(21)

4.11
Epoxidation of Alkenes Catalyzed by Porphyrin Cobalt Complexes

Metallaporphyrins proved to be efficient catalysts for oxidation reactions in organic chemistry [35]. Pozzi's group was the first to develop a synthesis of perfluoralkyl substituted tetraarylporphyrins such as **48** [36, 37]. Since one perfluoroalkyl chain on each aryl group was not sufficient to provide a good solubility of the ligand in perfluorinated solvents, the introduction of two "ponytails" was performed [37]. The cobalt complex Co-**48**, prepared from the porphyrin **48** and $Co(OAc)_2$, is a useful catalyst for the epoxidation of unfunctionalized olefins by dioxygen and 2-methylpropanal under fluorous biphasic conditions. With only 0.1 mol% of this catalyst, the oxidation of cyclooctene is completed after 3 h at rt using a solvent system of perfluorohexane and acetonitrile Eq. (22).

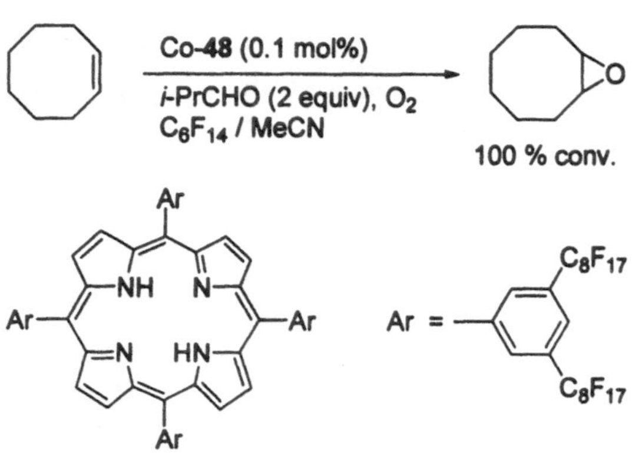

100 % conv.

(22)

Interestingly, the presence of two distinct phases did not hamper the oxidation of the alkene dissolved in the organic solvent. However, a high stirring rate (1300 min^{-1}) is required. Furthermore, the catalyst solution could be reused for a further reaction run leading to the same result. Terminal olefins such as 1-dodecene require a longer reaction time; after 14 h only 48% conversion was observed.

4.12
Epoxidation of Alkenes Catalyzed by Chiral Salen Manganese Complexes

From the work of Jacobsen and Katsuki, it is known that chiral manganese salen complexes are excellent catalysts for the asymmetric epoxidation of alkenes [38]. Substituted alkenes as well as terminal olefins and styrene derivatives are epoxidized in high yield and enantiomeric excess under homogeneous reaction conditions. Very recently, the first chiral salen complexes which are selectively soluble in perfluorinated solvents have been synthesized and their application in asymmetric synthesis has been investigated [39, 40].

The catalyst Mn-49 is obtained in six steps from the protected 3,5-diiodosalicylic acid (50) involving an Ullmann-coupling with perfluorooctyl iodide in the presence of copper to introduce the perfluoroalkyl chain. After transformation of the ester function into an aldehyde and demethylation the perfluoroalkylated salicylaldehyde 51 was treated with the chiral diamine 52 leading to the C_2-symmetric salen ligand 49 in 75% yield, which was converted into the corresponding manganese complex Mn-49 by refluxing with Mn(OAc)$_2$ Eq. (23).

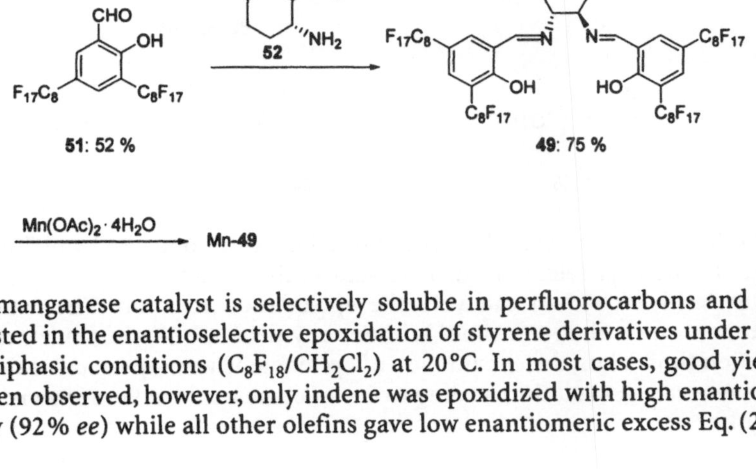

(23)

The manganese catalyst is selectively soluble in perfluorocarbons and has been tested in the enantioselective epoxidation of styrene derivatives under fluorous biphasic conditions (C$_8$F$_{18}$/CH$_2$Cl$_2$) at 20°C. In most cases, good yields have been observed, however, only indene was epoxidized with high enantioselectivity (92% ee) while all other olefins gave low enantiomeric excess Eq. (24).

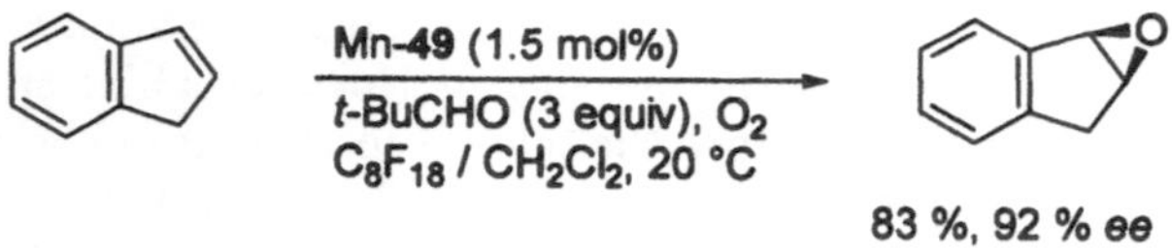

$$83\ \%,\ 92\ \%\ ee \tag{24}$$

4.13
Selenium Catalyzed Epoxidation of Olefins

Sharpless has shown that phenylselenic acid catalyzes the epoxidation of olefins with hydrogen peroxide or t-BuOOH [41]. However, the toxicity of selenium compounds precludes many applications of this catalytic epoxidation. To avoid contamination of the reaction products with selenium compounds polystyrene-bound phenylselenic acid has been used [42]. An alternative to the solid phase chemistry is the immobilisation of the selenium catalyst in the fluorous phase [43]. The arylbutylselenide 53 readily prepared from the corresponding aryl bromide 54 and lithium butylselenide is an excellent catalyst for the performance of epoxidation reactions in a fluorous biphase system. With 5 mol % of the catalyst 53, various polysubstituted and functionalized olefins 55 are epoxidized in a biphasic system of bromoperfluorooctane and benzene using hydrogen peroxide (60 % in water, 1.5–2.0 equiv) leading to epoxides 56 in good to excellent yields Eq. (25).

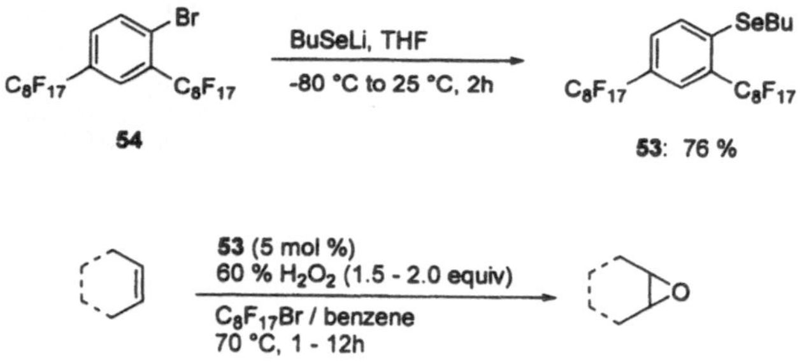

$$\tag{25}$$

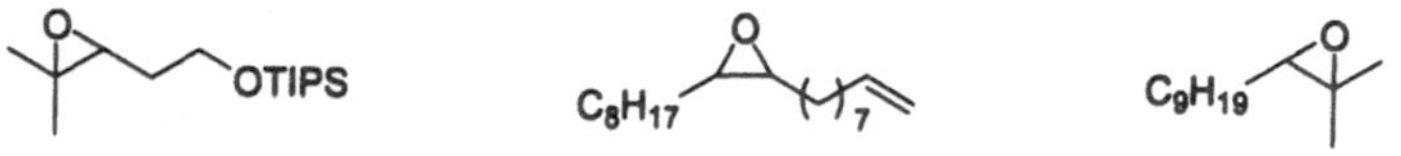

In this reaction the aryl butylselenide 53 is oxidized in situ by hydrogen peroxide to the aryl selenic acid. This acid itself is oxidized to the corresponding peracid wich then catalyzes the epoxidation reaction. The catalyst can be reused more than ten times without any decrease in yield nor increase in reaction time. Thus, the epoxidation of cyclooctene was repeated ten times with the same catalyst solution leading to cyclooctene oxide in 90–93 % yield within only 1 h.

5
Conclusion

Since the pioneer studies of Vogt and Kaim and the first publication of Horváth and Rábai in the field of fluorous biphase catalysis, this method has found many applications in organic chemistry, especially in oxidation reactions. Epoxidations of alkenes, even in an enantioselective manner, have been carried out in perfluorinated solvents as well as sulfide oxidation, hydroformylation, C-C-coupling reactions, etc. The fluorous biphase system combines the advantages of both homogeneous and heterogeneous catalysis: on the one hand, the homogeneous reaction conditions due to the miscibility of perfluorinated solvent with organic solvents at only higher temperature and, on the other hand, the facile separation of the mostly expensive catalyst which allows its reuse for further reaction runs. Furthermore, this method avoids the contamination of the product phase with traces of metals. In contrast to aqueous biphase catalysis, this system is also compatible with moisture sensitive compounds such as organometallics. These advantages also make this method interesting for industrial applications although perfluorinated solvents are still relatively expensive.

Acknowledgments. We thank the Deutsche Forschungsgemeinschaft (Schwerpunktprogramm "Sauerstofftransfer" and Leibniz-Programm) and the Fonds der Chemischen Industrie for generous financial support. We thank Elf-Atochem (France), Witco AG, BASF AG, Bayer AG, Chemetall GmbH, and Sipsy S.A. (France) for the generous gifts of chemicals. We also acknowledge the contributions of our coworkers I. Klement, H. Lütjens and F. Lhermitte.

6
References

1. Reichardt C (1978) Solvent Effects in Organic Chemistry. VCH, Weinheim
2. a) Cornils B, Herrmann W (eds) (1996) Applied Homogeneous Catalysis with Organometallic Compounds. VCH, Weinheim, b) Cornils B, Herrmann W (eds) (1998) Aqueous Phase Organometallic Catalysis. Wiley-VCH, Weinheim, c) Cornils B (1995) Angew Chem Int Ed Engl 34:1574
3. Vogt M (1991) PhD thesis, University of Aachen
4. Horváth IT, Rábai J (1994) Science 266:72
5. a) Cornils B (1997) Angew Chem Int Ed Engl 36:2057, b) Horváth IT (1998) Acc Chem Res 31:641
6. Wesseler EP, Iltis R, Clarc LC (1977) J Fluorine Chem 9:137
7. a) Riess JG (1995) New J Chem 19:893, b) Sadtler VW, Krafft MP, Riess JG (1996) Angew Chem Int Ed Engl 35:1976
8. Sharts CM, Reese HR (1978) J Fluorine Chem 11:637
9. a) Riess JG, Le Blanc M (1978) Angew Chem Int Ed Engl 17:621, b) Riess JG, Le Blanc M (1982) Pure Appl Chem 54:2383
10. Zhu DW (1993) Synthesis 953
11. Pereira SM, Savage P, Simpson GW (1995) Synth Commun 25:1023
12. Abraham MH (1960) J Chem Soc 4131
13. Klement I, Knochel P (1995) Synlett 1113
14. Klement I, Knochel P (1996) Synlett 1005
15. a) Brown HC, Dodson VH (1957) J Am Chem Soc 79:2302, b) Abraham MH, Davies AG (1959) J Chem Soc 429, c) Brown HC, Negishi E (1971) J Am Chem Soc 93:6682, d) Brown

HC, Midland MM, Kabalka GW (1971) J Am Chem Soc 93:1024, e) Tanahashi Y, Lhomme J, Ourisson G (1972) Tetrahedron 28:2655, f) Brown HC, Midland MM, Kabalka GW (1986) Tetrahedron 42:5523, g) Brown HC, Midland MM (1987) Tetrahedron 43:4059

16. a) Nozaki K, Oshima K, Utimoto K (1988) Tetrahedron Lett 29:6125, b) Barton DHR, Jang DO, Jaszberenyi JC (1990) Tetrahedron Lett 31, 4681, c) Köster R, Bellut H, Fenze W (1974) Liebigs Ann. 54

17. a) Horvárth IT, Kiess G, Cook RK, Bond JE, Stevens PA, Rábai J, Mozeleski E (1998) J Am Chem Soc 120:3133, b) Alvey LJ, Rutherford D, Juliette JJJ, Gladysz JA (1998) J Org Chem 63:6302

18. Rutherford D, Juliette JJJ, Rocaboy C, Horváth IT, Gladysz JA (1998) Catalysis Today 42:381

19. Juliette JJJ, Horváth IT, Gladysz JA (1997) Angew Chem int Ed Engl 36:1610

20. a) Stille JC (1986) Angew Chem Int Ed Engl 25:508, b) Miyaura N, Suzuki A (1995) Chem Rev 95:2457, c) Farina V, Krishnan B (1991) J Am Chem Soc 113:9585, d) Farina V, Kapadia S, Krishnan B, Wang C, Liebeskind LS (1994) J Org Chem 59:5905, e) Klement I, Rottländer M, Tucker CE, Majid TN, Knochel P (1996) Tetrahedron 52:7201

21. Betzemeier B, Knochel P (1997) Angew Chem Int Ed Engl 36:2643

22. a) Yoshino N, Kitamura M, Seto T, Shibata Y, Abe M, Ogino K (1992) Bull Chem Soc Jpn 65:2141, b) McLoughlin VCR, Thrower J (1969) Tetrahedron 25:5921

23. a) Betzemeier B, Knochel P unpublished results b) Bhattacharyya P, Gudmunsen D, Hope E, Kemmitt RDW, Paige DR, Stuart AM (1997) J Chem Soc: Perkin Trans 1 3609

24. Kainz S, Koch D, Baumann W, Leitner W (1997) Angew Chem Int Ed Engl 36:1628

25. Kling R, Sinou D, Pozzi G, Choplin A, Quignard F, Busch S, Kainz S, Koch D, Leitner W (1998) Tetrahedron Lett 39:9439

26. Klement I, Lütjens H, Knochel P (1997) Angew Chem Int Ed Engl 36:1454

27. a) Massyn C, Pastor R, Cambon A (1974) Bull Chem Soc Fr 5:975, b) Pedler AE, Smith RC, Tatlow JC (1971) J Fluorine Chem 1:433

28. a) Yamada T, Takai T, Rohde O, Mukaiyama T (1991) Chem Lett 1, b) Yamada T, Takai T, Rohde O, Mukaiyama T (1991) Chem Lett 5

29. Endo A, Kajitani M, Mukaida M, Shimizu K, Sato GP (1988) Inorg Chem Acta 150:25

30. Betzemeier B, Lhermitte F, Knochel P (1998) Tetrahedron Lett 39:6667

31. a) Tsuji J (1984) Synthesis 369, b) Tsuji J, Nagashima H, Hori K (1980) Chem Lett 257, c) Roussel M Mimoun H (1980) J Org Chem 45:5387

32. a) Schlingloff G, Bolm C (1998) Other Catalyzed Allylic Oxidation. In: Beller M, Bolm C (eds) Transition Metals For Organic Synthesis. Wiley-VCH, Weinheim, p 193, b) Grennberg H, Bäckvall JE (1998) Palladium Catalyzed Allylic Oxidation. In: Beller M, Bolm C (eds) Transition Metals For Organic Synthesis. Wiley-VCH, Weinheim, p 200

33. Vincent JM, Rabion A, Yachandra VK, Fish RH (1997) Angew Chem Int Ed Engl 36:2346

34. Pozzi G, Cavazzini M, Quici S (1997) Tetrahedron Lett 38:7605

35. Montanari F, Casella L (eds) (1994) Metalloporphyrins Catalyzed Oxidations, Kluver, Dordrecht

36. Pozzi G, Banfi S, Manfredi A, Montanari F, Quici S (1996) Tetrahedron 52:11879

37. Pozzi G, Montanari F, Quici S (1997) J Chem Soc Chem Commun 69

38. a) Jacobsen EN, (1993) Asymmetric Catalytic Epoxidation of Unfunctionalized Olefins. In: Ojima I (ed) Catalytic Asymmetric Synthesis. VCH, Weinheim, b) Katsuki T (1996) J Mol Cat 113:87

39. Pozzi G, Cinato F, Montanari F, Quici S (1998) J Chem Soc Chem Commun 877

40. Recent publication on enantioselective protonation carried out under FBS conditions: Takeuchi S, Nakamura Y, Ohgo Y, Curran DP (1998) Tetrahedron Lett 39:8691

41. a) Sharpless KB, Young MW (1975) J Org Chem 40:947, b) Hori T, Sharpless KB (1978) J Org Chem 43:1689, c) Kuwajima I, Shimizu M, Urate H (1982) J Org Chem 47:837, d) Syper L (1989) Synthesis 167

42. Taylor RT, Flood LA (1983) J Org Chem 48:5160

43. Betzemeier B, Lhermitte F, Knochel P (1999) Synlett 489

Benzotrifluoride and Derivatives: Useful Solvents for Organic Synthesis and Fluorous Synthesis

James J. Maul[1] · Philip J. Ostrowski[2] · Gregg A. Ublacker[3] · Bruno Linclau[4] · Dennis P. Curran[4]

[1] Occidental Chemical Corporation, Technology Center, Grand Island, NY 14072, USA.
  E-mail: jimmaul@aol.com.
[2] Occidental Chemical Corporation, Niagara Falls, NY 14303, USA.
  E-mail: phil_ostrowski@oxy.com.
[3] Occidental Chemical Corporation, Dallas, TX 75244, USA. *E-mail: gregg_ublacker@oxy.co.*
[4] Department of Chemistry, University of Pittsburgh, Pittsburgh, PA 15260 USA.
  E-mail: curran+@pitt.edu

Benzotrifluoride (BTF, trifluoromethylbenzene, α,α,α-trifluorotoluene, $C_6H_5CF_3$) and related compounds are introduced as new solvents for traditional organic synthesis and for fluorous synthesis. BTF is more environmentally friendly than many other organic solvents and is available in large quantities. BTF is relatively inert and is suitable for use as a solvent for a wide range of chemistry including ionic, transition-metal catalyzed and thermal reactions. It is especially useful for radical reactions, where it may replace benzene as the current solvent of choice for many common transformations. BTF and related solvents are also crucial components of fluorous synthesis since they can dissolve both standard organic molecules and highly fluorinated molecules. This chapter provides an overview of the reactivity and toxicological properties of BTF and analogs and then summarizes their recent uses as reaction solvents in both traditional organic and new fluorous synthesis.

Keywords. Benzotrifluoride, Reaction solvent, Organic synthesis, Fluorous synthesis, Green chemistry.

Topics in Current Chemistry, Vol. 206
© Springer-Verlag Berlin Heidelberg 1999

Abbreviations

BTF	benzotrifluoride (trifluoromethylbenzene, α,α,α-trifluoro-toluene
DBA	dibenzylidene acetone
3,4-DCBTF	3,4-dichlorobenzotrifluoride
TCE	*sym*-tetrachloroethane
DMF	*N,N*-dimethylformamide
DMSO	dimethyl sulfoxide
EDC	*N*-(3-dimethylaminopropyl)-*N'*-ethylcarbodiimide hydrochloride
HFMX	hexafluorometaxylene, 1,3-*bis*-(trifluoromethyl)benzene
HFPX	hexafluoroparaxylene, 1,4-*bis*-(trifluoromethyl)benzene
LAH	lithium aluminium hydride
MABTF	3-amino-benzotrifluoride
MCBTF	3-chlorobenzotrifluoride
MNBTF	3-nitrobenzotrifluoride
PCBTF	4-chlorobenzotrifluoride
PPTS	pyridinium *para*-toluenesulphonate
TFMBA	2-trifluoromethyl benzoic acid

1
Introduction

Benzotrifluoride (BTF, trifluoromethylbenzene, α,α,α-trifluorotoluene) (Fig. 1) is a clear, free-flowing liquid with a boiling point of 102°C, a melting point of

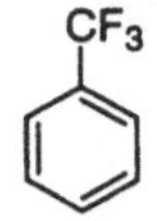

CF$_3$

Fig. 1   Benzotrifluoride (BTF)

−23 °C, and a density of 1.2 g/ml (25 °C). It has a characteristic odor resembling other aromatic solvents like toluene. BTF is slightly more polar than THF and ethyl acetate and slightly less polar than dichloromethane and chloroform.

BTF belongs to an important group of trifluoromethyl-substituted aromatic compounds, which have broad applications as intermediates or building blocks for crop protection chemicals, insecticides and pharmaceuticals, as well as dyes. Related higher boiling compounds that are produced in multimillion pound quantities include 4-chlorobenzotrifluoride (PCBTF) and 3,4 dichlorobenzotrifluoride (3,4-DCBTF).

While these materials are manufactured as intermediates, BTF and its analogs are relatively inert and hence, potentially useful as solvents for reactions and extractions (Sects. 3 and 4), as well as for non-chemical applications such as solvents for coatings and cleaning of surfaces (Sect. 2.5). The trifluoromethyl group on the aromatic ring is very stable to basic conditions even at elevated temperatures, and somewhat stable to aqueous acid conditions at moderate temperatures (Sect. 2.3.3). BTF has recently become available at low price. Moreover, its lower toxicity (see Sect. 2.6) and higher boiling point (which minimizes losses during evaporation) make it an ecologically suitable replacement for solvents like dichloromethane and benzene. Despite these favorable characteristics, BTF is relatively unknown as a solvent. However, our experience [1] and that of others is beginning to show that BTF is indeed suitable as solvent for many different reactions. In this chapter, we provide an overview of the features of BTF and related solvents and summarize their recent uses as reaction solvents in both traditional organic synthesis and new fluorous reactions. This information should prove helpful to others evaluating when and how to employ BTF as a solvent in organic synthesis.

2
General Introduction to Benzotrifluoride

2.1
Industrial Preparation of BTF

BTF is prepared industrially from toluene in two synthetic steps: 1) free radical perchlorination of the methyl group, followed by 2) fluorine/chlorine exchange of the three chlorine atoms with anhydrous hydrogen fluoride (Scheme 1). The chlorination step may be catalyzed by light of suitable wavelength (UV) and is conveniently carried out in the liquid phase. The fluoride/chloride exchange can be catalyzed by the presence of metal halide compounds, such as pentahalide (Cl, F) salts of antimony and molybdenum, and is effected under a variety of

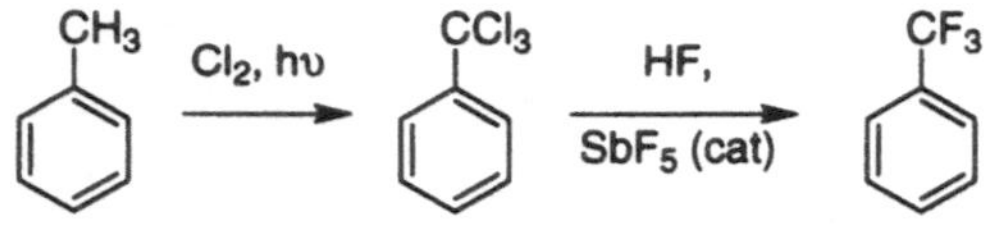

Scheme 1

conditions of temperature and pressure including liquid phase (high pressure and temperature) [2–4], liquid phase (ambient pressure) [5–9] or vapor phase (low pressure high temperature) [4].

2.2
Benzotrifluoride Analogs

4-Chlorobenzotrifluoride (PCBTF), 2,4-dichlorobenzotrifluoride (2,4-DCBTF) and 3,4-dichlorobenzotrifluoride (3,4-DCBTF) share major aspects of their syntheses with BTF. PCBTF and 2,4-DCBTF are manufactured from 4-chlorotoluene and from 2,4-dichlorotoluene as outlined in Scheme 1. 3,4-DCBTF is prepared by the electrophilic (metal halide catalyzed) chlorination of PCBTF (Scheme 2). Hexafluoroxylenes have been synthesized from the corresponding xylenes (*ortho*, *meta* and *para*) by sequences similar to that outlined in Scheme 1.

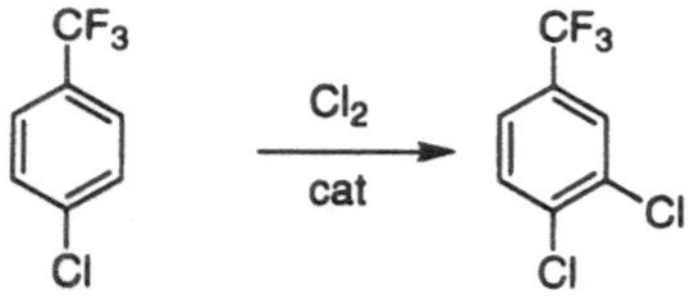

Scheme 2

2.3
The Chemistry of BTF

The chemistry of BTF is centered in three areas: aromatic electrophilic substitution as influenced by the CF_3 group, aromatic nucleophilic substitution as influenced by the CF_3 group, and the stability of the CF_3 group itself to hydrolytic conditions of acid and base.

2.3.1
Aromatic Electrophilic Substitution

Groups on aromatic rings are classified in aromatic electrophilic substitution reactions by their resonance and inductive effects. The trifluoromethyl group exerts no resonance effect and it exerts a negative Inductive (–I) effect (electron withdrawing) on the aromatic ring. Consequently, the CF_3 group deactivates the ring to aromatic electrophilic substitution and directs primarily toward the meta position. This is exemplified in Scheme 3, by the nitration of BTF to provide *m*-nitro-BTF (MNBTF). The subsequent reduction of MNBTF provides *m*-amino-BTF (MABTF), which is an important chemical intermediate (Sect. 2.4). For similar mechanistic reasons, chlorination of BTF with Cl_2 yields *m*-chloro-BTF (MCBTF) and chlorination of PCBTF yields 3,4-DCBTF (Scheme 2).

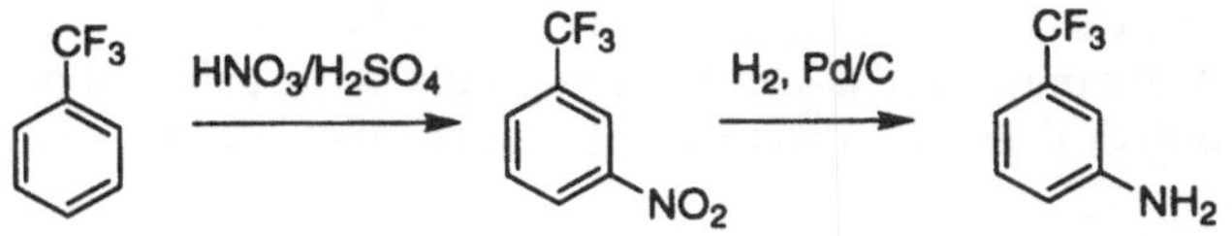

Scheme 3

2.3.2
Aromatic Nucleophilic Substitution

Aromatic nucleophilic substitution is promoted by functions with $-I$ and $-R$ (Resonance) effects that are ortho or para to the group undergoing aromatic nucleophilic substitution. The negative Inductive ($-I$) effect of the CF_3 group is usually insufficient, on its own, to effect such substitution of the chloride atom of PCBTF. However, with the assistance of one or two nitro-groups, the reaction is quite facile (Scheme 4) and of major commercial significance. The intermediate 4-chloro-3,5-dinitro-BTF is prepared by the nitration of PCBTF (aromatic electrophilic substitution) and the chlorine atom is readily replaced by substituted amines (aromatic nucleophilic substitution) to produce several herbicides. Aromatic nucleophilic substitution on 3,4-DCBTF with substituted phenols yields a family of herbicides known as diphenyl ethers (Sect. 2.4).

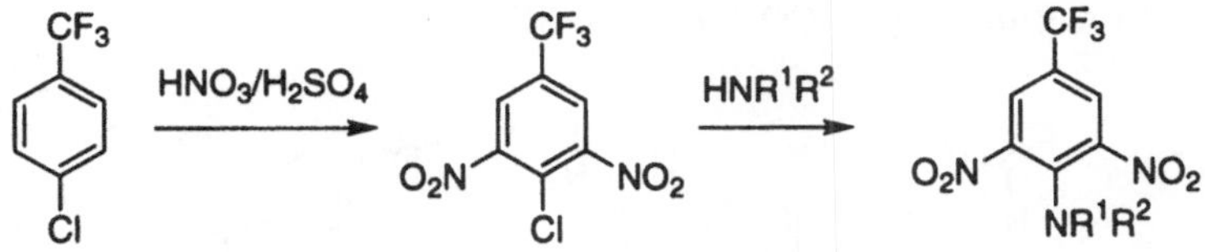

Scheme 4

2.3.3
Other Reactions

A very important consideration to the chemistry of BTFs is the stability of the CF_3 group toward acid or base. The CF_3 group of BTF is stable to strongly basic conditions. For example (Scheme 5), BTF can be treated with butyllithium/ potassium t-butoxide in tetrahydrofuran to prepare a lithiated aromatic compound, which yields 2-trifluoromethyl benzoic acid (TFMBA) upon treatment with CO_2 [10].

BTF is also reasonably stable towards acid. Under conditions of nitration with mixed acid (H_2SO_4/HNO_3), the CF_3 group of BTF (or PCBTF) is stable, but BTF will hydrolyze to benzoic acid after heating with 100% sulfuric acid [11] or with HBr [12, 13]. This hydrolytic stability allows BTFs to be considered a solvent for

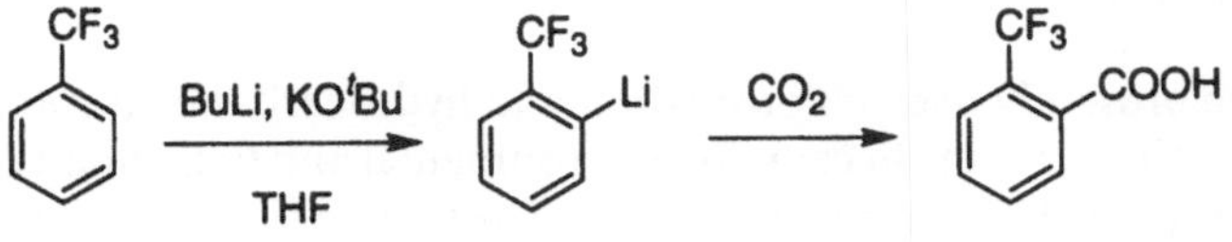

Scheme 5   TFMBA

reactions and extractions (Sects. 3 and 4) as well as in coatings, etc. (Sect. 2.5). Although BTF can be used as a solvent for many Lewis acid promoted reactions, the CF_3 group of BTF does react with strong Lewis-Acids such as $AlCl_3$ [14].

2.4
Uses of BTFs as Synthetic Intermediates in Chemical Synthesis

The use of BTFs as synthetic intermediates extends beyond agricultural products such as herbicides, but to enumerate each application is beyond the scope of the present work. Multi-million pound quantities of BTF, PCBTF, 2,4-DCBTF and 3,4-DCBTF are used in the synthesis of herbicides and major examples of each are presented here.

Agricultural products which are derived from BTF (MABTF) include the herbicides Fluometuran and Flurochloridone which are the respective products of Ciba and Zeneca Corporations (Fig. 2). Examples of herbicides which are derived from PCBTF include Trifluralin ($R = N(CH_2CH_2CH_3)_2$) and Ethalfluralin ($R = N(Et)CH_2C(CH_3) = CH_2$) sold by Dow Elanco Corporation (Fig. 2). Ishihara Corporation produces the fungicide Fluazinam (Fig. 2) from 2,4-DCBTF. Examples of the diphenyl ether herbicides (Fig. 2) which are derived from 3,4-DCBTF are: Oxyfluorfen ($R = Et$) by Rohm & Haas, Fomesafen ($R = C(O)NHSO_2CH_3$) by Zeneca and Acifluofen ($R = COOH$) by BASF.

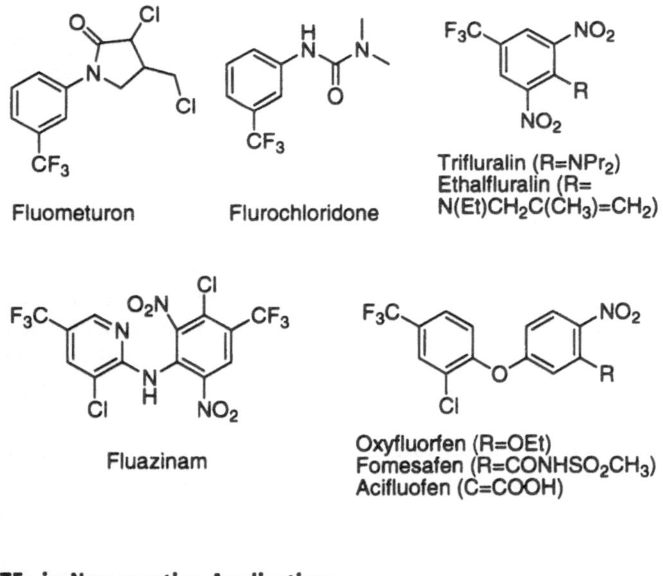

Fluometuron   Flurochloridone   Trifluralin (R=NPr$_2$)
                                          Ethalfluralin (R=
                                          N(Et)CH$_2$C(CH$_3$)=CH$_2$)

Fluazinam   Oxyfluorfen (R=OEt)
                   Fomesafen (R=CONHSO$_2$CH$_3$)
                   Acifluofen (C=COOH)

Fig. 2

2.5
Uses of BTFs in Non-reactive Applications

Many of the presently used chlorocarbon and hydrocarbon solvents have come under increasing scrutiny because of environmental and toxicological concerns. PCBTF, and 3,4-DCBTF were introduced by Occidental Chemical Corporation in 1992 for industrial solvent uses under the trademark of OXSOL [15]. BTF was added to the OXSOL family of solvents in 1996. In 1994, the EPA granted PCBTF an exemption from VOC (Volatile Organic Compound) regulation [16]. The major solvent use for these compounds involves PCBTF as an exempt solvent in paint and coating formulations. Consequently, paint formulators have been able to use PCBTF as a tool to comply with ever tightening VOC regulations [17–19]. Through 1998 the major coatings use area has been in urethane systems to include two component, moisture cure and alkyd resins. Additional use was found in epoxy, polyester, acrylic, silicone, ethyl silicate, phenolic varnish, vinyl butyral, melamine/urea formaldehyde, and nitrocellulose resin systems. Application areas include automotive refinishing, industrial, maintenance, metal furniture and appliances, wood furniture, marine coatings, aerospace, conformal coatings and concrete sealers. PCBTF can also be used as a non-VOC universal diluent at the spray gun for modern high solids coatings. Other minor applications include solvent use for metal cleaning, precision cleaning, adhesives, inks, dye carriers and uses as a functional fluid.

The considerable chemical stability of BTF suggests also its consideration as an alternative solvent for chemical reaction processes (Sects. 3 and 4). The relative stability of the BTFs in general (BTF, PCBTF and 3,4-DCBTF) as well as their desirable environmental and toxicological properties promote their consideration for a broad spectrum of applications as industrial solvents.

2.6
Toxicological and Environmental Properties of BTFs

2.6.1
Toxicology Summary of BTFs [20–23]

We provide here a toxicology summary for BTF (CAS #98–08–8), PCBTF (CAS #98–56–6) 3,4 DCBTF (CAS #328–84–7) and HFMX (hexafluoro meta-xylene, CAS # 402–31–3) with comparisons to benzene (CAS #71–43–2) and dichloromethane (CAS #75–09–2).

Both benzene and dichloromethane are well characterized toxicologically. As demonstrated by the brief acute toxicity summaries for both benzene and dichloromethane, the two chemicals seem to exhibit greater acute toxicity when administered to rats orally and more severe irritation to both rabbit eye and skin compared to BTF, PCBTF, 3,4-DCBTF, or HFMX. Dichloromethane and benzene have also been subjected to a number of subchronic and chronic bioassays. Benzene has been characterized as a human carcinogen while dichloromethane has caused cancer in laboratory animals. Dichloromethane and benzene have also exhibited a number of target organ effects as well as reproductive and teratogenic effects both in laboratory animals and humans. These data are too extensive to be summarized in this brief overview. Although the toxicity data generated for BTF, PCBTF, 3,4-DCBTF and HFMX are somewhat limited compared to benzene and dichloromethane, the results may indicate that these chemicals will not show the same chronic toxicity as these two chemicals.

BTF (CAS #98-08-8)
Acute Oral LD_{50} > 5000 mg/kg (rat) – practically non-toxic
Acute Inhalation LC_{50} (4 h) > 7958 ppm (rat) – practically non-toxic
Skin Irritation (Draize) 1.5/8.0 (rabbit) – slightly irritating
Eye Irritation (Draize) 0.3/110 – washed, 1.3/110 – unwashed (rabbit) –
no appreciable effect
Ames Test – Negative, with and without metabolic activation

No subacute/subchronic or chronic toxicological data are available for BTF.

PCBTF (CAS #98-56-6)
Acute Oral LD_{50} > 6650 mg/kg (rat) – practically non-toxic
Acute Dermal LD_{50} > 2700 mg/kg (rabbit) – practically non-toxic
Acute Inhalation LC_{50} (4 h) = 4479 ppm (rat) – slightly toxic
Skin Irritation (Draize) 0.3/8.0 – abraded, 0.2/8.0 – intact (rabbit) –
no appreciable effect
Eye irritation (Draize) 0.0/110 – washed, 2.0/110 – unwashed (rabbit) –
no appreciable effect
Ames Test – Negative, with and without metabolic activation

Inhaled or ingested PCBTF is largely exhaled unchanged. In a 90-day rat in-
halation study there were dose related liver effects. The "No Observable Effect
Level" (NOEL) for these effects was 50 ppm. If reversible hepatocyte hyper-
trophy is considered to be an adaptive response, the "No Observable Adverse
Effect Level" (NOAEL) for this study is 250 ppm. In a 28-day study, there were
changes in the male rat kidney that were not considered relevant to humans.
PCBTF was not neurotoxic at concentrations as high as 250 ppm for 90 days.
Central Nervous System (CNS) effects were observed in rats exposed to PCBTF
at or above 2822 ppm for 4 h. PCBTF is not considered to be a reproductive
hazard at dose levels as high as 45 mg/kg. Based on the results of genotoxicity
studies, PCBTF is not anticipated to be either mutagenic or oncogenic in mam-
malian systems.

3,4-DCBTF (CAS # 328-84-7)
Acute Oral LD_{50}=2900 mg/kg (rat) – slightly toxic
Acute Dermal LD_{50} >2000 mg/kg (rabbit) – practically non-toxic
Acute Inhalation LC_{50} (4 h) >1804 ppm (rat) – No mortality produced
Skin Irritation (Draize) 0.7/8.0 – (rabbit) – slightly irritating
Eye irritation (Draize) 1.5/110 – (rabbit) – slightly irritating
Ames Test – Negative, with and without metabolic activation

3,4-DCBTF has been subjected to a 14-day oral gavage, 28 day feeding, and
a modified 90 day oral gavage and reproductive study in rats. 3,4-DCBTF is
not considered to be a reproductive hazard at dose levels as high as 45 mg/kg.
The results of the 90-day study suggest that the liver and possibly the kidney
may be target organs. Increases in the liver weight and liver/body weight were
evident at the 45 mg/kg dose in males and an apparent dose response trend was
present throughout all the male groups. Similar trends were evident but less
striking in the female groups. The liver weight changes were not accompanied

by any notable abnormalities in cell size or structure. Therefore, the liver weight changes are probably due to an adaptive response of the liver to a foreign substance.

Significant increases were seen with male kidney weights at a 15 mg/kg effect level. Male kidney weight tended to increase with increased dosage. A corresponding trend was seen in female kidney weights but was not statistically significant. Histopathology revealed no abnormality in kidney cell size or structure. The cause of the increased kidney weight parameters is similar to that of the liver, i.e. metabolic induction.

HFMX   (CAS # 402–31–3) HexaFluoroMetaXylene
   Acute Oral LD_{50} > 5000 mg/kg (rat) – practically non-toxic
   Acute Dermal LD_{50} > 2000 mg/kg (rabbit) – practically non-toxic
   Acute Inhalation LC_{50} (4 h) – (rat) > 5710 ppm and < 17,129 ppm
   Skin Irritation (Draize) 0.3/8.0 – (rabbit) – no appreciable effect
   Eye irritation (Draize) 0.7/110 – (rabbit) – no appreciable effect
   Dermal Sensitization (Guinea Pig) – Buehler Assay – Negative, not a skin sensitizer

Benzene (CAS # 71–43–2)
   Acute Oral LD_{50} = 930 mg/kg (rat) – moderately toxic
   Acute Inhalation LC_{50} (7 h) = 10,000 ppm (rat) – practically non-toxic
   Skin Irritation (Draize) (rabbit) – moderately irritating
   Eye irritation (Draize) (rabbit) – moderately/severely irritating

Dichloromethane (CAS # 75–09–2)
   Acute Oral LD_{50} = 1600 mg/kg (rat) – moderately toxic
   Acute Inhalation LC_{50} (30 min) = 25,334 ppm (rat)
   Skin Irritation (Draize) (rabbit) – moderately/severely irritating
   Eye irritation (Draize) (rabbit) – mild/moderately irritating

2.6.2
Environmental Impact of BTFs

Use patterns and physical properties dictate that releases of benzotrifluoride products will primarily partition into air [24]. All of these chemicals have full atmospheric lifetimes on the order of one to two months. This short lifetime indicates that these products will not be implicated in ozone depletion and global warming. Final degradation products will be CO_2, H_2O, HF and HCl.

PCBTF received an exemption from VOC regulations based on the fact that its atmospheric hydroxyl radical reaction rate is slower than that of ethane [25]. A VOC is defined as "any compound of carbon, excluding carbon monoxide, carbon dioxide, carbonic acid, metallic carbides or carbonates, and ammonium carbonate, which participates in atmospheric photochemical reactions" [26]. A VOC exemption petition for BTF was filed with the EPA on March 11, 1997. Volatile organic compounds (VOCs) emission is controlled by regulation in efforts to reduce the tropospheric air concentrations of ozone.

Estimated full lifetimes and water solubility of BTF and several analogs are listed below:

Product	Atmospheric Lifetime	Water Solubility at 20 °C
BTF	32 days	250 ppm
PCBTF	66 days	29 ppm
3,4-DCBTF	40 days	12 ppm

Hydrolysis and biodegradation are not significant environmental removal processes for these compounds. Since the benzotrifluorides have low water solubility and moderate vapor pressures, accidental environmental release will result in volatilization into air. Calculations indicate that a spill of PCBTF into a model river will have a volatilization half-life of 3.95 h. PCBTF has a moderate potential for bioconcentration. Bioconcentrations were experimentally determined to be between 121.8 to 202.0 which is far below the concern threshold of 5000 [27]. Bioconcentration is a measurement of ratio of the rate constant of biological uptake in water, via non dietary routes, versus biological elimination [28].

2.6.3
Information on Disposal of BTFs

In large-scale applications, waste streams of BTFs may be safely reclaimed by distillation in an explosion-proof distillation unit or still. Most stills available today have reclaim efficiencies of 90 to 99%. Using a high efficiency, thin-film evaporation still, 95% or more of the BTF may be recovered from the spent solvent sludge. To further increase the amount of reclaimed solvent and to reduce the volume of waste that must be disposed of, stills can be equipped to employ steam stripping.

If no on-site reclaiming efforts are made, the still bottoms from a typical distillation unit will usually contain between 1 and 10% BTF, and they must be disposed of according to proper Resource Conservation and Recovery Act (RCRA) hazard classifications. Pure BTF and PCBTF have flash points less than 140 °F and qualify as D001 hazardous wastes, while pure 3,4-DCBTF is not regulated. These products do not contain any listed concentrations of compounds recognized by RCRA as hazardous wastes. In accordance with state and local regulations, their still bottoms may be added to other combustible products and incinerated as fuel oils, thereby avoiding costly hazardous waste disposal fees. The Heat of Combustion values are listed below:

Product	BTU/lb.
PCBTF	7700
DCBTF	4830
BTF	8060

On a small scale, laboratory quantities of the BTF's can be safely combined with other halogenated compounds for appropriate disposal [29].

3
Benzotrifluoride as a Solvent in Traditional Organic Synthesis

3.1
Introduction

As measured by E_T values (see Sect. 3.2), the polarity of BTF is intermediate between that of THF and ethyl acetate on one hand and dichloromethane and chloroform on the other hand. This suggests that BTF should be able to dissolve a wide variety of moderately polar organic compounds. However, BTF does not have significant Lewis-basic properties and does not form strong hydrogen bonds. Due to the trifluoromethyl group, BTF is more polar than benzene or toluene, so that relatively polar molecules that do not dissolve in common aromatic solvents do dissolve in BTF.

To demonstrate the potential of BTF as a general solvent for organic synthesis, some representative transformations of different reaction types are listed below according to reaction category. A few of these applications have been reported in the primary literature, but most have not. We have explored the use of BTF in two different ways. First, to probe the potential scope, we selected representative examples of an assortment of important reaction types and then conducted the reactions in both the literature solvent and in BTF. A selection of these comparative reactions is listed below. Second, since it quickly became apparent that BTF was a useful reaction solvent, we started to adapt it to our own research purposes. Selected examples of this type are also reported along with comparison yields. Most of the pairs of reactions were only conducted once, and thus small differences in yield should not be considered meaningful indications of the superiority of one solvent over another. We also include recent results from other groups. Examples are organized by reaction class with a short section on currently known limitations at the end.

3.2
Solvent Properties of BTF – Comparison with Other Common Solvents

BTF has unique properties that encourage its consideration as a solvent for synthetic chemistry. BTF (bp 102 °C) forms an azeotrope (bp 80 °C, 90 % BTF) with water which can be used to drive dehydration reactions [30]. Since BTF is heavier than water, a reverse Dean-Stark trap must be used. Commercial BTF (distilled) contains less than 50 ppm water, and this low water content suggests that commercial BTF could be used directly for many applications. The use of phosphorus pentoxide as drying agent is also recommended [31], but we found that over time, decomposition of BTF took place, resulting in black precipitation. We instead typically dried BTF over anhydrous potassium carbonate followed by distillation; however, we have not measured the water content of the BTF before and after drying so the effectiveness of this procedure is not known.

The melting point of BTF (–29 °C) is high by solvent standards and this can limit some types of low-temperature reactions. The melting point can be lowered by adding other solvents, but the effect is modest (Table 1). The freezing

Table 1. Estimated freezing point depression of BTF when mixed with other solvents

	CH_2Cl_2	THF	Toluene	Benzene	Et_2O
m.p. solvent[a]	−97	−108	−93	5	−118
5%[b]	−33	−36	−34	−36	−35
10%	−36	−39	−37	−38	−40
20%	−40	−47	−42	−43	−41
33%	−45	−50	−47	−47	−43
50%	−56	−65	−55	−28	−56

[a] all melting points are in °C.
[b] 5% of the solvent mixed with BTF.

Table 2. Comparison of BTF with other solvents

	Boiling Point (°C)	Melting Point (°C)	Vapor Pres. mm Hg (25°C)	Density (25°C)	Molecular Weight	Polarity (Dipole Moment) Debye Units	E_T^N (Ref 32)	Flash Point (°C TCC)
BTF	102	−29	40	1.18	146.11	2.9	0.241	12
DCM	40	−96.7	430	1.32	84.94	1.57	0.309	None
THF	66	−108	162	0.88	72.11	1.74	0.207	−17
Et_2O	34	−116	534	0.71	74.12	1.14	0.117	−40
Benzene	80	5.5	95	0.87	78.11	0.0	0.111	−11
Toluene	111	−95	28	0.86	92.14	0.43	0.099	4
PCBTF	139	−36	8	1.34	180.56	1.2	TBD	43[a]
3,4-DCBTF	174	−12	2	1.47	215.00	1.5	TBD	77
$CHCl_3$	61	−64	194	1.47	119.39	1.2	0.259	None
CCl_4	77	−23	114	1.58	153.84	0	TBD	None
HFMX	116	−34	18	1.34	214.11	2.5	TBD	26
HFPX	116	3	18	1.38	214.11	0.6	TBD	21

[a] TOC fire point 97°C.

points in Table 1 were determined by the naked eye upon slow cooling, so they are estimates that should not be regarded as highly accurate.

A comparison of the physical properties of several BTFs with those of common organic solvents is presented in Table 2 [32].

3.3
Thermal Reactions

BTF and its mono- and dichloro derivatives are very useful for conducting thermal reactions (Scheme 6), as illustrated by the sulfur dioxide extrusion reaction [33] in BTF (6.1) and the Cope rearrangement [34] in 3,4-dichloro BTF (6.2). In the latter case, the products were isolated after acid-base extraction. One limitation for the use of BTF and derivatives in thermal reactions is that strong aqueous acidic reagents cannot be used, since BTF is prone to hydrolysis at ele-

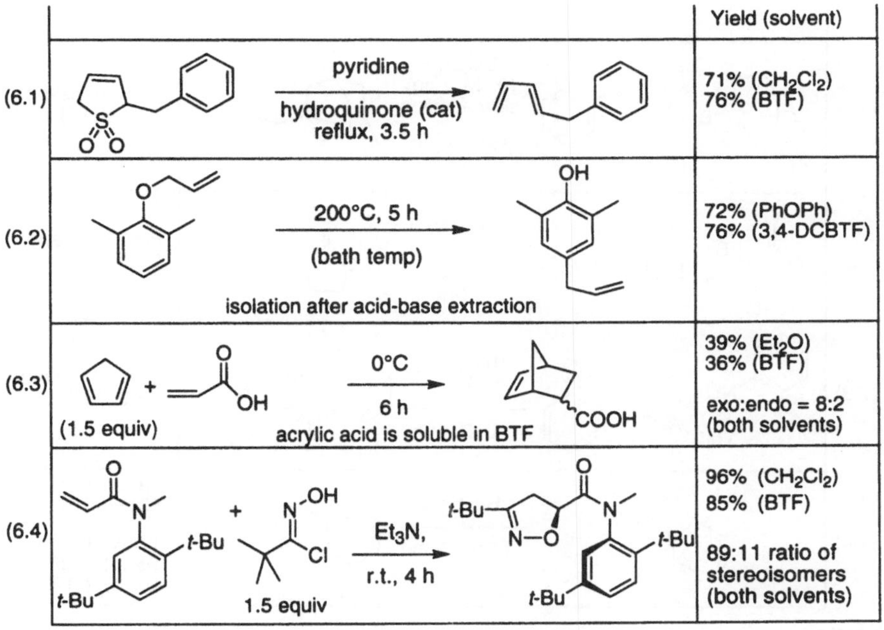

			Yield (solvent)
(6.1)		pyridine → hydroquinone (cat) reflux, 3.5 h	71% (CH$_2$Cl$_2$) 76% (BTF)
(6.2)		200°C, 5 h (bath temp) isolation after acid-base extraction	72% (PhOPh) 76% (3,4-DCBTF)
(6.3)	+ (1.5 equiv)	0°C 6 h acrylic acid is soluble in BTF	39% (Et$_2$O) 36% (BTF) exo:endo = 8:2 (both solvents)
(6.4)	t-Bu, t-Bu + 1.5 equiv	Et$_3$N, r.t., 4 h	96% (CH$_2$Cl$_2$) 85% (BTF) 89:11 ratio of stereoisomers (both solvents)

Scheme 6

vated temperatures. A Diels-Alder reaction [35] (6.3) and a nitrile oxide cyclo-addition reaction [36] (6.4), both at room temperature, proceeded well in BTF. In the nitrile oxide cycloaddition, the ratio of diastereomers is independent of the solvent used. Although we did not conduct a thermal reaction in PCBTF, it seems likely that this solvent is also suitable for reactions at high temperature.

3.4
Radical Reactions

Despite its toxicity, benzene is still often the solvent of choice for radical reactions because of its reluctance to undergo radical addition or hydrogen abstraction. Only in particular cases, alternatives such as toluene, cyclohexane, CH$_2$Cl$_2$ and even THF can be used. Our experiments suggest that BTF should replace benzene as a general solvent for many radical reactions (Scheme 7). The Giese reaction [37] (7.1), a typical allylation [38] (7.2) and a sophisticated isonitrile tandem cyclization reaction [39] (7.3) proceed equally well in benzene and BTF. The yield of the Giese reaction with a poor radical acceptor such as styrene [40] remains the same by exchanging benzene for BTF. There is no limitation on the initiation procedure (AIBN, light). Addition of tributyltin hydride to a triple bond [41] (7.4) in BTF proceeds with somewhat lower yield than in THF, with comparable E/Z ratios.

The reduction of alkyl halides using a fluorous tin hydride in BTF is a very general type of reaction that is discussed in Sect. 4.3.1. These results clearly

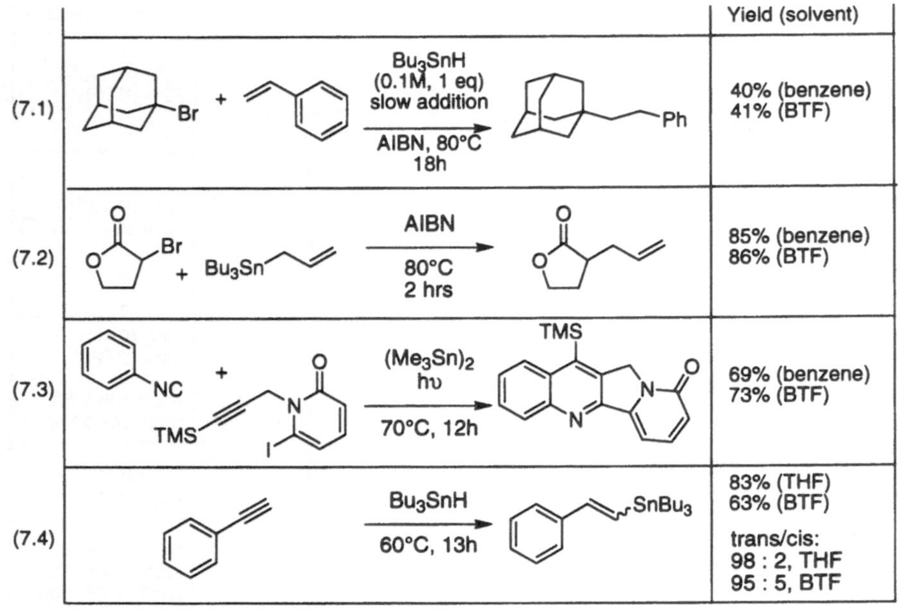

				Yield (solvent)
(7.1)	adamantyl-Br + styrene	Bu₃SnH (0.1M, 1 eq) slow addition → AIBN, 80°C 18h	adamantyl-CH₂CH₂-Ph	40% (benzene) 41% (BTF)
(7.2)	(α-bromo-γ-butyrolactone) + Bu₃Sn-allyl	AIBN 80°C 2 hrs	(α-allyl-γ-butyrolactone)	85% (benzene) 86% (BTF)
(7.3)	PhNC + TMS-alkynyl-propargyl-N-pyridinone-I	(Me₃Sn)₂ hυ 70°C, 12h	TMS-quinoline-indolizine fused product	69% (benzene) 73% (BTF)
(7.4)	phenylacetylene	Bu₃SnH 60°C, 13h	Ph-CH=CH-SnBu₃	83% (THF) 63% (BTF) trans/cis: 98 : 2, THF 95 : 5, BTF

Scheme 7

demonstrate that BTF is quite resistant towards radicals. Its low toxicity and higher polarity (which should solubilize a larger variety of compounds) make it an ideal replacement for benzene in radical reactions.

3.5
Lewis-Acid Reactions

BTF is known to react with strong Lewis-Acids such as $AlCl_3$. [42] However, milder Lewis-Acids do not readily react with BTF. Zinc chloride catalyzed Friedel-Crafts acylation (8.1) leads to better yields in refluxing BTF compared to *sym*-tetrachloroethane. [43] The deactivating trifluoromethyl group is presumably responsible for the inertness of BTF towards aromatic substitution under these conditions. Titanium tetrachloride has successfully been used for Sakurai [44] (8.2), Mukaiyama-aldol [45] (8.3) and Diels-Alder [46] (8.4) reactions. Lectka and coworkers [47] reported an efficient catalytic enantioselective imino-ene reaction using a copper catalyst in BTF (8.5), which gave better results in BTF compared to traditional solvents like THF or dichloromethane. They suggested that the aromatic character of BTF has a favorable influence for the enantioselectivity. Trimethylaluminum and dimethylaluminum-perfluorophenolate were successfully used for the cleavage of acetals derived from C_2-symmetric diols [48] (8.6), although the cleavage was not very stereoselective. The yield in BTF was quantitative and the reaction was faster than in THF. Coloration of the reaction mixture during the preparation of the pentafluorophenolate complex suggests that this Lewis-Acid does interact with BTF to some extent.

In a very recent paper, Mikami and coworkers [80] report Friedel-Crafts and Diels-Alder reactions catalyzed by the fluorinated Lewis acid $Yb(N(SO_2C_4F_9)_2)_3$. BTF is both stable to the reaction conditions and is successful in dissolving both the substrates and the fluorinated Lewis acid (see Sect. 4).

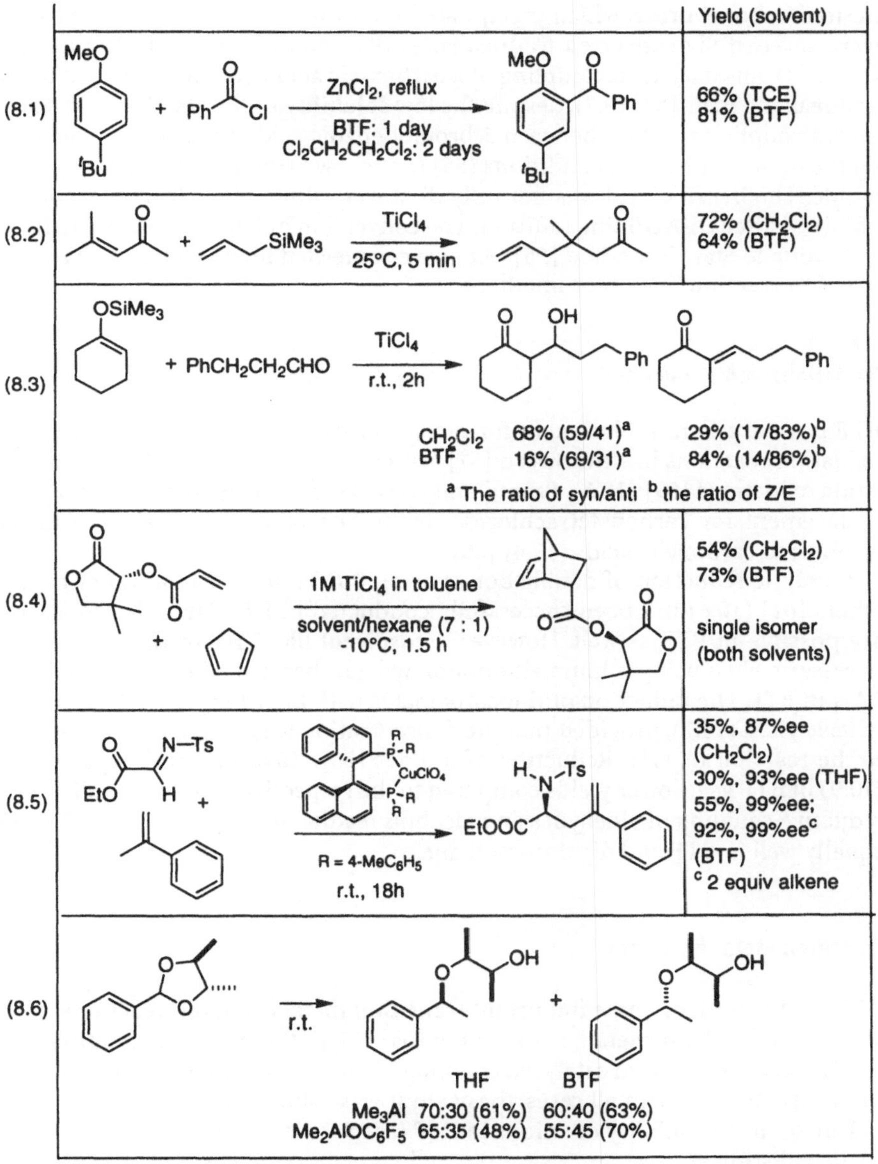

		Yield (solvent)
(8.1)	$ZnCl_2$, reflux / BTF: 1 day / $Cl_2CH_2CH_2Cl_2$: 2 days	66% (TCE) 81% (BTF)
(8.2)	$TiCl_4$ 25°C, 5 min	72% (CH_2Cl_2) 64% (BTF)
(8.3)	$TiCl_4$ r.t., 2h	CH_2Cl_2 68% (59/41)[a] 29% (17/83%)[b] BTF 16% (69/31)[a] 84% (14/86%)[b] [a] The ratio of syn/anti [b] the ratio of Z/E
(8.4)	1M $TiCl_4$ in toluene / solvent/hexane (7:1) / -10°C, 1.5 h	54% (CH_2Cl_2) 73% (BTF) single isomer (both solvents)
(8.5)	$R = 4\text{-}MeC_6H_5$ r.t., 18h	35%, 87%ee (CH_2Cl_2) 30%, 93%ee (THF) 55%, 99%ee; 92%, 99%ee[c] (BTF) [c] 2 equiv alkene
(8.6)	r.t.	THF BTF Me_3Al 70:30 (61%) 60:40 (63%) $Me_2AlOC_6F_5$ 65:35 (48%) 55:45 (70%)

Scheme 8

3.6
Functional Group Transformations

Standard acylation, [49] silylation [50] and tosylation [51] (9.1) reactions of alcohols can be conducted in BTF. The conversion of primary alcohols to iodides with iodine and triphenylphosphine [52] (9.2; 9.3), however, does not seem to work very well. Reaction of sulfonyl isocyanates with anilines (9.4) led to the desired sulfonyl ureas, which precipitated out of the reaction mixture. Ketones were successfully converted to dioxolanes (9.5) and imines [53] (9.6) using a reverse Dean-Stark trap. Coupling of a carboxylic acid with an aniline led to the amide using EDC [54] (9.7), despite the low solubility of the starting acid. However, a coupling reaction between 3-bromobenzoic acid and p-methoxybenzylamine under Mitsunobu conditions [55] (not shown) did not lead to the desired amide. The benzoic acid was not fully dissolved, although a clear solution was obtained after DEAD/PPh$_3$ addition. The conversion of ketones to vinyl triflates according to Stang's method [56] (9.8) gave somewhat lower yields in BTF compared to reaction in dichloromethane.

3.7
Oxidations and Reductions

BTF is an excellent replacement for dichloromethane in some important mild oxidation reactions like the Swern [57] (10.1), Dess-Martin [58] (10.2) and peroxide oxidations [59] (10.3). Prasad and coworkers reported that BTF is a good replacement for carbon tetrachloride in the Sharpless oxidation of primary alcohols to carboxylic acids (10.4) [60].

Catalytic reduction of double bonds using Pd/C (10.5) or removal of benzyl ethers [61] (10.6) has been successfully conducted in BTF. Hydride reductions are possible in BTF as well. However, a cosolvent like THF or diethyl ether is necessary when using lithium aluminum hydride, because of the insolubility of LAH in BTF. The reduction of thiocarbamates with LAH [62] in BTF proceeded in 95% yield (10.7), provided that the concentration was 4 times less compared to the reaction in THF. Reduction of iodides [53] (10.8) and carboxylic esters (10.9) in BTF gave lower yields compared to the respective reactions in THF. The reductive coupling of aldehydes and alcohols leading to ethers [63] (10.10) goes equally well in BTF and dichloromethane.

3.8
Transition Metal Reactions

BTF can be used for some important transition metal catalyzed reactions, such as the Grubbs olefin-metathesis reaction [64] (11.1), the Petasis olefination [65] (11.2) and the palladium catalyzed coupling of vinyl tin compounds and sulfonyl chlorides [66] (11.3). In all cases, the catalyst was soluble in BTF.

Due to the insolubility of anions in BTF (see Sect. 3.8), a palladium catalyzed allylic substitution reaction using the malonate anion [67] could not be tried. However, an intramolecular version using tosylcarbamate as nucleophile (11.4)

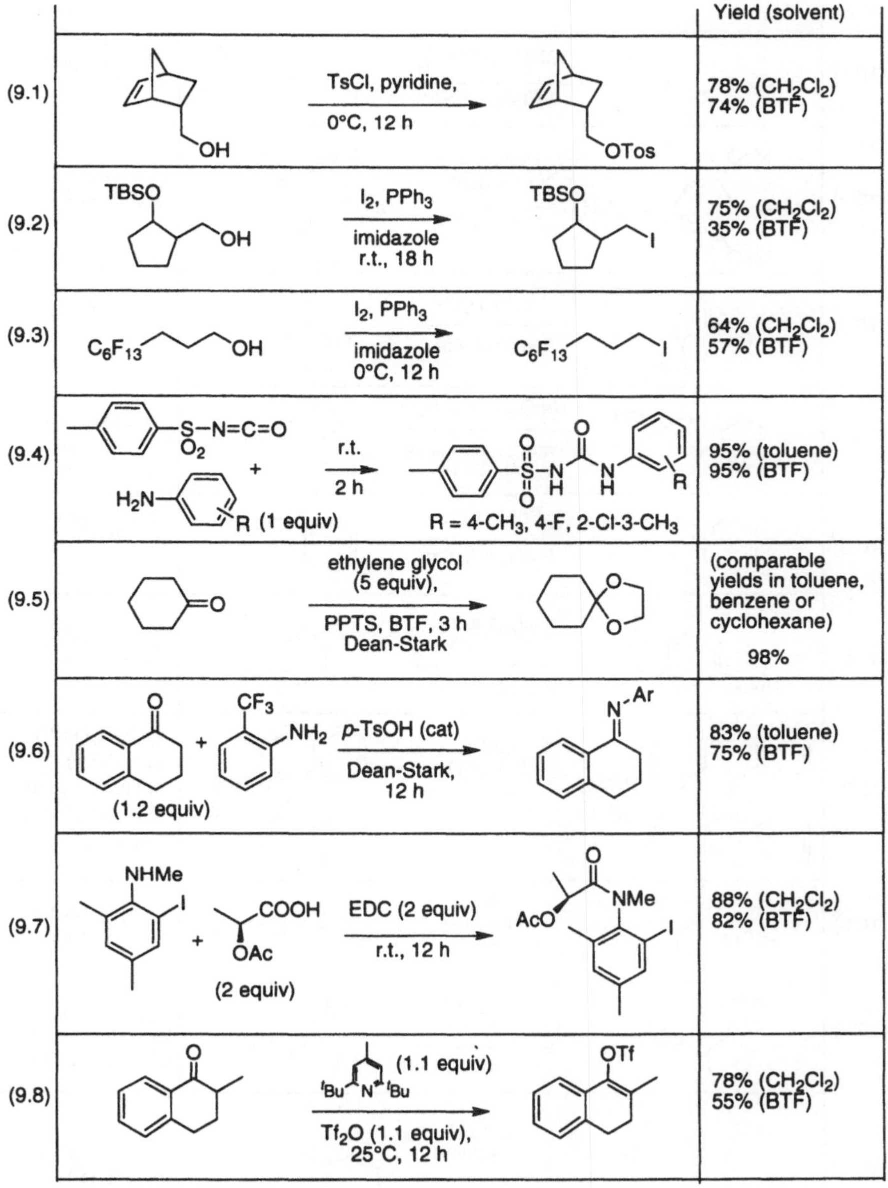

				Yield (solvent)
(9.1)		TsCl, pyridine, 0°C, 12 h		78% (CH$_2$Cl$_2$) 74% (BTF)
(9.2)		I$_2$, PPh$_3$ imidazole r.t., 18 h		75% (CH$_2$Cl$_2$) 35% (BTF)
(9.3)	C$_6$F$_{13}$\~\~\~OH	I$_2$, PPh$_3$ imidazole 0°C, 12 h	C$_6$F$_{13}$\~\~\~I	64% (CH$_2$Cl$_2$) 57% (BTF)
(9.4)	+ H$_2$N— R (1 equiv)	r.t. 2 h	R = 4-CH$_3$, 4-F, 2-Cl-3-CH$_3$	95% (toluene) 95% (BTF)
(9.5)		ethylene glycol (5 equiv), PPTS, BTF, 3 h Dean-Stark		(comparable yields in toluene, benzene or cyclohexane) 98%
(9.6)	+ (1.2 equiv)	p-TsOH (cat) Dean-Stark, 12 h		83% (toluene) 75% (BTF)
(9.7)	+ (2 equiv)	EDC (2 equiv) r.t., 12 h		88% (CH$_2$Cl$_2$) 82% (BTF)
(9.8)		(1.1 equiv) ^{t}Bu–N–^{t}Bu Tf$_2$O (1.1 equiv), 25°C, 12 h		78% (CH$_2$Cl$_2$) 55% (BTF)

Scheme 9

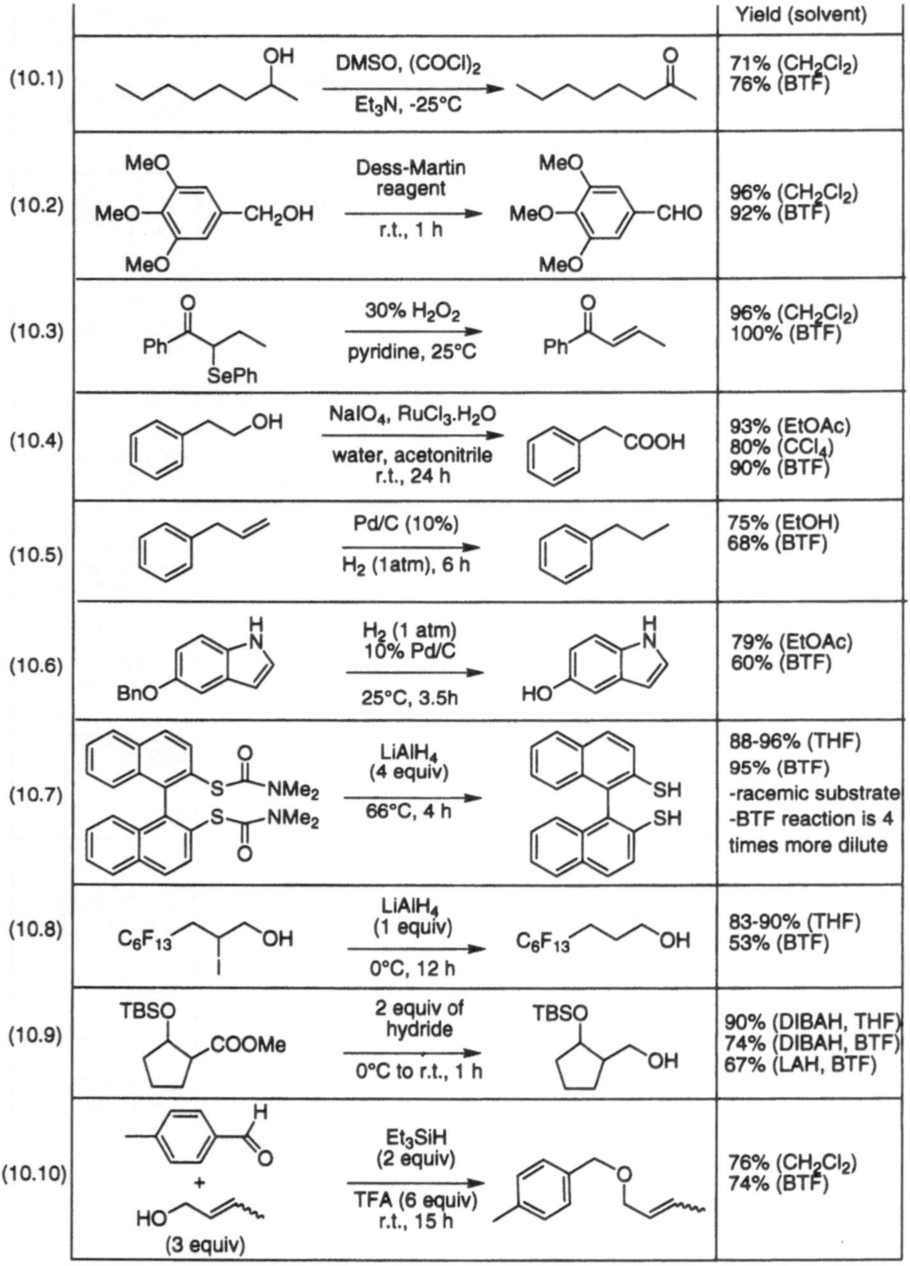

			Yield (solvent)
(10.1)	![hept-2-ol] OH	DMSO, (COCl)$_2$ / Et$_3$N, -25°C	![heptan-2-one] O	71% (CH$_2$Cl$_2$) 76% (BTF)
(10.2)	MeO, MeO, MeO —CH$_2$OH	Dess-Martin reagent / r.t., 1 h	MeO, MeO, MeO —CHO	96% (CH$_2$Cl$_2$) 92% (BTF)
(10.3)	Ph—C(=O)—CH(SePh)—CH$_2$CH$_3$	30% H$_2$O$_2$ / pyridine, 25°C	Ph—C(=O)—CH=CH—CH$_3$	96% (CH$_2$Cl$_2$) 100% (BTF)
(10.4)	Ph—CH$_2$CH$_2$OH	NaIO$_4$, RuCl$_3$.H$_2$O / water, acetonitrile r.t., 24 h	Ph—CH$_2$COOH	93% (EtOAc) 80% (CCl$_4$) 90% (BTF)
(10.5)	Ph—CH$_2$CH=CH$_2$	Pd/C (10%) / H$_2$ (1atm), 6 h	Ph—CH$_2$CH$_2$CH$_3$	75% (EtOH) 68% (BTF)
(10.6)	BnO— indole	H$_2$ (1 atm) 10% Pd/C / 25°C, 3.5h	HO— indole	79% (EtOAc) 60% (BTF)
(10.7)	binaphthyl bis-S-C(=O)NMe$_2$	LiAlH$_4$ (4 equiv) / 66°C, 4 h	binaphthyl bis-SH	88-96% (THF) 95% (BTF) -racemic substrate -BTF reaction is 4 times more dilute
(10.8)	C$_6$F$_{13}$—CH$_2$—CHI—CH$_2$OH	LiAlH$_4$ (1 equiv) / 0°C, 12 h	C$_6$F$_{13}$—CH$_2$CH$_2$CH$_2$OH	83-90% (THF) 53% (BTF)
(10.9)	TBSO, cyclopentane—COOMe	2 equiv of hydride / 0°C to r.t., 1 h	TBSO, cyclopentane—CH$_2$OH	90% (DIBAH, THF) 74% (DIBAH, BTF) 67% (LAH, BTF)
(10.10)	4-Me-C$_6$H$_4$—CHO + HO—CH$_2$—CH=CH—CH$_3$ (3 equiv)	Et$_3$SiH (2 equiv) / TFA (6 equiv) r.t., 15 h	4-Me-C$_6$H$_4$—CH$_2$—O—CH$_2$—CH=CH—CH$_3$	76% (CH$_2$Cl$_2$) 74% (BTF)

Scheme 10

led to good results. In this case, the catalyst was soluble in BTF, although the preparation took much longer compared to the preparation in THF because the starting Pd(dba)$_3$-CHCl$_3$ complex is only sparingly soluble in BTF. Other transition metal catalyzed reactions such as the Pauson-Khand reaction, [68] the Sharpless asymmetric epoxidation and the Jacobsen epoxidation [69] did not seem to work well in BTF (examples not shown).

3.9
Phase Transfer Reactions

BTF is a suitable solvent for conducting heterogeneous reactions with phase transfer catalysis. The reaction of benzyl chloride with sodium cyanide (in toluene and BTF) or potassium cyanide (in acetonitrile and BTF) using two different phase transfer catalysts [70] gave similar yields (12.1; 12.2). Aromatic nucleophilic substitution of chloro-2,4-dinitro benzene using potassium fluoride and 18-crown-6 in acetonitrile [71] or BTF (12.3) also gave similar results. The reaction in BTF went approximately two times faster. Aliphatic nucleophilic substitution of octyl bromide with the same reagents but elevated temperature (83 °C) was very slow and incomplete both in acetonitrile and BTF (12.4). In BTF, however, no elimination product was observed. The same reaction conducted in PCBTF at reflux temperature (145 °C), led only to recovery of starting material.

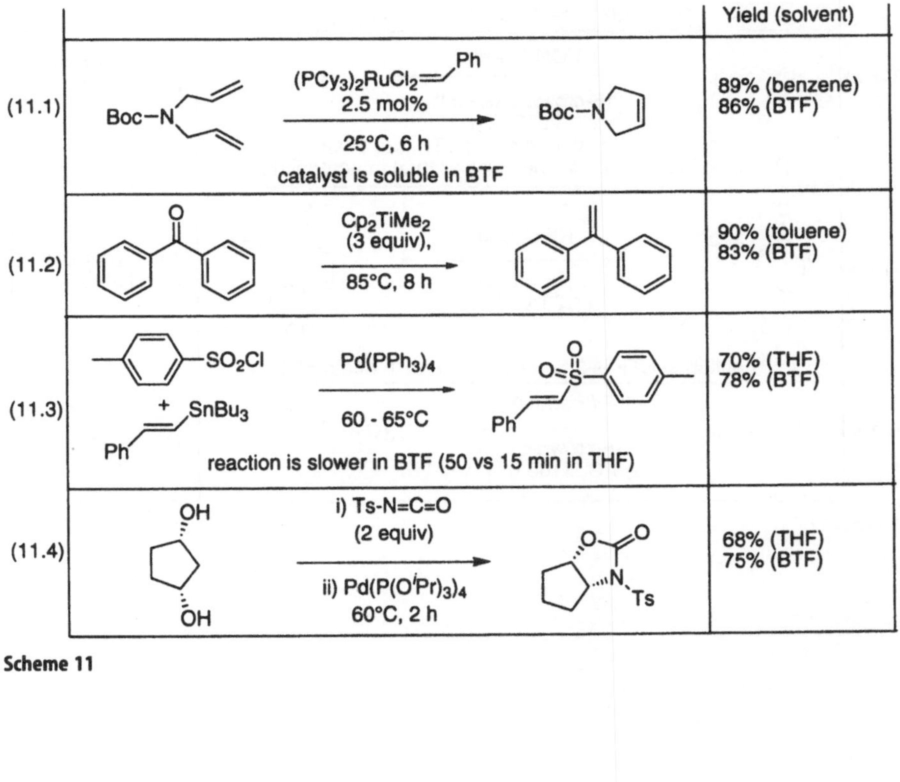

				Yield (solvent)
(11.1)	Boc—N (diallyl)	(PCy$_3$)$_2$RuCl$_2$= Ph 2.5 mol% 25°C, 6 h catalyst is soluble in BTF	Boc—N (ring)	89% (benzene) 86% (BTF)
(11.2)	benzophenone	Cp$_2$TiMe$_2$ (3 equiv), 85°C, 8 h	1,1-diphenylethylene	90% (toluene) 83% (BTF)
(11.3)	—⟨⟩—SO$_2$Cl + Ph SnBu$_3$	Pd(PPh$_3$)$_4$ 60 - 65°C reaction is slower in BTF (50 vs 15 min in THF)	O=S⟨⟩— Ph	70% (THF) 78% (BTF)
(11.4)	cyclopentane diol (OH, OH)	i) Ts-N=C=O (2 equiv) ii) Pd(P(O^{i}Pr)$_3$)$_4$ 60°C, 2 h	cyclic carbamate N-Ts	68% (THF) 75% (BTF)

Scheme 11

3.10
Limitations of BTF

There is no general solvent that is useful for all reactions, and BTF naturally has its limitations. In addition to the limitations posed by the freezing point, boiling point and chemical stability mentioned before, BTF is not very Lewis-basic and therefore is not a good substitute for reactions that require solvents like ethers, DMF, DMSO, etc. Not surprisingly, ions are not readily dissolved in BTF and many types of "anionic" reactions do not work well in BTF. For example, attempted deprotonations of esters and ketones with LDA in BTF were not successful. Reaction of diethyl malonate with NaH (5 equiv) and reaction with MeI[72] (6 equiv) in BTF was very heterogeneous and yielded 60% of the dimethylated product, compared to 89% in THF. No reaction was observed if the same malonate anion was used as a nucleophile in a Pd-catalyzed allylic substitution reaction in BTF (see 3.7). Wittig reactions also did not work very well in BTF. The ylid of ethyl triphenyl phosphonium bromide [73] was formed only slowly in BTF, and the characteristic deep red color was never obtained.

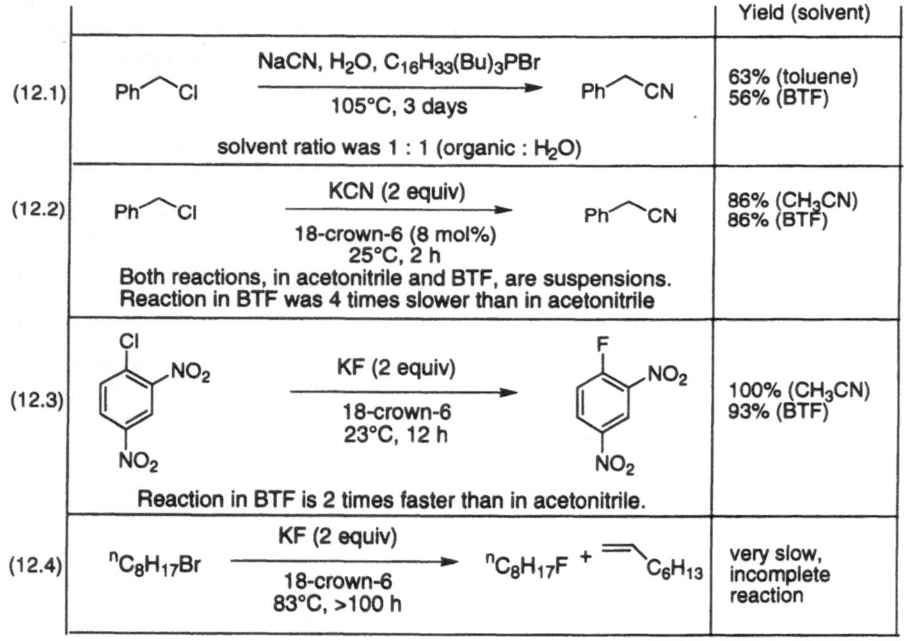

		Yield (solvent)
(12.1)	Ph–Cl $\xrightarrow[\text{105°C, 3 days}]{\text{NaCN, H}_2\text{O, C}_{16}\text{H}_{33}\text{(Bu)}_3\text{PBr}}$ Ph–CN solvent ratio was 1 : 1 (organic : H$_2$O)	63% (toluene) 56% (BTF)
(12.2)	Ph–Cl $\xrightarrow[\substack{\text{18-crown-6 (8 mol\%)} \\ \text{25°C, 2 h}}]{\text{KCN (2 equiv)}}$ Ph–CN Both reactions, in acetonitrile and BTF, are suspensions. Reaction in BTF was 4 times slower than in acetonitrile	86% (CH$_3$CN) 86% (BTF)
(12.3)	Cl–C$_6$H$_3$(NO$_2$)$_2$ $\xrightarrow[\substack{\text{18-crown-6} \\ \text{23°C, 12 h}}]{\text{KF (2 equiv)}}$ F–C$_6$H$_3$(NO$_2$)$_2$ Reaction in BTF is 2 times faster than in acetonitrile.	100% (CH$_3$CN) 93% (BTF)
(12.4)	^{n}C$_8$H$_{17}$Br $\xrightarrow[\substack{\text{18-crown-6} \\ \text{83°C, >100 h}}]{\text{KF (2 equiv)}}$ ^{n}C$_8$H$_{17}$F + C$_6$H$_{13}$	very slow, incomplete reaction

Scheme 12

3.11
Evaluation of Other Uses of BTF: Extraction, Chromatography

BTF is not miscible with water and hence, it can be used for extraction purposes. Like dichloromethane, it is heavier than water. We observed, however, that phase separation between water and BTF is sometimes difficult. BTF can be used as solvent for chromatography as well. However, it has a typical intense aromatic odor, which is not convenient when using large quantities. In addition, because of the high boiling point, the solvent removal process is relatively slow (compared to solvents like ether and dichoromethane) and loss of relatively volatile products can occur. Because of these practical disadvantages, we have limited the use of BTF as a chromatography solvent to the purification of fluorous compounds (see 4.0).

4
BTF as a Solvent in Fluorous Synthesis

4.1
Introduction: Fluorous Synthesis

A number of fluorous methods have recently been introduced to facilitate the separation of organic reaction components, and these techniques are potentially useful all the way from large scale organic reactions to small scale parallel and combinatorial synthesis [74]. All these techniques rely on the ability of "fluorous" (highly fluorinated) molecules to partition in the fluorous phase in a liquid-liquid (or solid-liquid) extraction between an organic solvent and a fluorous solvent (or fluorous solid phase [75]). A key feature of all fluorous techniques is that at the workup stage, all desired (*or* all undesired) products that are labeled with a fluorous tag can be quickly separated from non-tagged products by using simple workup techniques. [76]

4.2
Solubility Issues in Fluorous Synthesis

The possibility to run reactions in a homogeneous fashion is one of the attractive features of solution phase synthesis techniques over the conceptually related solid phase synthesis techniques. However, the incorporation of fluorous chains to permit molecules to partition into a fluorous phase naturally begins to alter the solubility properties of the resulting fluorinated organic molecules. Indeed, molecules with very large fluorous domains can have little or no solubility in many common organic solvents. Thus, solubility and selection of a reaction solvent are crucial considerations in designing fluorous reactions or reactions sequences.

The reaction solvent problem can be addressed in several ways. First it is possible to use mixtures of fluorocarbon and organic solvents. Many of these mixtures are biphasic, and thus reactions conducted in them are heterogeneous, but homogenous blends can sometimes be obtained by heating. A problem with this approach is that fluorocarbons are very non-polar and are extraordinarily poor solvents for most organic molecules. A second approach is to select organic solvents that have some capability to dissolve fluorous compounds, and ethers (especially THF) are very useful in this regard. But perhaps the most general approach is to select a lightly fluorinated organic solvent. Experience shows that these "hybrid" solvents do indeed have the capability to dissolve both organic and fluorous reactions components. Among the lightly fluorinated solvents that have been used so far in fluorous synthesis, BTF is by far the most important because it serves as a suitable replacement for most chlorocarbon and hydrocarbon solvents (which might not dissolve the fluorous reaction component). It does not serve as a replacement for Lewis basic solvents like THF or DMF, but these solvents can often be used directly or BTF can simply be added as a cosolvent to aid in dissolution of the fluorous reaction component. The selected examples that are shown below serve only as representatives of the many successful fluorous reactions that we have conducted in BTF.

4.3
Examples of Fluorous Synthesis in BTF

4.3.1
Radical Reactions Using Fluorous Tin Hydride [77]

One of the first applications of fluorous synthesis was the use of a fluorous tin hydride 1 (Scheme 13) in radical reactions. The fluorous chains on the tin enable removal by fluorous-organic extraction, but at the same time, the solubility of 1 in benzene and other organic solvents is dramatically decreased. Consequently, reduction of 2 using 1 in benzene gave only low yields of reduced product. However, BTF is able to solubilize both the organic substrate and the fluorous reagent 1, and reduction of adamantyl bromide in BTF gave adamantane 3 in 90% yield (Scheme 13).

$$\text{Ad-Br} + (C_6F_{13}CH_2CH_2)_3SnH \xrightarrow[\text{reflux}]{\text{AIBN}} \text{Ad-H} + (C_6F_{13}CH_2CH_2)_3SnBr$$

1.2 equiv   90%

2   **1**   **3**

Scheme 13

A catalytic procedure using $NaCNBH_3$ for regeneration of the tin hydride has also been developed. In this case, t-BuOH is added as a cosolvent to solubilize the borohydride. The ease of the fluorous extraction procedure contrasts sharply with the difficult removal of the organic tributyltin derivatives and is further illustrated by the successful application of the fluorous tin hydride in a parallel synthesis using the Giese reaction (Scheme 14). This catalytic procedure using BTF as the reaction solvent is now among the most convenient for conducting tin hydride radical reactions. As an added bonus, the tin reagents are readily recovered from the fluorous phase in both the stoichiometric and catalytic procedures.

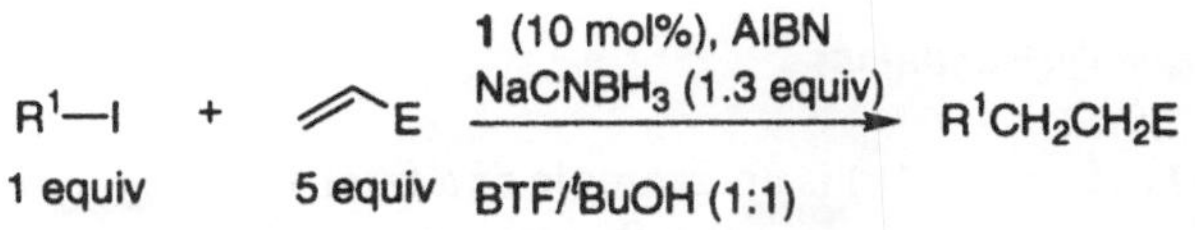

$$R^1\!-\!I \;+\; \diagup\diagdown E \xrightarrow[\text{BTF/}^t\text{BuOH (1:1)}]{\substack{\textbf{1} \text{ (10 mol\%), AIBN} \\ \text{NaCNBH}_3 \text{ (1.3 equiv)}}} R^1CH_2CH_2E$$

1 equiv   5 equiv

R^1 : E	CN	CO_2Me	COMe
$C_{15}H_{31}$	72%	92%	67%
$c\text{-}C_6H_{11}$	75%	65%	75%
Ad	89%	94%	78%

Scheme 14

4.3.2
Fluorous Synthesis of Isoxazolines [78]

The power of fluorous strategies in multitstep synthesis is illustrated in the synthesis of isoxazolines (Scheme 15). The substrate **6** is made fluorous through the (necessary) protection of the hydroxyl functionality of **5** as the silyl ether. This reaction was done in THF but could be performed in BTF as well (see 3.5). Workup was effected by three-phase extraction (water/dichloromethane/perfluorohexanes). Cycloaddition reaction of **7** with a 10-fold excess of nitrile oxide leads to fluorous isoxazoline **8**. Fluorous-organic extraction separates the fluorous isoxazoline from the excess nitrile oxide **7**, furoxan dimer byproduct **9**, and any other excess reagents or reagent byproducts. This reaction was performed in BTF since the starting silyl ether **6** was not completely soluble in dichloromethane. Finally, removal of the fluorous tag leads to the isoxazoline **10** as the only product in the organic phase. The sequence shows that fluorous synthesis is very effective for obtaining pure compounds using fast extraction procedures, and that BTF is an essential solvent for conducting reactions with fluorous substrates. The generality of the cycloaddition reaction in BTF was evident as over a dozen successful examples were conducted.

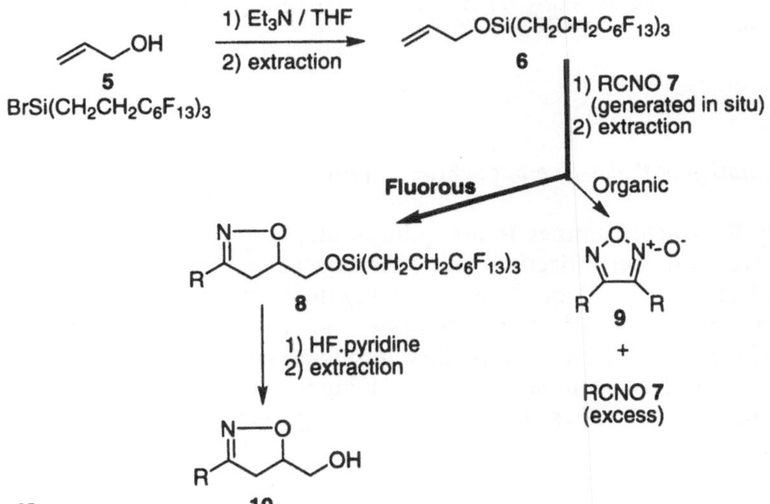

Scheme 15

4.3.3
Fluorous Tin Azide Cycloadditions

The tin azide 11 (Scheme 16) is an example of a fluorous reagent that, unlike the fluorous tin hydride 1, transfers the fluorous tag to the product. This type of "phase switch" is very powerful; only the nitrile that reacts with 11 is "switched" to the fluorous phase whereas any unreacted nitrile or impurities are left behind in the organic phase. Cycloaddition reaction of 11 with excess nitrile leads to fluorous tetrazoles 12. Again, BTF is a unique reaction solvent that serves to dissolve all reactants. Purification is effected in the usual way via fluorous-organic extraction. Deprotection yields the tetrazoles 13 in good yields and purities.

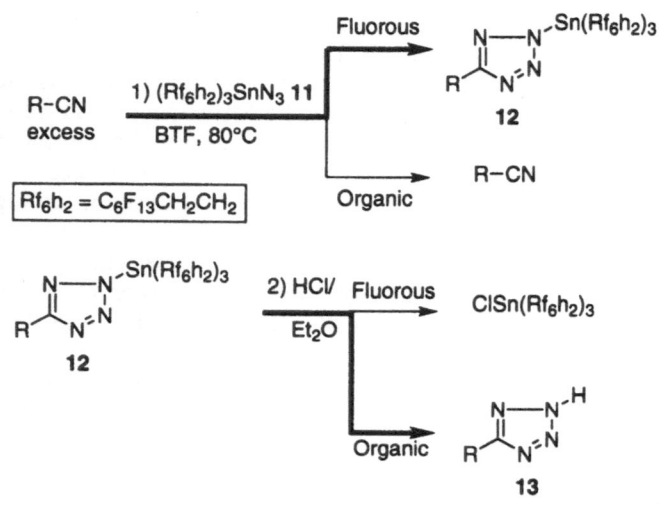

Scheme 16 Yields: R $= CH_3$, 83%; $PhCH_2$, 77%; p-$CH_3C_2H_4$, 61%; p-$MeOC_6H_4$, 72%

4.3.4
Derivatization of Highly Fluorous Substrates [79]

Highly fluorinated amines 14 are quite insoluble in dichloromethane or even THF. However, derivatization under homogeneous conditioning is possible using BTF as solvent since both the highly fluorous substrates 14 and organic reagents like p-toluenesulfonyl chloride and dansyl chloride are soluble in BTF. The BTF is removed before the fluorous-organic extraction. This example shows that BTF is useful not only in applying fluorous reagents to organic synthesis but also in synthesizing the fluorous reagents themselves.

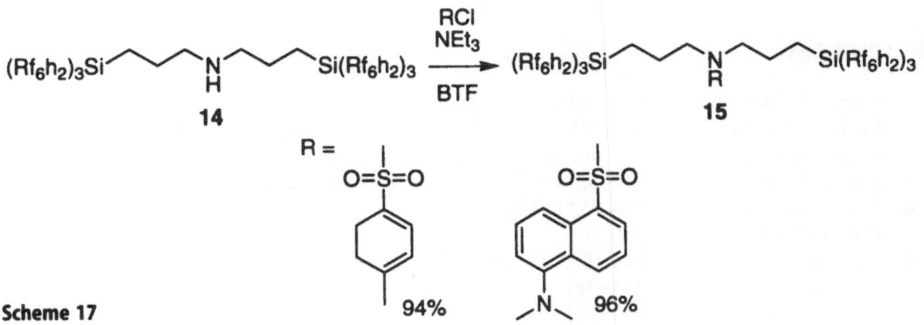

Scheme 17

5
Conclusions

The use of BTF as a solvent for organic reactions has only been studied over the last three years, but its characteristics and potential are beginning to emerge. BTF is readily available in highly pure form in bulk quantities at moderate prices. Its physical properties and favorable toxicity and environmental properties may render it advantageous to substitute BTF or related solvents for chloro- or hydrocarbon solvents in a diverse range of reactions. However, BTF is not Lewis basic and there is probably only a niche role for BTF as a replacement (or additive) for ether, alcohol, amide and related solvents.

While it is instructive to view BTF as a potential substitute for other types of solvents, chemical common sense and experience teach us that no one solvent is an exact substitute for another. Indeed, some applications where BTF is advantageous are already beginning to emerge. For example, in our group BTF has displaced benzene as the solvent of first choice for many kinds or radical reactions. And the unique features of BTF and related lightly fluorinated solvents for fluorous synthesis techniques cannot be substituted for by any traditional organic solvent.

Although the use of non-traditional methods to conduct organic reactions (solventless, supercritical CO_2, etc.) is an area that is increasing in importance, the vast majority of organic reactions are still conducted in an organic solvent. However, environmental, safety toxicity and cost concerns have been gradually contracting the number of acceptable organic solvents over the past few decades. The early work on BTF and related solvents shows very good potential, and it would appear that a rare opportunity has arisen for the organic community to expand its stable of commonly useful solvents.

Acknowlegements. Bruno Linclau is indebted to the Belgian American Educational Foundation for a fellowship. Bruno Linclau and Dennis P. Curran thank Oxychem for generous supplies of BTF and related solvents and Oxychem, NIH and NSF for research funding. They also acknowledge the contributions of the many coworkers who ran comparative reactions in BTF and organic solvents. Included among these are: David Bom, Christine Chen, Daniel Christen, Dr. Mark Ebden, Dr. Matthias Frauenkron, Ana Gabarda, Giovanna Gualtieri, Gregory Hale, Brian Haney, Ulrich Iserloh, Sun-Young Kim, SooKwang Lee, Weidong Liu, Zhiyong Luo, Dr. Lena Ripa, Alexey Rivkin, Dr. Ashvani Singh, Maria Christina Solimini, and Prof. Seiji Takeuchi.

6
References

1. Ogawa A, Curran DP (1997) J Org Chem 62:450
2. Barbour AK, Belf LF, Buxton MW (1963) In: Stacey M, Tatlow JC, Edwards AG (eds) Advances in Fluorine Chemistry, vol 3. Butterworths, London, p 81
3. Brown JH, Suchling CW, Wholley WB (1949) J Chem Soc, Supp Issue S 95
4. Lademann R et al, US 3,966,832 (1978)
5. Robota S, US 3,859,372 (1975)
6. Baxamusa J, Robota S, US 4,183,873 (1980)
7. Sendlak LP, US 4,129,602 (1978)
8. Sendlak LP, US 4,130,594 (1978)
9. Ramanadin A, Seigneurin S, US 4,462,937 (1984)
10. Schlosser M, Katsoulos G, Takagishi S (1990) Synlett 747
11. Lefave GM (1949) J Am Chem Soc 71:4148
12. Simon JH, McArthur RE (1947) Ind Eng Chem 39:366
13. Hudlicky M (1976) Chemistry of Organic Fluoride Compounds. Ellis Harwood, p 274
14. a) Henne AL, Newman MS (1938) J Am Chem Soc 60:1697; b) Ramchandani RK, Wakharkar RD, Sudalai A (1996) Tetrahedron Lett 37:4063
15. OXSOL is a registered trademark of Occidental Chemical Corporation
16. 40 CFR 51.100
17. Nagy G, Tramontana D (1995) Am Paint Coatings J 37
18. Hare CH (1997) Paint and Coatings Industry, XII, 10:202
19. Hare CH (1998) Modern Paint Coatings, 1:30
20. Registry of Toxic Effects of Chemical Substances (RTECS) compiled by the National Institute for Occupational Safety and Health of the U S Department of Health and Human Services, 1998, *http://www.tomescps.com* (accessed January 1999)
21. Hazardous Substance Data Bank (HSDB) compiled by the National Library of Medicine, 1998, *http://www.tomescps.com* (accessed January 1999)
22. (1996) Sax's Dangerous Properties of Industrial Materials, 9th edn. Van Nostrand Reinhold, New York
23. Knaak JB, Smith LW, Fitzpatrick RD, Olson JR, Newton PE (1998) Inhalation Toxicology 10:65
24. SIDS Initial Assessment Report, OECD, 1996 – >99% to air
25. 59 FR 50693–96, October 5, 1994
26. 40 CFR 51.100
27. 63 FR 53421, October 5, 1998
28. Manahan SE (1992) Toxicological Chemistry. Ann Arbor, MI, 138
29. National Research Council (1995) Prudent Practices in the Laboratory Handling and Disposal of Chemicals. National Academy Press, Washington DC
30. Yaws CL (1994) Handbook of Vapor Pressure, vol 2. Gulf, Houston, p 269
31. Perrin DD, Armarego WLF, Perrin DR (1980) Purification of Laboratory Chemicals, 2nd edn. Pergamon Press, New York
32. E_T^N values are an empirical parameter of solvent polarity which are derived by measurement of the long-wave UV/ Visible absorption band of the negative solvachromatic Pyridinium-N-phenoxide betaine dyes in the solvent being studied. Higher values are an indication of greater solvent polarity. The above values were taken from the text by Reichardt C (1990) Solvents and Solvent Effects in Organic Chemistry, 2nd edn. VCH Publishing, Weinheim, p 365
33. Gómez AM, López JC, Fraser-Reid B (1993) Synlett 10:943
34. a) Kistyakowsky GB, Tichenor RL (1942) J Am Chem Soc 64:2302; b) Tarbell DS, Kincaid JF (1940) J Am Chem Soc 62:728
35. Diels O, Alder K (1928) Liebigs Ann Chem 460:98
36. Curran DP, Hale GR, Geib SJ, Balog A, Cass QB, Degani ALG, Hernandes MZ, Freitas LCG (1997) Tetrahedron:Asymmetry 8:3955

37. Giese B (1983) Angew Chem Int Ed Engl 22:753
38. Hanessian S, Léger R, Alpegioni M (1992) Carbohyd Res 228:145
39. Curran DP, Liu H, Josien H, Ko SB (1996) Tetrahedron 52:11385
40. Giese B, Kretzschmar G (1983) Chem Ber 116:3267. See also ref. 37
41. a) Kikukawa K, Umekawa H, Wada F, Matsuda T (1988) Chem Lett 881; b) Labadie JW, Stille JK (1983) J Am Chem Soc 105:6129
42. a) Henne AL, Newman MS (1938) J Am Chem Soc 60:1697; b) Ramchandani RK, Wakharkar RD, Sudalai A (1996) Tetrahedron Lett 37:4063
43. Kulka M (1954) J Am Chem Soc 76:5469
44. a) Sakurai H, Hosomi A, Hayashi J (1984) Org Synth 62:86 b) Fleming I, Dunogués J, Smithers R (1989) Org React 37:57
45. Mukaiyama T, Banno K, Narasaka K (1974) J Am Chem Soc 96:7503
46. Poll T, Sobczak A, Hartmann H, Helmchen G (1985) Tetrahedron Lett 26:3905
47. Drury WJ III, Ferraris D, Cox C, Young B, Lectka T (1998) J Am Chem Soc 120:11006.
48. Ishihara K, Hanaki N, Yamamoto H (1983) J Am Chem Soc 115:10695.
49. a) Höfle G, Steglich W, Vorbruggen H (1978) Angew Chem Int Ed Engl 17:569; b) Höfle G, Steglich W (1972) Synthesis 619.
50. Chandhary SK, Hernandez O (1979) Tetrahedron Lett 20:99.
51. Henkel JG, Spurlock LA (1973) J Am Chem Soc 95:8339
52. Sigiyama K, Hirao A, Nakahama S (1996) Macromol Chem Phys 197:3149
53. Strekowski L, Wydra RL, Harden DB, Honkan VA (1990) Heterocycles 31:1565
54. Kitagawa O, Izawa H, Sato K, Dobashi A, Taguchi T (1998) J Org Chem 63:2634
55. Mitsunobu O (1980) Synthesis 1
56. a) Azord-Hossain M (1997) Tetrahedron Lett 38:49. b) Stang PJ, Trepkow W (1980) Synthesis 983
57. Okuma K, Swern D (1978) Tetrahedron 34:1651
58. a) Dess DB, Martin JC (1983) J Org Chem 48:4155; b) Ireland RE, Liu L (1993) J Org Chem 58:2899; c) Dess DB, Martin JC (1991) J Am Chem Soc 113:7277
59. a) Reich HJ, Wollowitz S (1993) Org React 44:1; b) Reich HJ, Renga JM, Reich IL (1975) J Am Chem Soc 97:5434
60. Prasad M, Lu Y, Kim H-Y, Hu B, Repic O, Blacklock TJ, submitted for publication
61. Ek A, Witkop B (1954) J Am Chem Soc 76:5579
62. Fabbri D, Delogu G, De Lucchi O (1993) J Org Chem 58:1748
63. Josien H, Ko SB, Bom D, Curran DP (1998) Chem Eur J 4:67
64. Fu GC, Nguyen ST, Grubbs RH (1993) J Am Chem Soc 115:9856
65. Perasis NA, Bzowej EI (1990) J Am Chem Soc 112:6392
66. Labadie SS (1989) J Org Chem 54:2496
67. Trost BM, Van Vranken DL (1993) J Am Chem Soc 115:444
68. Chung YK, Lee BY, Jeong N, Hudecek M, Pauson PL (1993) Organometallics 12:220
69. Jacobsen EN, Zhang W, Muci AR, Ecker JR, Deng L (1991) J Am Chem Soc 113:7063
70. Starks CM, Liotta C (1978) Phase Transfer Catalysis, Principles and Techniques. Academic Press, New York, p 110
71. Liotta CL, Harris HP (1974) J Am Chem Soc 96:2250
72. Reynolds KA, O'Hagan D, Gani D, Robinson JA (1988) J Chem Soc Perkin Trans I 3195
73. Vedejs E, Meier GP, Snoble KAJ (1981) J Am Chem Soc 103:2823
74. a) Studer A, Hadida S, Ferrito R, Kim S-Y, Jeger P, Wipf P, Curran DP (1997) Science 275:823; b) Curran DP (1996) Chemtracts-Organic Chemistry 9:75; c) Horvath, IT (1998) Acc Chem Res 31:641; d) Cornils B (1997) Angew Chem Int Ed 36:2057
75. a) Curran DP, Hadida S, He M (1997) J Org Chem 62:6714; b) Kainz S, Luo ZY, Curran DP, Leitner W (1998) Synthesis 142
76. Curran DP (1998) Angew Chem Int Ed 37:117
77. Curran DP, Hadida S (1996) J Am Chem Soc 118:2531
78. Studer A, Curran DP (1997) Tetrahedron 53:6681
79. Linclau B, Singh AK, Curran DP (1999) J Org Chem, in press
80. Nishikido J, Nakajima H, Saeki T, Ishii A, Mikami K (1998) Synlett 1347

Reactions in Supercritical Carbon Dioxide (scCO$_2$)

Walter Leitner

Max-Planck-Institut für Kohlenforschung, Kaiser-Wilhelm-Platz 1,
D-45470 Mülheim an der Ruhr, Germany.
E-mail: leitner@mpi-muelheim.mpg.de

The present contribution highlights recent developments in the application of compressed (liquid or supercritical) carbon dioxide as a solvent for chemical reactions. After a brief introduction to the basic physical properties of scCO$_2$, some practical aspects of the use of compressed gases are discussed. A survey of successful applications of compressed and particularly supercritical CO$_2$ in organic synthesis is provided with an emphasis on metal-catalyzed reactions.

Keywords. Environmentally benign synthesis, Catalysis, „Green" solvents, Supercritical fluids, Carbon dioxide.

Topics in Current Chemistry, Vol. 206
© Springer-Verlag Berlin Heidelberg 1999

Abbreviations

acac	acetylacetonate
ADMET	acyclic diene metathesis
BARF	tetrakis[3,5-bis(trifluoromethyl)phenyl]borate
BINAP	2,2'-bis(diphenylphosphino)-1,1'-binaphtyl
BINAPHOS	2-(diphenylphosphino)-1,1'-binaphten-2'yl-1,1'-binaphten-2,2'-diylphosphite
CESS	catalysis and extraction using supercritical solutions
cod	*cis,cis*-1,5-cyclooctadiene
d_c	critical density
dimcarb	dimethylammonium carbamate
DMSO	dimethysulfoxide
DMF	*N,N*-dimethylformamide
dppb	1,4-bis(diphenylphosphino)butane
dppe	1,2-bis(diphenylphosphino)ethane
DuPHOS	1,2-bis(2,5-dialkylphospholano)benzene)
FTIR	Fourier transform infrared (spectroscopy)
hfacac	hexafluoroacetylacetonate
nbd	norbornadiene
NMR	nuclear magnetic resonance (spectroscopy)
p_c	critical pressure
PPA	polyphenylacetylene
RCM	ring closing metathesis
ROMP	ring opening metathesis polymerization
sc	supercritical (with compounds)
SCF	supercritical fluid
SFE	supercritical fluid extraction
T_c	critical temperature
Tos	tosylate, $p\text{-}CH_3C_6H_4SO_3^-$
tpp	triphenylphosphine
UV-Vis	ultraviolet-visible (spectroscopy)
$z\text{-}H^xF^y$	substituent of type $(CH_2)_x(CF_2)_yF$ in position(s) z of an aryl ring

1
Introduction

Supercritical fluids (SCFs) have fascinated researchers ever since the existence of a critical temperature was first noted more than 175 years ago [1]. Although initial studies focussed mainly on the physical properties of supercritical phases, their chemical reactivity was of interest from the beginning, too [2]. In fact, several well established industrial processes for the production of bulk chemicals occur under temperatures and pressures beyond the critical data of the reaction mixture, the Haber-Bosch process and the high pressure polymerization of ethylene being just the most outstanding examples. Based on the pioneering work of Kurt Zosel at the Max-Planck-Institut für Kohlenforschung in Mülheim [3], SCFs have also been used commercially for more than two decades as solvents for separation processes in food industry [4]. Supercritical fluid extraction (SFE) using supercritical carbon dioxide (scCO$_2$) is part of the current state-of-the-art for the industrial production of decaffeinated coffee and of hops aroma. In sharp contrast, the interest in the use of SCFs as reaction media for complex organic syntheses has sparked mainly during the last five to ten years and it is just fair to refer to SCFs as "modern solvent systems" in this context [5].

The unique physico-chemical properties of the supercritical state, as briefly outlined in Sect. 2, make SCFs in general highly attractive reaction media for chemical synthesis. Probably the most important incentive for the use of SCFs in organic chemistry comes, however, from in an increasing demand for environmentally and toxicologically benign processes for the production of high-value chemicals. Water and carbon dioxide are clearly the most attractive solvents for such applications of "Green Chemistry" [6]. In practical systems, the choice of an SCF for a synthetic application will also depend on the critical data, which must not be too drastic in order to keep equipment and process costs to a minimum. The present chapter will therefore focus on the use of compressed (supercritical or liquefied) CO$_2$, because it best meets the criteria of ecological and economical constraints. The reader should be aware, however, of the rich and exciting chemistry which is carried out using other SCFs, especially in supercritical alkanes and in the more polar solvents scCHF$_3$ and scH$_2$O. We will further restrict our discussion mainly to applications of CO$_2$ in organic syntheses using homogeneous or heterogeneous metal catalysts. Related areas like free-radical polymerization, the synthesis of inorganic or organometallic compounds, enzymatic reactions, and reactions that are used mainly as probes for near-critical phenomena are outside the scope of the present review. For additional information on these topics and for further reading, we recommend the thematic issue of *Chemical Reviews* on SCFs [7] and a recent monograph on their use in chemical synthesis [5].

2
Physico-Chemical Properties of Supercritical Carbon Dioxide as a Solvent

The schematic representation of the phase diagram of pure carbon dioxide in Fig. 1 shows the aggregation state of CO$_2$ as a function of pressure and temperature. The solid, liquid, and gaseous state are separated by the melting, sublimation, and evaporation curve, respectively. These three states are in equilibrium at the triple point. The critical point marks the high-temperature end of the evaporation line and is characterized by the critical temperature $T_c = 31.1$ °C and the critical pressure $p_c = 73.8$ bar [8]. No distinct liquid or vapor phase can exist beyond the critical point and the new supercritical phase has properties which are often reminiscent of both states. Similar to gases, scCO$_2$ has a very low surface tension, low viscosity, and high diffusion rates. Furthermore, scCO$_2$ forms fully homogeneous single phase mixtures with many reaction gases over a wide range of composition. On the other hand, the density of scCO$_2$ can easily be

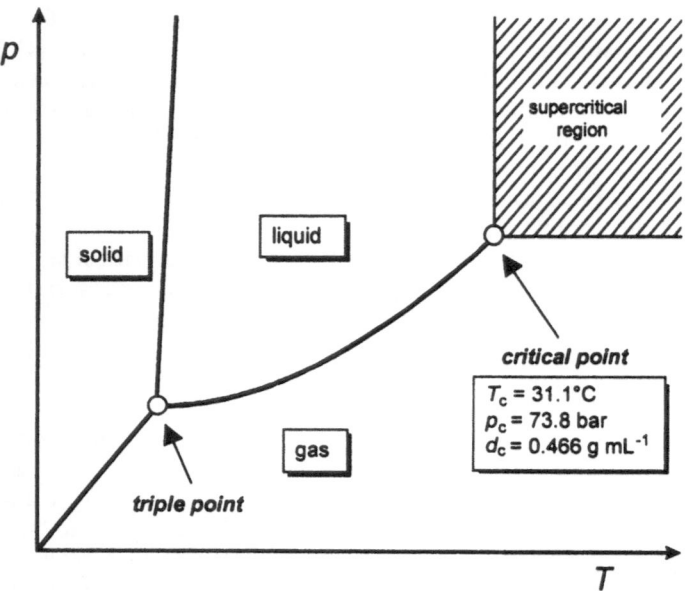

Fig. 1. Schematic phase diagram of carbon dioxide

adjusted to liquid-like values and the solubility of liquid or solid material in these media can be orders of magnitude higher than predicted from the ideal gas law.

The bulk density of CO_2 at the critical point ($d_c = 0.466$ g mL^{-1} [8]) is the mean value of the densities of the liquid and the gaseous phase just before entering the supercritical region. It must be noted, however, that the local density of an SCF is subject to large fluctuations and may differ considerably from the bulk density, especially around solute molecules [9]. Many attempts have been made to uti-lize "chemical probes" like radical reactions or Diels-Alder additions to investigate these so-called clustering or augmentation phenomena [10]. For most practical applications, however, it seems more important that the high compressibility of the supercritical phase allows the bulk density of $scCO_2$ to be varied continuously from very low to liquid-like values with relatively small variations in temperature and/or pressure. For example, the density of $scCO_2$ at 37 °C is only 0.33 g mL^{-1} at 80 bar, but it rises to 0.80 g mL^{-1} at 150 bar [8].

Variation in solvent density corresponds to changing the amount of CO_2 in a reactor of constant volume, and hence the chemical potential and the mole fraction of a solute can be varied at constant molar (mole per volume) concentration. Obviously, such changes may have a strong impact on chemical equilibria and reaction rates, which in turn determine yields and selectivities of synthetic processes [11]. In addition, a number of solvent properties of the fluid phase are directly related or change in parallel with the density. Accordingly, such properties can be "tuned" in $scCO_2$. Variation of the so-called "solvent power" is the most obvious application and discussed in more detail below.

The dielectric constant of an SCF varies also in parallel with density, and this change may directly influence chemical reactions. It must be noted, however, that the polarity of pure CO_2 varies only in a fairly limited regime between approximately 1.0 and 1.6, indicating that pure $scCO_2$ is always a fluid of very low polarity. The polarity can be substantially increased with even small amounts of polar co-solvents, and reagents may of course play this role as well.

The phase behavior of the reaction mixture is of paramount importance for a chemical reaction in $scCO_2$. In order to exploit the physical properties of the supercritical state most effectively during a reaction, one will generally try to define conditions where a single homogeneous phase is present. On the other hand, it may be desirable to achieve controlled separation during work-up or to establish two-phase systems leading, for example, to catalyst immobilization. It is therefore mandatory to carry out chemical reactions in $scCO_2$ using window-equipped reactors that allow for the visual control of the reaction mixture (Sect. 3). Furthermore, the expected phase behavior should be already consider-ed during the design of a process to provide at least a rough estimate of pro-mising reaction conditions.

Unfortunately, the simple picture shown in Fig. 1 is only valid for pure CO_2 and the phase behavior of mixtures is much more complicated [4]. It is not suf-ficient to adjust temperature and pressure above the critical values of pure CO_2 to assure a single-phase reaction mixture in the presence of other components! The phase behavior of mixtures is a function of composition and the actual phase diagram can vary considerably even for seemingly similar components. Reaction solutions are mixtures of at least three (substrate, product and sol-vent) but in most cases more components of variable composition and a full description of the phase behavior is not practical in most cases. Fortunately, most exploratory studies towards chemical synthesis are carried out under fairly dilute conditions (substrate concentrations $\ll 10$ mol %), and the problem can therefore often be simplified to getting an idea of the solubility of the various components in $scCO_2$.

The "solvent power" of a fluid phase is of course related to its polarity, but depends also strongly on the bulk density of the SCF directly [3, 12]. As a result, the solubility of a substrate in $scCO_2$ can be varied considerably with pressure and temperature, whereby higher density corresponds generally to higher solu-bility. Appreciable solubilities in pure $scCO_2$ usually require densities of $d > d_c$ and some researchers refer to reaction phases as "supercritical" only if they ful-fill this additional criterion. The solvent properties at the critical point show, however, a rapid but gradual onset rather than a sharp discontinuity. Naturally, the structure of the solute plays also an important role for its solubility in $scCO_2$ with three factors being most important. Firstly, the solubility increases with increasing vapor pressure of the substrate. Secondly, compounds with low po-larity are better soluble than very polar compounds or salts. Finally, some spe-cific groups like perfluoro or polysiloxane substituents are known to result in a high affinity to compressed CO_2 beyond simple polarity arguments. These "CO_2-philic" substituents can lead to quite dramatic solubility enhancements allowing for example the use of $scCO_2$ as a solvent for high molecular weight polymers [13] and for metal ions [14]. They play also a major role in the design of surfactants and dispersion agents used to generate microemulsions or emulsions of $scCO_2$ with otherwise immiscible organic liquids or even water [15]. Most recently, such homogeneous water/$scCO_2$ mixtures have been shown to allow simple inorganic [16] and organic [17] reactions to be carried out under mild conditions in SCFs providing an aqueous microenvironment.

A number of potential benefits can be associated with the use of $scCO_2$ in organic synthesis and the following brief summary is intended to provide guidelines for the selection of interesting target reactions. The environmental and toxicological benefits upon replacement of potentially hazardous organic solvents are obvious and may be of particular importance in the production of biologically active fine chemicals like pharmaceuticals, food additives, cosmetics, or agro-chemicals. The physical properties discussed above open a wide range of fascinating applications, offering for example density as an additional reaction parameter. This may be of particular importance for processes that are highly sensitive to small variations in the reaction environment. The selective and variable solvent power allow for enhanced separation schemes which seems especially attractive for the recovery of homogeneous catalysts in integrated processes of catalysis and extraction using supercritical solution (CESS process, [18, 23]). The high miscibility with reactant gases excludes problems arising from slow gas/liquid diffusion which are often encountered in two-phase reaction systems. The combination of fast diffusion with comparably good solution ability is particularly attractive for heterogeneous catalysis, where reaction rates and catalyst lifetime are often limited by mass transfer and coking, respectively [19].

Chemical interactions of CO_2 with substrates, products or catalysts can also play a major role in defining rate and selectivity of a given reaction. This chemical influence must not necessarily be positive, making it even more important to remember that CO_2 does not always provide an "inert" medium. For example, hydrogen carbonate and protons are generated in the presence of water (pH $\approx$ 3), carbamic acids or carbamates are formed with basic N-H functionalities, and the coordination ability and reactivity towards various transition metal centers is well established [20]. One of the appealing prospects of the chemical reactivity of CO_2 is its simultaneous use as solvent *and* C_1 building block in metal-catalyzed processes.

3
Basic Equipment and Reactor Design

Working with $scCO_2$ or other compressed gases involves the use of high pressure equipment and rigorous safety precautions have to be taken [21]. The legal requirements vary from country to country and depend on the size of the equipment as well as the maximum pressures and temperatures. In general, no special buildings or autoclave chambers are required for laboratory scale equipment (< 250 mL) at pressures below 300 bar. Regardless of the volume, no manipulation apart from the opening or closing of sampling or addition valves should ever be carried out on systems under pressure. Furthermore, it must be assured that the operator is never directly exposed to any pressurized equipment. Various designs of strong polycarbonate shields or covers can be used for this purpose.

A variety of different reactor systems has been described for the use of $scCO_2$ in batch-wise and continuos operation. In principle, all systems consist of a central high pressure autoclave which is connected to various additional components via valves and tubing. Commercial pumps and/or compressors can be used to deliver the CO_2 and to add or recirculate fluids. Devices to measure, record, and control pressure and temperature (wall temperature and inside temperature) should be installed to allow at least a minimum of reaction control. Exploratory studies of chemical synthesis involve mostly batch-type procedures, and some typical reactors used in our laboratories are shown in Fig. 2.

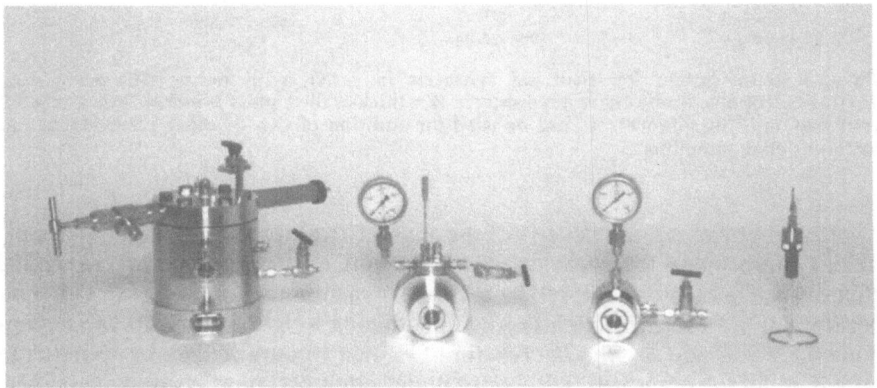

Fig. 2. High pressure reactors and spectroscopic cells as used at the Max-Planck-Institut für Kohlenforschung for exploratory studies on chemical synthesis in $scCO_2$

The reactor size varies from approx. 4 mL to 225 mL and all designs allow for visual control of the reaction mixtures. The larger reactors are basically modifications of standard stainless steel autoclaves and similar equipment is commercially available from manufactures of high pressure autoclaves. The design used at the Mülheim laboratories [3, 22] has two opposite thick-walled glass windows and various fittings allow direct coupling to online spectroscopy (e.g. reflectance FTIR, [23]) or gas chromatographic techniques [24]. A complete setup is schematically depicted in Fig. 3. The smaller reactors shown in Fig. 2 have a volume of 25 mL and 10 mL, respectively. This design and similar cells can be used directly for spectroscopic monitoring (e.g. FTIR, UV-Vis) if a suitably transparent window material (sapphire, CaF_2, diamond, etc.) is chosen. The smallest reactor on the right of Fig. 2 is a 5-mm high pressure NMR tube, manufactured from a sapphire single crystal and a pressure head of non-magnetic steel [25]. The sapphire crystals are commercially available to fit into 5 mm and 10 mm standard NMR probes and the design is used routinely in our laboratory under appropriate and rigorous safety precautions.

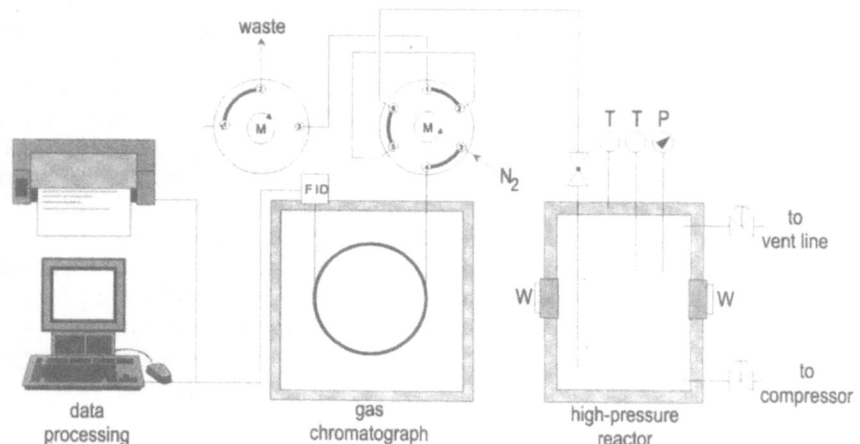

Fig. 3. Reactor system for chemical synthesis in scCO$_2$ with online GC monitoring. T = thermocouple; P = pressure transducer; W = thick-walled glass window. The ports "to vent line" and "to compressor" can be used for addition of CO$_2$ or other components and for additional sampling

As outlined above, the density of the supercritical reaction mixture is of particular importance for chemical synthesis and must therefore be controlled effectively. This can be achieved by accurate measurements of pressure and temperature or – more conveniently – by introducing weighed amounts of all components including CO$_2$ into the reactor of known volume. It seems important to note that partial pressures (and hence molar amounts) of gases in supercritical reaction mixtures cannot be simply deduced from pressure differences during their introduction. The total pressure of a mixture of CO$_2$ and H$_2$ (10 bar) prepared at room temperature and heated to 80 °C is substantially different from the pressure generated by heating the same amount of CO$_2$ to 80 °C and then increasing the pressure by 10 bar with H$_2$. This problem can be overcome with suitable calibration curves or by introducing known volumes (rather than pressures) of reactant gases using pumps of the syringe-type.

In summary, the equipment required to carry out reactions in scCO$_2$ is nowadays readily available and various levels of investigations from fast screening to in situ spectroscopy can be envisaged, just like for synthesis in conventional solvents. Safety issues are of course of highest priority, but are readily met on a laboratory scale. Under these constraints, and if some basic thermodynamic properties of compressed gases are considered, chemical synthesis in scCO$_2$ can be readily integrated in the toolbox of modern organic synthesis.

4
Applications of Supercritical Carbon Dioxide as a Reaction Medium for Chemical Synthesis

4.1
Hydrogenation Reactions

4.1.1
Heterogeneously Catalyzed Hydrogenation Reactions

The high miscibility with H_2, the smaller number of phase boundaries (SCF/catalyst vs gas/liquid/catalyst), and the high diffusion rates were recognized relatively early as potentially useful properties of SCFs in hydrogenation reactions using heterogeneous noble metal catalysts. A substantial amount of research in this field has been and still is being carried out in industrial laboratories or in cooperation with the chemical industry. Therefore, the information in the published literature reflects only a limited fraction of the most recent developments.

The Pd-catalyzed hydrogenation of fatty acids and their derivatives has been successfully performed using supercritical propane (scC_3H_8) [26] and $scCO_2$ [27]. In contrast, the enantioselective hydrogenation of ethyl acetate using a cinchona-modified Pt-catalyst required the use of an alkane solvent, because CO_2 lead to rapid catalyst deactivation [28]. Heterogeneous hydrogenation in SCFs was also studied intensively as part of a commercial vitamin synthesis [29] and the hydrogenation kinetics of a not fully specified α,β-unsaturated ketone using a Pd/Al_2O_3 catalyst in $scCO_2$ were reported [30].

Most recently, a detailed survey was published on conversion rates and selectivities of the Pd-catalyzed hydrogenation of various organic substrates in $scCO_2$, including the reduction of $C=C$, $C=O$, $C=N$ and NO_2 functional groups [31]. The increased reaction rates in the presence of compressed CO_2 were found to allow the use of surprisingly small flow reactor systems. The chemoselectivity with multifunctional substrates could be adjusted nicely by small variations in the reaction parameters. The hydrogenation of acetophenone (Scheme 1) provides an illustrative example.

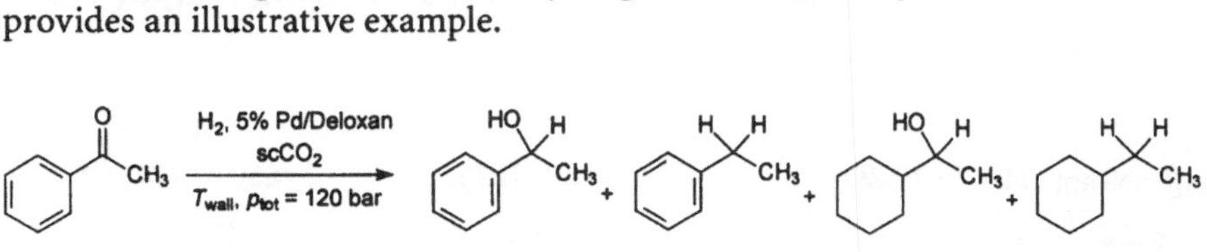

$T_{wall} = 90°C$, H_2 : sub. = 2 : 1	90%	7.5%	0%	0%
$T_{wall} = 180°C$, H_2 : sub. = 5 : 1	14%	41%	28%	17%
$T_{wall} = 300°C$, H_2 : sub. = 6 : 1	0%	4.5%	0%	89.5%

Scheme 1

4.1.2
Homogeneously Catalyzed Hydrogenation Reactions

The hydrogenation of CO_2 to formic acid or its derivatives is an interesting approach to the use of CO_2 as a raw material in chemical synthesis [32, 33]. The equilibrium between CO_2 and H_2 to give HCOOH is catalyzed efficiently by soluble rhodium catalysts in organic solvents [34] or aqueous solution [35]. The addition of an amine is necessary to shift the equilibrium to the side of the desired product [34b]. The first highly efficient catalytic system for this reaction was operating in a mixture of DMSO and NEt_3 as the reaction medium and was based on the rhodium catalyst 1 formed in situ from a suitable precursor and the chelating phosphine ligand 1,4-bis(diphenylphosphino)butane (dppb) (Scheme 2) [34a]. Attempts to use this catalyst without DMSO employing CO_2 in the supercritical state as the solvent and reagent failed, however [33]. In contrast, the ruthenium complex $[Ru(PMe_3)_4Cl_2]$ (2) proved to be highly efficient under supercritical conditions and the rates (given as turnover frequencies, TOF = mole product per mole metal per hour) were remarkably higher than with the same catalyst in a liquid solvent [36]. The key to the success was the use of a trialkyl rather than an arylphosphine, providing sufficient solubility of the active species in the initially homogeneous reaction mixture. Indeed, the solubility of 2 in a supercritical mixture of CO_2, H_2 and NEt_3 was experimentally verified [37]. Replacing the aryl groups of dppb by cyclohexyl rings in the rhodium complexes 3 lead to a catalyst that could also be used in $scCO_2$ [38]. The difference in reaction rates using 3 in $scCO_2$ or DMSO was found to be less pronounced than that between $scCO_2$ and THF in the ruthenium case.

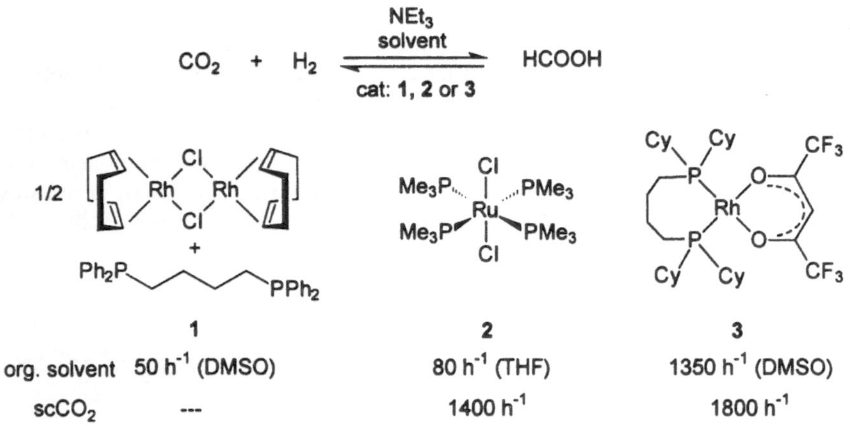

Scheme 2

In presence of ruthenium catalysts, the formation of HCOOH under supercritical conditions was coupled to subsequent condensation with MeOH or Me_2NH to give methyl formate [39] or DMF [40], respectively. The phase behavior of the system utilizing the secondary amine differs from that of the other reaction media, because CO_2 and Me_2NH spontaneously form liquid dimethylammonium carbamate (dimcarb) under the reaction conditions. Hence, a liquid phase is observed at any stage of the reaction owing to the presence of dimcarb initially and of water at a later stage [37]. In this particular case, the conversion to DMF occurs with outstanding efficiency regardless whether the $scCO_2$-soluble catalyst **2** [40] or the $scCO_2$-insoluble complex [Ru(dppe)₂Cl₂] (dppe = 1,2-bis(diphenylphosphino)ethane) [41] is used. The ruthenium catalyst can also be tethered to a solid support by sol-gel techniques to provide a highly efficient heterogenized catalyst [42].

The hydrogenation of $scCO_2$ demonstrates nicely the importance of the phase behavior for homogeneous catalysis. Although a single homogeneous phase is not always a necessary prerequisite for efficient catalysis, a reasonable degree of solubility is crucial for catalysts to be used under fully homogenous supercritical conditions. The chiral rhodium catalyst **4** containing a chiral ligand of the DuPHOS-type has been successfully adapted for efficient asymmetric homogenous hydrogenation of dehydroamino acid derivatives under homogeneous conditions in $scCO_2$ [43]. The solubility problems associated with the ionic nature of the complex were overcome by the use of lipophilic counterions like triflate ($CF_3SO_3^-$) or the BARF-anion (BARF = tetrakis[3,5 bis(trifluorome-thyl)phenyl]borate, Scheme 3). It is important to note that the enantioselectivities in $scCO_2$ are at least comparable to those obtained in conventional solvents for most substrates and even considerably higher for β,β-disubstituted double bonds.

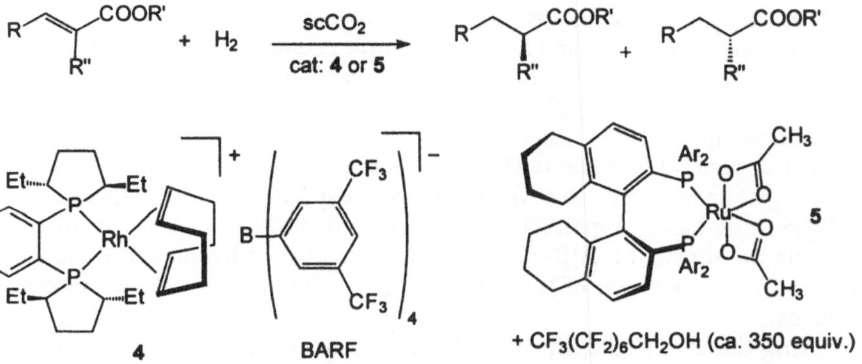

Scheme 3

As already apparent from the studies on CO_2 hydrogenation, the large class of catalytically important arylphosphine ligands and their metal complexes do not generally exhibit sufficient solubility for catalysis under homogeneous conditions in $scCO_2$. This is further exemplified with neutral chiral ruthenium catalysts for asymmetric hydrogenation of α,β-unsaturated carboxylic acids (Scheme 3) [44]. The well known catalyst [Ru(BINAP)(OAc)₂] (BINAP=2,2'-bis-(diphenylphosphino)-1,1'-binaphtyl) cannot be used in $scCO_2$ for solubility reasons. Partial hydrogenation of the ligand's naphtyl moieties as in complex **4** increases the solubility slightly, but only the addition of a perfluorinated alcohol as co-solvent finally results in high reaction rates and selectivites in $scCO_2$.

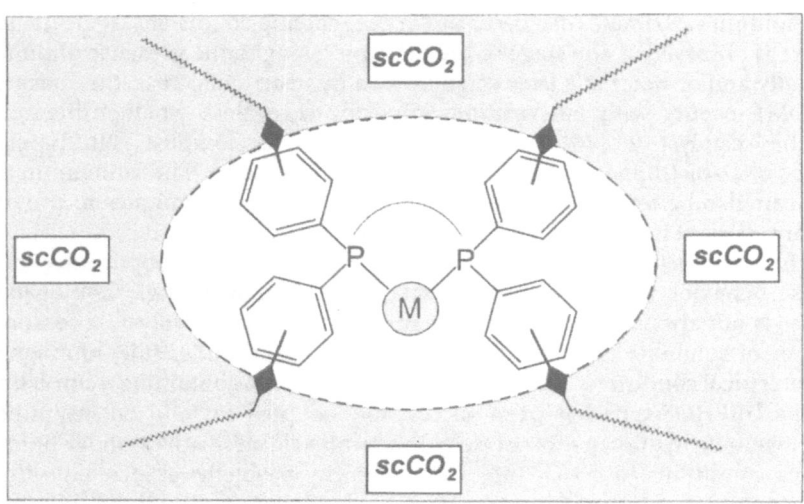

Fig. 4. A general concept for solubilizing catalytically relevant ligands for the use in $scCO_2$, exemplified for a bidentate aryl phosphine ligand. Perfluoroalkyl chains or other solubilizers (*waved line*) are fixed at the periphery of the ligand. The position of the substitution and the use of appropriate spacers (*bold line*) are used to „tune" the stereo-electronic properties of the ligand

Our concept to ameliorate the low solubility of catalytically relevant ligands is schematically depicted in Fig. 4 for a complex containing a bidentate chelating arylphosphine [45]. The catalytically active fragment is made available in $scCO_2$ by the fixation of "CO_2-philic" solubilizers in the periphery of the ligand. In a first approach to demonstrate this strategy, we used perfluoroalkyl chains of general formula $(CH_2)_x(CF_2)_yF$ as solubilizers and introduced them into typical arylphosphines like triphenylphosphine (tpp) or 1,2bis(diphenylphosphino)ethane (dppe). The perfluorinated derivatives are designated by putting the shortcut z-H^xF^y in front of the established ligand acronym, whereby z gives the position of the substituent(s) in the aryl ring relative to phosphorus.

Using the chemical shift of the [103]Rh-nucleus in rhodium phosphine complexes as a sensitive probe for electronic or geometric variations in coordination compounds [46], it was demonstrated that long chains with $x = 2$ and $y = 6$ or 8 in *meta* or *para* position relative to phosphorus ($z = 3$ or 4) exhibit only a very small electron-donating effect on the metal center, comparable to methyl groups in the same position [45]. As initially shown by UV-Vis spectroscopy, the solubility of the complexes increased dramatically compared to the insoluble parent compounds, allowing even NMR spectroscopic investigations of complexes like [(3-H^2F^6-dppe)Rh(hfacac)] (6; hfacac = hexafluoroacetylacetonate) in $scCO_2$ solution [47]. The catalytic activity of the new ligands in homogeneous $scCO_2$ solution was first demonstrated for the hydroformylation reactions described in Sect. 4.2 [45]. As exemplified in Scheme 4, complex 6 could also be used efficiently as a catalyst precursor for homogeneous hydrogenation of

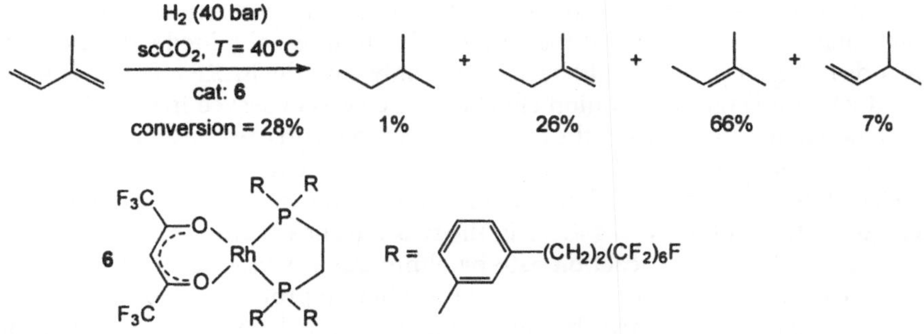

Scheme 4

C = C double bonds in scCO₂ [47]. Isoprene was selectively reduced to a mixture of iso-pentenes with very little over-reduction to iso-pentane.

Following the same strategy, we and other researchers are now synthesizing and using catalysts with ligands bearing H²F⁶-substituents or other permutations of x, y, and z for various applications in scCO₂ [48] and further examples of the approach will be discussed throughout this chapter. Closely related are efforts to synthesize perfluorinated stoichiometric reagents for use in chemical synthesis in scCO₂. A H²F⁶-substituted tin hydride has been prepared and applied to various radical reactions including reductive dehalogenation in scCO₂ [49]. A perfluorinated anthraquinone derivative has been devised as "H₂-carrier" for an scCO₂-based alternative process for the production of H₂O₂ [50].

The enantioselective hydrogenation of imines is a powerful approach to the synthesis of chiral secondary amines. The series of complexes 7 (Scheme 5) either with or without the 3-H²F⁶-substituent and containing various anions was synthesized to investigate the influence of the ligand substitution pattern and the anion on the catalytic efficiency in scCO₂ [23]. As expected, the substitution in the ligand increased the solubility of the complexes, but had very little impact on the enantioselectivity. The anion, however, did not only exhibit an effect on

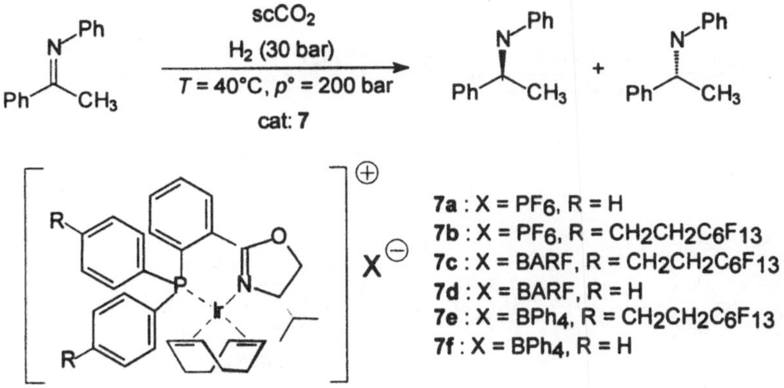

Scheme 5

solubility, but determined also largely the asymmetric induction of the individual catalysts. High ee's (80–81%) comparable to those obtained with the same ligands in liquid solvent could be achieved only with the BARF anion. It is noteworthy that no significant anion effect on the ee was observed in CH_2Cl_2.

Detailed investigations of the course of reaction with catalyst **7d** revealed that a much lower catalyst loading was required for efficient catalysis in $scCO_2$ than in CH_2Cl_2 under otherwise identical conditions. It was shown that this effect was due to a different kinetic behavior in the two solvent systems, rather than to an increase in the overall reaction rate by eliminating mass-transfer limitations. Furthermore, it was demonstrated that the reaction product could be effectively separated from the catalyst by SFE, using the same solvent for the reaction and the purification step (CESS process). A single catalyst load was used in seven subsequent catalytic runs with almost constant enantioselectivity and a decrease of activity was noticeable only after the fourth run.

4.2
Hydroformylation

The hydroformylation of propene using $[Co_2(CO)_8]$ as a catalyst is one of the very early example of a reaction utilizing a homogeneous organometallic catalyst in $scCO_2$ [51]. The high solubility of the precatalyst and the active intermediates in this system is largely due to the high vapor pressure of homoleptic carbonyl and hydrido carbonyl complexes. The study was mainly devoted to the elucidation of the reactivity of catalytically active species by ^{59}Co-NMR spectroscopy in $scCO_2$, exploiting the fact that linewidths of quadrupolar nuclei can be considerably reduced in SCFs. However, the potential benefits of using $scCO_2$ as the medium for the catalytic process were also explicitly addressed and an enhanced selectivity for the linear n-aldehyde at comparable catalytic activity was noted. A recent kinetic study of the reaction provided further information and also revealed some typical limitations of the cobalt hydroformylation catalyst, in particular the need for high catalyst loading (propene/Co = ca. 4) and high reaction temperatures ($\geq 75\,°C$, generally $> 90\,°C$) [52].

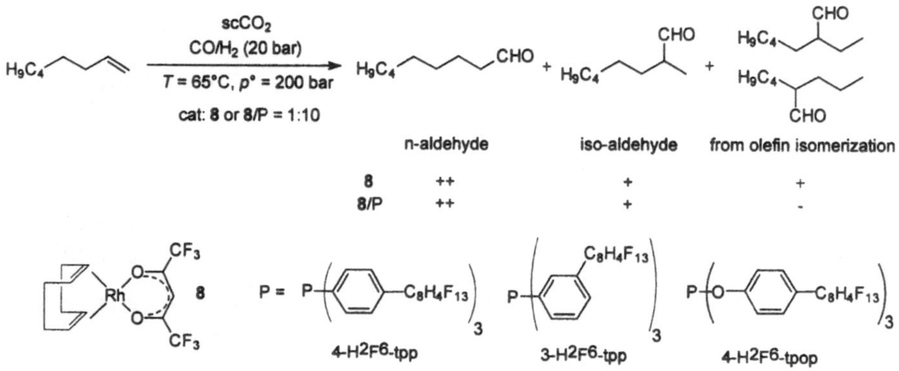

Scheme 6

Highly active unmodified rhodium catalysts for the hydroformylation of various olefins in scCO$_2$ are formed under mild conditions from [(cod)Rh(hfacac)] (8; cod = cis,cis-1,5-cyclooctadiene) and a number of other simple rhodium precursors [24]. Especially for internal olefins, the rate of hydroformylation is considerably higher than using the same catalysts in conventional liquid solvents under otherwise identical conditions. A detailed study of the hydroformylation of 1-octene (Scheme 6) using the online GC setup shown in Fig. 3 revealed a network of competing isomerization and hydroformylation when 8 was used without additional modifiers. As a result, the regioselectivity for the desired linear n-aldehyde varied considerably with conversion. At 60% conversion, the product aldehydes contained almost 80% of nonanal, whereas only 58% linear aldehyde were present in the final product mixture.

If complex 8 was used in the presence of H^{2}F^6-substituted "CO$_2$-philic" arylphosphines or arylphosphites, a soluble, ligand-modified rhodium catalyst was formed [24, 45]. Using, 3-H^{2}F^6-tpp at a P/Rh-ratio of 10:1, a total number of catalytic turnover (TON) of 2000 was readily achieved with an initial TOF = 500 h^{-1}. This value is considerably lower than for the unmodified system in scCO$_2$ (TOF = 1350 h^{-1}), but still well in the range of useful practical application. In fact, the initial TOF with 3-H^{2}F^6-tpp in scCO$_2$ was found to be slightly higher than with the unsubstituted tpp ligand in toluene, although the opposite trend might have been expected from the electronic properties of the two ligands. As in conventional solvents, the modified catalytic system lead to a high and constant regioselectivity of up to 85% for the n-aldehyde with only very little – if any – isomerization or other side reactions. Using the CESS approach, the catalysts could be recycled five times without any noticeable decrease in activity or selectivity, and the rhodium content in the product was as low as 1 ppm.

More recently, perfluoroalkyl substituted aryl phosphine ligand lacking the ethylene spacer (4-H^{0}F^6-tpp and 4-H^{0}F^1-tpp) have also shown very promising results for the homogeneous hydroformylation of long chain terminal olefins in scCO$_2$ [53, 54]. Furthermore, trialkylphosphines like PEt$_3$ have been used successfully to form homogenous catalysts for hydroformylation in scCO$_2$ [55]. These ligands differ considerably from the triarylphosphines in their electronic and steric properties, leading to somewhat different regioselectivities and partial reduction of the aldehydes to the corresponding alcohols.

The outstanding levels of enantiocontrol asymmetric in the hydroformylation of vinyl arenes using rhodium catalysts with (R,S)-BINAPHOS have stimulated attempts to apply such catalysts also in scCO$_2$ [56, 57]. The results were found to depend strongly on the phase behavior of the reaction mixture, because the low solubility of the chiral phosphine/phosphite ligand in scCO$_2$ prevented an efficient chirality transfer with this ligand under fully homogeneous conditions [57]. In contrast, the perfluoroalkyl-substituted derivative (R,S)-3-H^{2}F^6-BINAPHOS gave excellent results in the enantioselective hydroformylation of styrene in the presence of compressed CO$_2$, including conditions where a single homogeneous phase was present (Scheme 7) [58]. Independent of the phase behavior, the reaction rates and the ee's compared well with those observed with the established catalyst in benzene solution. Most remarkably, the regioselectivity for the chiral branched aldehyde over the achiral linear byproduct was significantly higher with the perfluoroalkyl derivative compared to the use of BINAPHOS under the established conditions. Control experiments revealed that this high regioselectivity was mainly due to the ligand substitution pattern rather than the change of solvent.

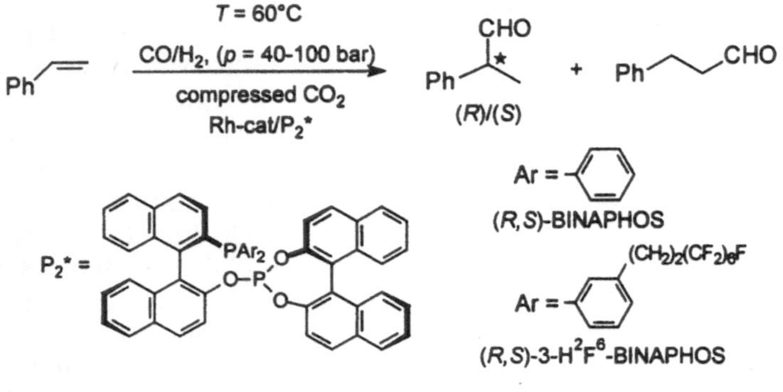

Scheme 7

4.3
Rhodium-Catalyzed Polymerization of Phenylacetylene

Polymerization in compressed (liquid or supercritical) CO_2 is a field of great current interest and radical polymerization has been studied in quite some detail [13, 59]. These studies demonstrate that control of the morphology and the microstructure of the resulting polymers is possible under certain conditions. Furthermore, the use of a solvent that is a gas under ambient conditions allows the preparation of solvent free polymers and $scCO_2$ can be used to extract residues of monomer or low molecular weight oligomers. In contrast to radical and cationic polymerization, very few reports have focused up to now on transition metal-catalyzed polymer synthesis in compressed CO_2. Noticeable exceptions are the ring-opening metathesis polymerization (Sect. 4.5) and the copolymerization of epoxides and CO_2 to give polycarbonates (Sect. 4.7).

The polymerization of phenylacetylene using the complex [(nbd) Rh(acac)] **9** (nbd = norbornadiene, acac = acetylacetonate) as a catalyst precursor was found to occur with considerably higher rates in compressed (liquid or supercritical) CO_2 than in conventional solvent such as THF or hexane (Sche-

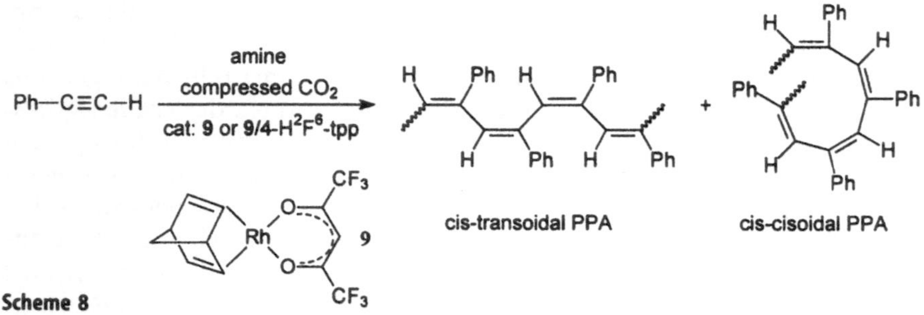

cis-transoidal PPA   cis-cisoidal PPA

Scheme 8

me 8) [60]. The high rate in CO$_2$ compared to THF was rather unexpected, because complex 9 is highly soluble in THF whereas it is very poorly soluble in pure CO$_2$. Furthermore, the polymerization occurred as a precipitation polymerization in CO$_2$, whereas the THF-system remained homogeneous up to very high conversions. In CO$_2$, polyphenylacetylene (PPA) with *cis-transoid* and *cis-cisoid* main chains was obtained, whereas the *cis-transoid* polymer was formed almost exclusively in THF.

A ligand modified soluble catalyst was formed from complex 9 and 4-H^{2}F^6-tpp in situ in compressed CO$_2$. The use of the "CO$_2$-philic" ligand resulted in a somewhat increased amount of the *cis-transoid* polymer in the product mixture. The molecular weight distribution remained fairly broad, although living polymerization systems have been described with a Rh/tpp catalyst in THF solution [61]. The broad distribution in CO$_2$ might be related to the heterogeneous reaction system: Although the reaction mixture with 9/4-H^{2}F^6-tpp was initially homogeneous, the polymer precipitated very rapidly leading to a heterogeneous polymerization process.

4.4
C-C Coupling Reactions

Palladium-catalyzed coupling reactions between aryl halides or triflates and vinylic substrates have emerged as very powerful tools to build up carbon skeletons and have found numerous applications in synthesis of biologically active natural or non-natural compounds. Consequently, there have been several recent attempts to use scCO$_2$ as a solvent for various processes of this type, including the Heck-coupling shown in Scheme 9 as a common test reaction. The use of perfluorinated compounds to generate soluble palladium catalysts was again a major theme.

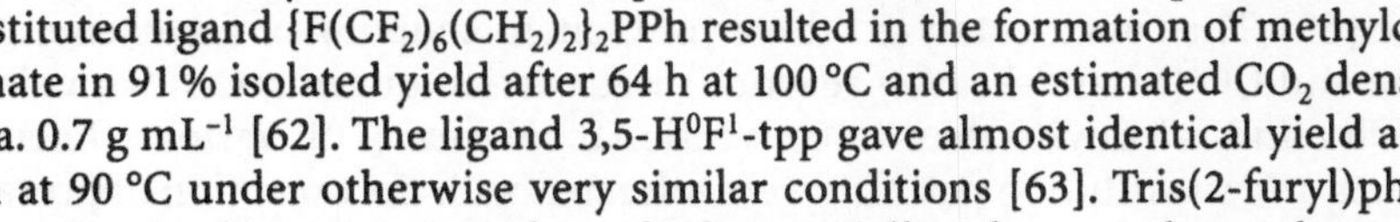

Scheme 9

An in situ catalyst formed from palladium acetate and the perfluoroalkyl-substituted ligand {F(CF$_2$)$_6$(CH$_2$)$_2$}$_2$PPh resulted in the formation of methylcinnamate in 91% isolated yield after 64 h at 100 °C and an estimated CO$_2$ density of ca. 0.7 g mL^{-1} [62]. The ligand 3,5-H^{0}F^1-tpp gave almost identical yield after 24 h at 90 °C under otherwise very similar conditions [63]. Tris(2-furyl)phosphine also lead to active catalysts [63], especially when used together with fluorinated Pd-sources like palladium trifluoroacetate [64]. Most of the above systems lost their activity if metallic palladium was formed under non-optimum reaction conditions. However, the use of a commercial heterogeneous Pd/C catalyst was reported to give up to 85% yield of methyl cinnamate in the presence of CO$_2$ at temperatures and pressures beyond the critical data [65]. The CO$_2$ density in the latter experiments can be estimated to ca. 0.22 g mL^{-1} and was hence considerably lower than the densities employed with the metal complexes.

The enantioselective nickel-catalyzed co-dimerization of substituted styrenes and ethylene ("hydrovinylation", Scheme 10) is the first example of an asymmetric C-C coupling reaction occurring efficiently in compressed (liquid or supercritical) CO_2 [66]. The chiral information was provided by the phosphorous ligand 10, and its Ni-complex was both activated and solubilized using the BARF anion. The new procedure provides a completely chlorine-free environmentally friendly procedure for enantioselective hydrovinylation giving comparable or even slightly higher chemo-, regio- and stereoselectivities than CH_2Cl_2 at temperatures above 0 °C. The optimum selectivities in CH_2Cl_2 using aluminium sesquichloride as activator are still somewhat higher, but require very low reaction temperatures of –60 °C and below.

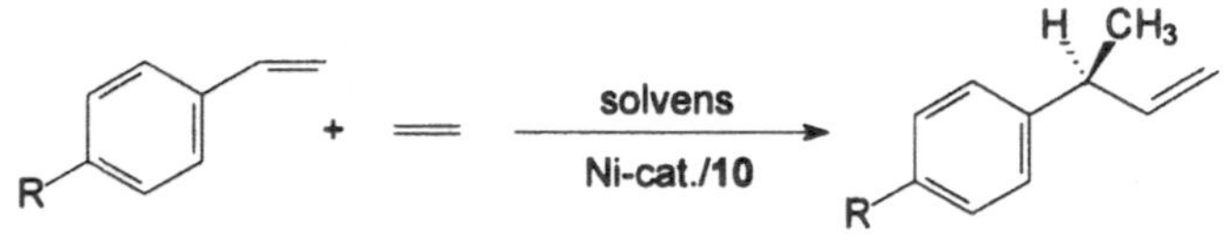

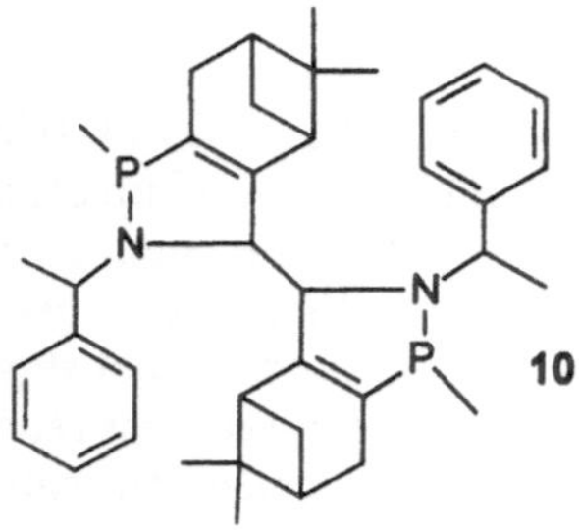

	23°C	35-40°C
CH_2Cl_2	70% sel. 85% ee	57% sel. 81% ee
CO_2	71% sel. 86% ee	77% sel. 83% ee

Scheme 10

The co-trimerization of two molecules of an alkyne with one molecule CO_2 to form 2-pyrones was recognized very early as an interesting possible approach to use CO_2 simultaneously as a solvent and a reagent [67]. The best results were obtained under homogeneous conditions using a soluble in situ catalyst consisting of [Ni(cod)$_2$] and PMe$_3$ [68, 69]. Oxygen transfer from CO_2 to the phosphine ligand under concomitant formation of inactive nickel carbonyl complexes was revealed as a major deactivation pathway limiting the catalyst's lifetime [69]. The cobalt-catalyzed three component coupling of an alkene, an alkyne and CO (Khand-Pauson reaction) was carried out in scCO$_2$ with similar activity and selectivity than in conventional solvents [70]. Various functional groups and substitution patterns were tolerated with the exception of protected secondary amines, which showed insufficient solubility.

A remarkable synthesis of oxalate salts using a supercritical mixture of CO_2 and CO under very drastic conditions (400 bar, 380 °C) in the presence of solid $Cs_2(CO_3)$ was reported [71]. Friedel-Crafts-type alkylations on zeolites [72] or other solid acid catalysts [73] have been studied using scCO$_2$ as the medium. Performing the reaction in the supercritical state was found to be superior to either liquid or gas phase processes. Using scCO$_2$ lead to enhanced catalyst lifetimes because of reduced coking and allowed the use of miniature flow-type reactors for continuous operation. The selectivity towards the monoalkylation product could be carefully controlled and was increased compared to conventional reaction conditions.

4.5
Olefin Metathesis

The ring opening metathesis polymerization (ROMP) has been studied in supercritical and liquid CO_2 using simple ruthenium salts [74] and the well-defined carbene complexes **11** and **12** as initiators (Scheme 11) [75]. With $[Ru(H_2O)_6](Tos)_2$ ($Tos = p\text{-}CH_3C_6H_4SO_3^-$) as the initiator, polymerization of norbornene occurred with moderate rates in the presence of MeOH as a co-solvent [74]. An interesting effect of the solvent mixture and the total pressure on the tacticity and the *cis*-content of the norbornenamer was noted and interpreted in terms of the geometric properties of the transition state [76]. In contrast, the molybdenum complex **11** proved highly soluble in the reaction mixture without any additives and norbornene was polymerized with high yields [75]. The molecular weights, their distribution, and the stereochemistry at the double bond of the norbornenamer were almost identical to material produced in CH_2Cl_2. High polymerization activity in pure CO_2 was also observed with Ru-catalysts **12**, although most of the initially charged complex remained insoluble as judged by visual inspection [75]. Under optimum conditions, the resulting polymer could be isolated as clean white material without further purification. The bulk amount of the initially charged complex was located in a relatively small area and could be removed mechanically without the need for the conventional redissolving/precipitation procedure [77].

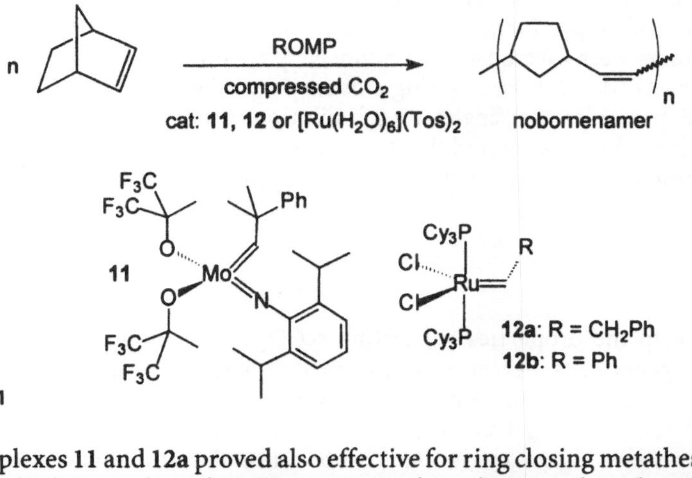

Scheme 11

Complexes **11** and **12a** proved also effective for ring closing metathesis (RCM) in scCO₂ [75]. A number of products varying from five-membered up to sixteen-membered carbo- and heterocycles could be synthesized using this approach. Owing to the poor solubility of **12a**, the products could be readily separated with recovery of an active catalyst in a CESS process. Furthermore, the RCM of the secondary amine shown in Scheme 12 occurred readily in $scCO_2$ to provide the natural product epilachnen, whereas the N-H functionality had to be protected during the same conversion in conventional solvents like CH_2Cl_2. The reversible formation of the carbamic acid prevented the deactivation of **12a** by the N-H group, demonstrating for the first time that $scCO_2$ can be used as a temporary protecting group and a solvent during catalytic reactions.

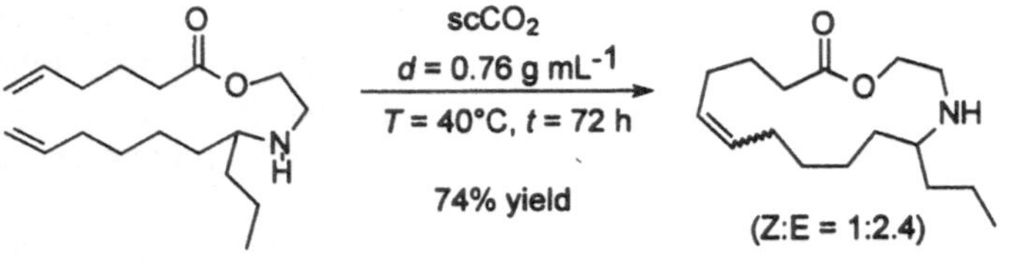

Scheme 12

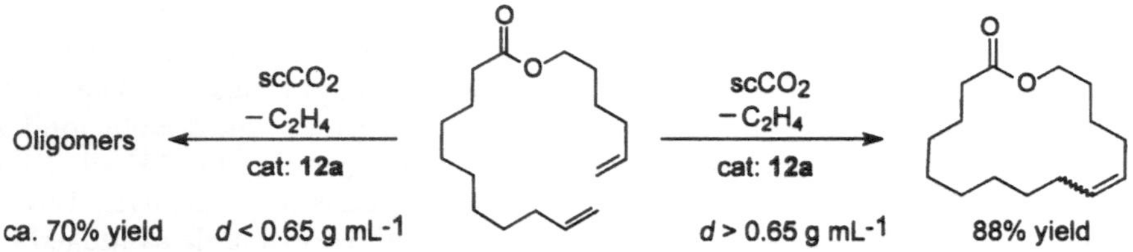

Scheme 13

A remarkable influence of the density of the reaction medium on the product distribution was observed in the RCM approach to the sixteen-membered macrocycle shown in Scheme 13. At densities above 0.65 g mL^{-1}, the macrocycle was formed by intramolecular ring closure in excellent yields without appreciable amounts of by-products. At the same molar concentration but lower densities, oligomers were formed almost exclusively via intermolecular acyclic diene metathesis (ADMET). Favoring ring closure over oligomerization with density in the SCF resembles the Ziegler-Ruggli dilution principle in conventional solvents and it has been suggested that the influence of the density on the chemical potential might be responsible for this striking similarity.

4.6
Oxidation

In addition to the properties suggesting $scCO_2$ as an interesting medium for catalysis in general, its inertness towards oxidation seems particularly attractive for oxidation reactions. In contrast to processes in liquid solution, byproducts from oxidation of the solvent can never be formed. Furthermore, the use of $scCO_2$ may allow higher oxidant/substrate ratios due to expanded explosion limits as compared to liquid phase or gas-phase oxidations. At the same time, the better heat transport capacities of $scCO_2$ relative to low density gaseous mixtures may also prove beneficial for process design of highly exothermic oxidation reactions. These favorable property profile has stimulated a rapidly growing number of studies of metal-catalyzed oxidation reactions in scCO$_2$.

The metal-catalyzed oxidation of olefins was investigated using inorganic and organic peroxides as oxidants. Cyclohexene was converted to adipic acid in a two-phase scCO$_2$/H$_2$O system using Na(IO$_4$) as the oxidant, but rapid deactivation of the ruthenium catalyst occurred [78]. Several papers have demonstrated almost simultaneously that epoxides can be obtained by metal-catalyzed oxidations in good yields under anhydrous conditions using ^{t}BuOOH as the oxidant, whereas diols were the main products when water was present [69, 79, 80]. The approach has been extended to diastereoselective [81] and even enantioselective epoxidations [80]. The promising results of these pioneering studies suggest that the design of special catalysts for selective oxidations in scCO$_2$ may prove rewarding in the future.

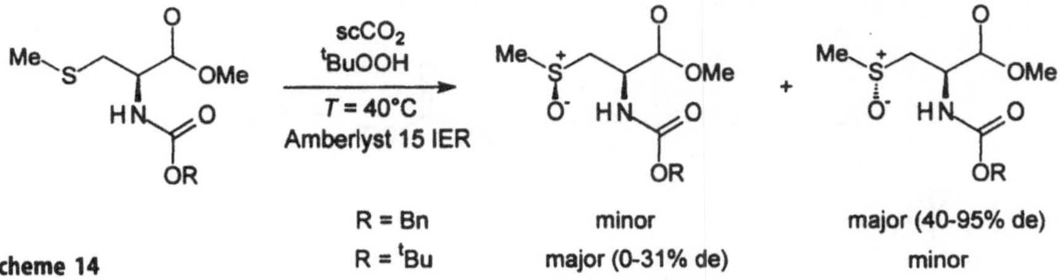

Scheme 14

R = Bn
R = ^{t}Bu

minor
major (0-31% de)

major (40-95% de)
minor

A remarkable effect of the reaction medium has been observed in the heterogeneously catalyzed diastereoselective oxidation of sulfur compounds as shown in Scheme 14 [82]. The reaction, which was completely non-selective in conventional solvents, could be optimized up to 95% de in scCO$_2$ whereby a dramatic dependence on pressure (and hence bulk density) was observed. Although no fully satisfactory explanation is yet available for these results, they seem to support again the potential to use the density of scCO$_2$ as an additional parameter for the optimization of organic syntheses.

The selective oxidation of organic substrates using molecular O$_2$ is a formidable challenge of modern synthetic chemistry and scCO$_2$ seems to hold considerable promise in this area. The rhodium-catalyzed aerobic oxidation of THF to γ-butyrolactone could be achieved in scCO$_2$ with TONs up to 130 and TOFs of ca. 15 h^{-1} for the desired product [83]. Investigations of catalytic [84] and non-catalytic [85] free-radical aerobic oxidations of alkanes in scCO$_2$ have been reported. Iron-group complexes of perfluorinated porphyrin ligands have been used for the oxidation of cyclic alkenes in supercritical CO$_2$ and improved selectivities compared to liquid phase reactions were noted in some cases [86]. Using aldehydes as co-oxidants (Mukayama-conditions), a variety of olefins including

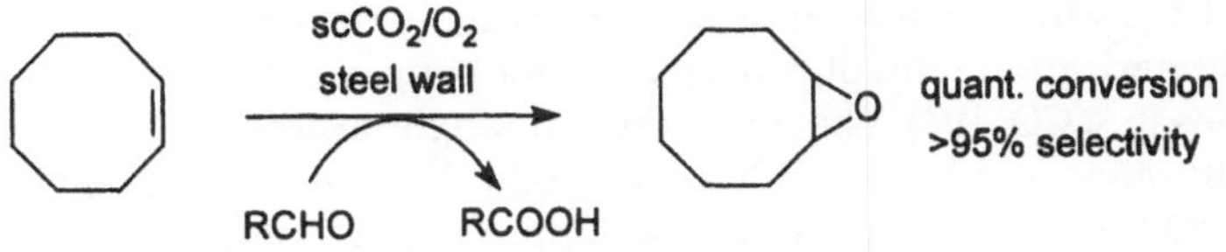

Scheme 15

scCO₂/O₂
steel wall

RCHO   RCOOH

quant. conversion
>95% selectivity

cyclooctene could be epoxidized very efficiently in $scCO_2$ in the absence of any metal complex catalyst (Scheme 15) [87]. The reaction was shown to be effectively initiated by stainless steel components of the reactor walls.

4.7
Synthesis of Organic Carbonates

The simultaneous use of CO_2 as a solvent and C1 building block has been already addressed in a number of examples throughout this chapter. This concept has also found attention for the synthesis of organic carbonates, which are already produced using CO_2 as a feedstock or for which CO_2 would be an environmentally benign alternative raw material [20]. Depending on the starting material, dialkyl or diaryl carbonates, cyclic carbonates, or polycarbonates can be envisaged as potential reaction products (Scheme 16).

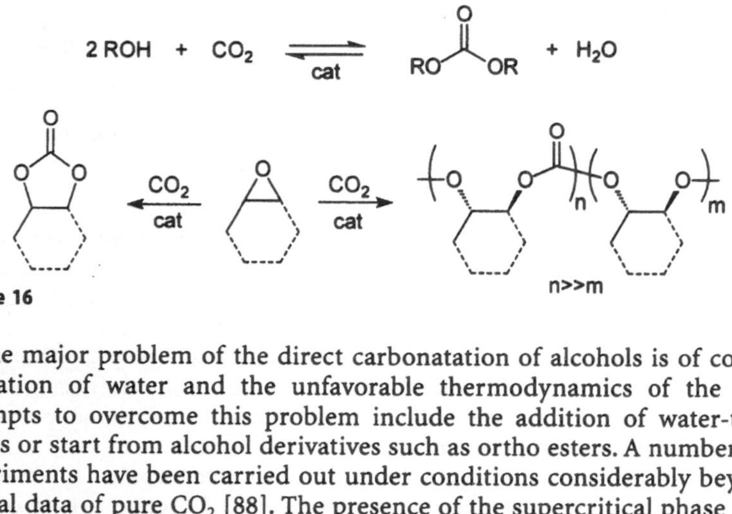

Scheme 16

The major problem of the direct carbonatation of alcohols is of course the formation of water and the unfavorable thermodynamics of the process. Attempts to overcome this problem include the addition of water-trapping agents or start from alcohol derivatives such as ortho esters. A number of such experiments have been carried out under conditions considerably beyond the critical data of pure CO_2 [88]. The presence of the supercritical phase as a solvent was explicitly addressed in the synthesis of glycerol carbonate from glycerol and CO_2 [89]. The tin-catalyzed conversion of trimethyl ortho ester to dimethyl carbonate in $scCO_2$ occurred with up to ca. 30 catalytic turnovers, whereby the highest yields and selectivities were observed in the vicinity of the critical pressure of pure CO_2 [90].

The copolymerization of propylene oxide and CO_2 was effectively initiated using a heterogeneous zinc glutarate catalyst [91]. Interestingly, a more recently developed molecular catalyst showed no activity under heterogeneous supercritical conditions, although it was highly active when dissolved in a liquid reaction mixture [92]. A fluorinated zinc carboxylate catalyst has been developed to allow high polymerization rates of > 400 g polymer per g of zinc under homogeneous conditions in $scCO_2$ [93]. The polymers had weight-averaged molecular weights in the 10^5 range and broad molecular weight distributions. The polymer contained > 90% carbonate linkages under optimized reaction conditions.

5
Summary and Outlook

Supercritical carbon dioxide has emerged as a highly promising reaction medium for organic synthesis and research activities in this field have been prolific during the last decade. Especially in combination with heterogeneous or homogeneous catalysis, the technology offers considerable potential for ecologically benign synthesis on a commercial scale. The potential benefits like increased reaction rates and increased or different chemo-, regio- and stereoselectivity are expected to stimulate efforts to add scCO₂ also to the toolbox of modern laboratory-scale synthesis. This progress should be facilitated as the necessary equipment becomes more readily available to the non-specialist's laboratory. At present, we are far from fully understanding how to exploit the unique properties most efficiently in general terms, but some trends and patterns begin to emerge. Detailed mechanistic studies including in situ spectroscopic techniques will help to rationalize the sometimes remarkable effects of the supercritical phase. The design of "CO₂-philic" reagents and catalysts will remain at the core of the developments and only a very small number of pos-sible modifications have been applied as yet. The requirement for a close interplay and for fruitful mutual stimulation between fundamental research in chemical synthesis, in catalysis, in spectroscopy, and in chemical engineering is probably one of the most fascinating aspects of this interdisciplinary and fast-developing field of chemical science.

6
References

1. Cagniard de LaTour C (1822) Ann Chim Phys 21:127
2. Jessop PG, Leitner W (1999) A Brief History of Chemical Synthesis in SCFs. In Jessop PG, Leitner W (eds) Chemical Synthesis in Supercritical Fluids. Wiley-VCH, Weinheim, p 13
3. Zosel K. (1978) Angew. Chem Int Ed Engl 17:702
4. McHugh M, Krukonis VJ (1994) Supercritical Fluid Extraction, 2nd edn. Butterworth-Heinemann, Boston
5. Jessop PG, Leitner W (eds) (1999) Chemical Synthesis in Supercritical Fluids. Wiley-VCH, Weinheim
6. Anastas PT, Warner JC (1998) Green Chemistry: Theory and Practice. Oxford University Press, Oxford
7. Noyori R. (ed) (1999) Chemical Reviews 99
8. Span R, Wagner W (1996) J Phys Chem Ref Data 25:1509
9. Tucker SC, Maddox MW, (1998) J. Phys. Chem. B 102:2437
10. Brennecke JF, Chateauneuf FE (1999) Chemical Reviews 99:433
11. Clifford AA (1999) Physical Properties of SCFs as Related to Chemical Reactions. In: Jessop PG, Leitner W (eds) Chemical Synthesis in Supercritical Fluids. Wiley-VCH, Weinheim, p 54
12. Giddings JC, Myers MN, McLaren L, Keller RA (1968) Science 162:67
13. DeSimone JM, Guan Z, Elsbernd CS (1992) Science 257:945
14. a) Lin YH, Brauer RD, Laintz KE, Wai CM (1993) Anal Chem 65:2549; b) Yazdi AV, Beckman EJ (1996) Ind Eng Chem Res 35:3644; c) Smart NG, Carleson T, Kast T, Clifford AA, Burford MD, Wai CM (1997) Talanta 44:137

15. a) Harrison K, Goveas J, Johnston KP, O'Rear EA (1994) Langmuir 10:3536; b) Yazdi AV, Lepilleur C, Singley EJ, Liu W, Adamsky FA, Enick RM, Beckman EJ (1996) Fluid Phase Equilib 117:297 c) Johnston KP, Harrison KL, Clarke MJ, Howdle SM, Heitz MP, Bright FV, Carlier C, Randolph TW (1996) Science 271:624
16. Clarke MJ, Harrison KL, Johnston KP, Howdle SM (1997) J Am Chem Soc 119:6399
17. a) Jacobsen GB, Lee CT, daRocha SRP, Johnston KP (1999) J Org Chem 64:1201–1206; b) Jacobsen GB, Lee CT, Johnston KP (1999) J Org Chem 64:1207–1210
18. Koch D, Kainz S, Janssen E, Leitner W (1999) manuscript in preparation.
19. a) Tiltscher H, Wolf H, Schelchshorn J (1984) Ber Bunsen-Ges Phys Chem 88:897; b) Fan L, Yan S, Fujimoto K, Yoshii K (1997) J Chem Eng Jpn 30:923; c) Subramaniam B, Clark MC (1998) Ind Eng Chem Res 37:1243
20. Leitner W (1996) Coord Chem Rev 153:257
21. Fink R, Beckman EJ (1999) High-Pressure Equipment Design. In: Jessop PG, Leitner W (eds) Chemical Synthesis in Supercritical Fluids. Wiley-VCH, Weinheim, p 67
22. For a detailed description see Koch D (1998) PhD thesis, Max-Planck-Insitut für Kohlenforschung/Universität Jena, Verlag Mainz, Aachen
23. Kainz S, Brinkmann A, Leitner W, Pfaltz A (1999) J Am Chem Soc 121:6421
24. Koch D, Leitner W (1998) J Am Chem Soc 120:13398
25. a) Roe DC (1985) J Magn Res 63:388; b) Gaemers S, Luyten H, Ernsting JM, Elsevier CJ (1999) Magn Reson Chem 37:1999
26. Härröd M, Macher MB, Högberg J, Møller P (1997) Proceedings of the 4th Italian Conference on Supercritical Fluids and their Applications, Capri, Italy, p 319
27. Tacke T, (1995) Chem.-Anlagen Verfahren, 28:18
28. Minder B, Mallat T, Pickel KH, Steiner K, Baiker A (1995) Catal Lett 34:1
29. Pickel KH, Steiner K (1994) Proceedings of the 3rd International Symposium on Supercritical Fluids, Strasbourg, France, p 25
30. Devetta L, Giovanzana A, Canu P, Bertucco A, Minder J (1999) Catalysis Today 48:337
31. a) Hitzler MG, Poliakoff M (1997) Chem Commun 1667; b) Hitzler MG, Smail FR, Ross SK, Poliakoff M (1998) Org Process Res Dev 2:137
32. Jessop PG, Ikariya T, Noyori R (1995) Chem Rev 95:259
33. Leitner W (1995) Angew Chem Int Ed Engl 34:2207
34. a) Graf E, Leitner W (1992) J Chem Soc Chem Commun 623; b) Leitner W, Dinjus E, Gaßner F (1994) J Organomet Chem 475:257
35. a) Gaßner F, Leitner W (1993) J Chem Soc Chem Commun 1465; b) Leitner W, Dinjus E, Gaßner F (1998) CO_2 Chemistry. In: Cornils B, Herrmann WA (eds) Aqueous-Phase Organometallic Catalysis. Wiley-VCH, Weinheim, p 486
36. Jessop PG, Ikariya T, Noyori R (1994) Nature 368:231
37. Jessop PG, Hsiao Y, Ikariya T, Noyori R (1996) J Am Chem Soc 118:344
38. Koch D, Leitner W (1996) unpublished results
39. Jessop PG, Hsiao Y, Ikariya T, Noyori R (1995) J Chem Soc Chem Commun 707
40. Jessop PG, Hsiao Y, Ikariya T, Noyori R (1994) J Am Chem Soc 116:8851
41. Kröcher O, Köppel RA, Baiker A (1997) Chem Commun 453
42. Kröcher O, Köppel RA, Baiker A, (1996) Chem Commun 1497
43. Burk MJ, Feng S, Gross MF, Tumas W (1995) J Am Chem Soc 117:8277
44. Xiao J, Nefkens SCA, Jessop PG, Ikariya T, Noyori R (1996) Tetrahedron Letters 37:2813
45. Kainz S, Koch D, Baumann W, Leitner W (1997) Angew Chem Int Ed Engl 36:1628
46. Leitner W, Bühl M, Fornika R, Six C, Baumann W, Dinjus E, Kessler M, Krüger C, Rufinska A (1999) Organometallics 18:1196
47. Kainz S, Koch D, Leitner W (1997) Homogeneous Catalysis in Supercritical Carbon Dioxide: A "Better Solution"?. In: Werner H, Schreier W (eds) Selective Reactions of Metal-Activated Molecules. Vieweg, Wiesbaden, p 151
48. The ligand 4-H^2F^6-tpp has been also successfully employed in fluorous biphasic catalysis: Kling R, Sinou D, Pozzi G, Choplin A, Quignard F, Busch S, Kainz S, Koch D, Leitner W (1998) Tetrahedron Lett. 39:9439; see the relevant chapter in this volume on the FBS approach

49. Hadida S, Super MS, Beckman EJ, Curran DP (1997) J Am Chem Soc 119:7406
50. Super MS, Enick RM, Beckman EJ (1997) J Chem Eng Data 42:664
51. a) Rathke JW, Klingler RJ, Krause TR (1991) Organometallics 10:1350; b) Klingler RJ, Rathke JW (1994) J Am Chem Soc 116:4772
52 Guo Y, Akgerman A (1997) Ind Eng Chem Res 36:4581
53. Banet A, Paige DR, Stuard AM, Chadbond IR, Hope EG, Xiao J (1998) Proceedings of the 11th International Symposium on Homogeneous Catalysis, St. Andrews, UK, P252
54. Palo DR, Erkey C (1998) Ind Eng Chem Res 37:4203
55. Bach I, Cole-Hamilton DJ (1998) Chem Commun 1463
56. Ojima I, Tzamarioudaki M, Chuang CY, Iula DM, Li Z (1998) Catalytic Carbonylations in Supercritical Carbon Dioxide. In: Herkes FE (ed) Catalysis of Organic Reactions. Marcel Dekker, New York, p 333
57. Kainz S, Leitner W (1998) Catal Lett 55:223
58. Franció G, Leitner W (1999) Chem Commun in press
59. Canelas DA, DeSimone JM (1997) Adv Polym Sci 133:103
60. Hori H, Six C, Leitner W (1999) Macromolecules 32: 3178
61. Kishimoto Y, Eckerle P, Miyatake T, Ikariya T, Noyori R (1994) J Am Chem Soc 116:12131
62. Carroll MA, Holmes AB (1998) Chem Commun 1395
63. Morita DK, Pesiri DR, David SA, Glaze WH, Tumas W (1998) Chem Commun 1397
64. Shezad N, Oakes RS, Clifford AA, Rayner CM (1999) Tetrahedron Lett 40:2221
65. Cacchi S, Fabrizi G, Gasparrini F, Villani C (1999) Synlett 345
66. Wegner A, Leitner W (1999) Chem Commun in press
67. Reetz MT, Könen W, Strack T (1993) Chimia 47:493
68. Dinjus E, Fornika R, Scholz M (1996) Organic Chemistry in Supercritical Fluids. In: van Eldik R, Hubbard CD (eds) Chemistry under Extreme or Non-Classical Conditions. Wiley, New York, p 219
69. Kreher U, Schebesta S, Walther D (1998) Z anorg allg Chem 624:602
70. Jeong N, Hwang SH, Lee YW Lim JS (1997) J Am Chem Soc 119:10549
71. Kudo K, Ikoma F, Mori S, Komatus K, Sugita N (1995) J Chem Soc Chem Commun 633
72. Gao Y, Shi YF, Zhu ZN, Yuan WK (1997) Proceedings of the 4th International Symposium on Supercritical Fluids, Sendai, Japan, p 531
73. Hitzler MG, Smail FR, Ross SK, Poliakoff M (1998) Chem Commun 1998 359
74. a) Mistele CD, Thorp HH, DeSimone JM (1995) Polym Prepr 36:507; b) Mistele CD, Thorp HH, DeSimone JM, (1996) J. M. S. – Pure Appl Chem A33:953
75. Fürstner A, Koch D, Langemann K, Leitner W, Six C (1997) Angew Chem Int Ed Engl 36:2466
76. Hamilton JG, Rooney JJ, DeSimone JM, Mistele CD (1998) Macromolecules 31:4387
77. Six C (1997) PhD thesis, Max-Planck-Insitut für Kohlenforschung/Universität Jena, Verlag Mainz, Aachen
78. Morgenstern DA, LeLacheur RM, Morita DK, Borkowsky SL, Feng S, Brown GH, Luan L, Gross MF, Burk MJ, Tumas W (1996) Supercritical Carbon Dioxide as a Substitute Solvent for Chemical Synthesis and Catalysis. In: Anastas PT, Williamson TC (eds) Green Chemistry. ACS Symp Ser 626, American Chemical Society, Washington DC, p 132
79. Haas GR, Kolis JW (1998) Organometallics 17:4454
80. Pesiri DR, Morita DK, Glaze W, Tumas W (1998) Chem Commun 1015
81. Haas GR, Kolis JW (1998) Tetrahedron Lett. 39:5923
82. Oakes RS, Clifford AA, Bartle KD, Pett MT, Rayner CM (1999) Chem Commun 247
83. Loeker F, Koch D, Leitner W (1998) Supercitical Carbon Dioxide as an Innovative Medium for Oxidation. In: Emig G, Kohlpaintner C, Lücke B (eds) Selective Oxidations in Petrochemistry, DGMK, Hamburg, p 209
84. a) Dooley K, Knopf F (1987) Ind Eng Chem Res 26:1910; b) Wu XW, Oshima Y, Koda S, (1997) Chem Lett 1045
85. a) Ochiogrosso RN, McHugh MA (1987) Chem Eng Sci 42:2478; b) Srinivas P, Mukhopadhyay M (1994) Ind Eng Chem Res 33:3118
86. Birnbaum ER, Le Lacheur RM, Horton AC, Tumas W (1999) J Mol Catal A: Chem 139:11

87. Loeker F, Leitner W (1999) submitted for publication
88. a) Ruf M, Chell FA, Walz R, Vahrenkamp H (1997) Chem Ber/Recueil 130:101; b) Fang S, Fujimoto K (1996) Appl Catal A: General 142:L1
89. Vieville C, Yoo JW, Pelet S, Mouloungui Z (1998) Catal Lett 56:245
90. Sakakura T, Saito Y, Okano M, Choi JC, Sako T (1998) J Org Chem 63:7095
91. Darensbourg DJ, Stafford NW, Katsurao T (1995) J Mol Catal A: Chem. 104:L1-L4
92. Darensbourg DJ, Zimmer MS (1999) Macromolecules 32:2137
93. Super M, Berluche E, Costello C, Beckman, EJ (1997) Macromolecules 30:368

Modern Solvent Systems in Industrial Homogeneous Catalysis

Boy Cornils

Hoechst AG, D-65926 Frankfurt/M, Germany

Water can be used very advantageously as the solvent in homogeneous catalytic reactions and thus gives rise to a specific class of aqueous biphasic reactions among two-phase reactions. This new use of water has two very positive consequences: an advantageous effect on the selectivity of the homogeneous catalytic reactions and the opportunity, for the first time, to "immobilize" the homogeneous catalyst by means of the "liquid support" water, then to separate it simply and immediately from the reaction products after its task is complete and thus to return it to the catalysis cycle. Apart from the industrial use of hydroformylation (oxo process), fine chemicals are also being increasingly prepared by the aqueous, homogeneously catalyzed biphasic technique.

Keywords: Hydroformylation, Oxo process, Aqueous biphasic homogeneous catalysis.

1
Introduction

1.1
Motivation for Work

Water cannot be regarded as a "modern solvent" in *all* cases, although there are indications of a renaissance in its use in organic chemistry [1–4]. This inter-

Topics in Current Chemistry, Vol. 206
© Springer-Verlag Berlin Heidelberg 1999

esting development has been influenced by, inter alia, the introduction of water as a solvent in aqueous biphasic homogeneous catalysis where it has revolutionized the process methodology and does indeed represent a "modern solvent".

Until about 20 years ago, homogeneously catalyzed reactions were the domain of organic phases and solvents. One reason for this was that the catalysts employed – apart from acid/base systems, usually sensitive and frequently thermally unstable organometallics [5–7] – were almost always hydrolytically labile, and the combination of homogeneous catalytic systems with water seemed illogical and either had little to recommend it or harbored no promise at all. On the contrary, in hydroformylation (oxo process), which has for decades been the most important application of homogeneous catalysis (1980 production: about 5 million metric tons per year [8], 1998: over 6 million metric tons per year [9]), the catalyst was usually decomposed by means of aqueous reactants and the catalyst cycle was thus deliberately interrupted (brief description in [10]). This is all the more surprising since there were indications of particular (positive) effects of an interaction between water and homogeneous organometallic catalysts [11]. Even earlier, in the new, cobalt-catalyzed oxo process, it had been recognized that during the hydroformylation significant amounts of water led to advantageous effects (the use of water-soluble catalyst precursors as in the BASF process [12] or the early Ruhrchemie process [13], the yield-increasing cleavage of initially formed formic esters by means of aqueous sodium formate solutions [14], etc).

All these observations were empirical, individual results of unsystematic experiments. Since water had been judged, as mentioned, to be incompatible with the metal carbonyl catalysts of the oxo process, this solvent was not a seriously considered alternative. This paper points the way to the introduction of water as a future-oriented solvent for industrial homogeneous catalysis. Applications of phase transfer catalysis will not be considered here (since they require additional, cost-increasing phase transfer agents), but the emphasis will be placed on aqueous biphasic homogeneous catalysis and its status and possibilities.

1.2
Background

Compared to the "classical" oxo process (classical in respect of the organic phase usually employed), the use of water was a revolutionary development following a report by Manassen entitled "Heterogenizing Catalysts" in 1973 [15]. This expressed the vision of a biphasic process in words for the first time: "A heterogenized catalyst allows much greater versatility (...). (...). However, the use of two immiscible liquid phases, one containing the catalyst and the other containing the substrate, must be considered. The two phases can be separated by conventional means and high degrees of dispersion can be obtained through emulsification. This ease of separation may be particularly advantageous in situations where frequent catalyst regeneration or reactivation is required."

This quotation indicates that the extraordinarily important advantage, in terms of process technology, of a catalyst phase which is immiscible with the reaction products was the actual impetus for the biphasic catalysis process outlined by Manassen. It is therefore possible to separate the reaction products from the homogeneous catalyst phase by simple phase separation (decantation) immediately after the reaction is complete and to recirculate the catalyst. This provides a very elegant solution to the age-old problem of homogeneous catalysis, namely the separation of the reaction products from the catalyst which is, by definition, homogeneously dissolved in them (cf. Fig. 1).

This concept allows the important advantages of homogeneous over heterogeneous catalysts to be utilized for the first time. Particular mention should be made here of the significantly greater variability of homogeneous catalysts which enables them to be tailored to the requirements of the desired reaction by means of steric and/or electronic modification (cf. Table 1).

It is the opportunity to understand the mode of action of homogeneous catalysts by modeling under realistic conditions and to comprehend their structure/activity relationships by varying parameters that makes homogeneous catalysis so interesting to the modern researcher. Biphasic catalysis makes the decisive disadvantage of homogeneous catalysis obsolete, something which has been unsuccessfully attempted for decades by "heterogenizing", i.e. by anchoring the intrinsically homogeneous catalysts on a matrix (support) having a heterogeneous structure and thus immobilizing them [16]. These attempts were doomed to failure because the strong and continually changing stresses on the bond between the catalyst metal and support lead to weakening of this bond and thus, finally, to leaching of the metal (cf. Fig. 2, [17]).

Biphasic catalysis, not anchoring on solid supports, was recognized as providing the fundamental means for combining all the advantages of homogeneous

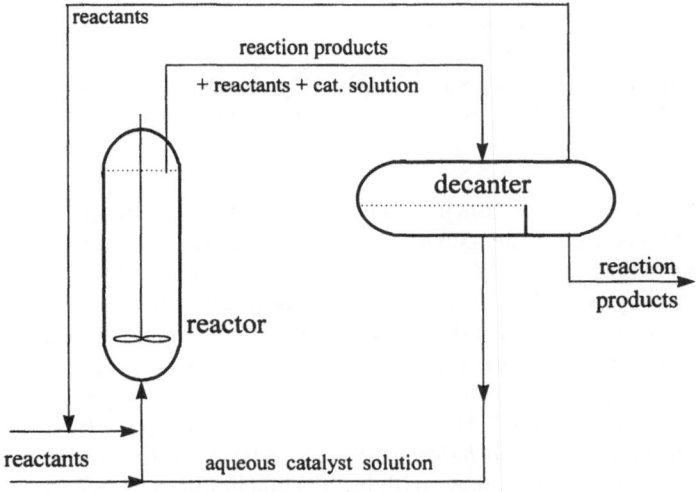

Fig. 1. Biphasic concept: the reactor (*left*) is supplied with the reactants; in the phase separator (*right*), the reaction products and the catalyst phase are separated. The catalyst phase is returned by a short route to the reactor

Table 1. Homogeneous vs. heterogeneous catalysis: the advantages and disadvantages

	Homogeneous catalysis	Heterogeneous catalysis
Activity (relative to metal content)	High	Variable
Selectivity	High	Variable
Reaction conditions	Mild	Harsh
Service life of catalysts	Variable	Long
Sensitivity toward catalyst poisons	Low	High
Diffusion problems	None	High
Catalyst recycling	Expensive	Not necessary
Variability of steric and electronic properties of catalysts	Possible	Not possible
Mechanistc understanding	Plausible under random conditions	More or less impossible

$$Co_2(CO)_7 \text{ support} + H_2 \rightleftharpoons 2 \ HCo(CO)_3 \text{ support}$$

$$HCo(CO)_3 \text{ support} \rightleftharpoons HCo(CO)_2 \text{ support} + CO$$

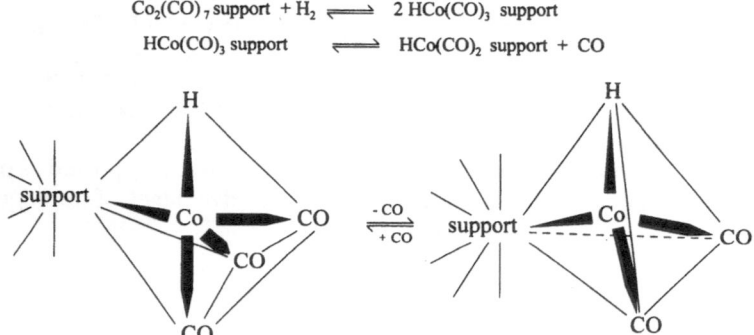

Fig. 2. Alternation between supported trigonal-bipyramidal hydridotricarbonylcobalt and tetrahedral hydridodicarbonylcobalt during the course of the catalytic cycle

catalysis. To avoid confusion, it should be emphasized at this point that the catalysis of the reactions under consideration here (e. g. hydroformylation) is *homogeneous* in nature even in the case of modification of the biphasic process. This can easily be proven by the fact that *all* criteria of a homogeneously catalyzed reaction are met (their catalysts are, inter alia, molecularly dispersed "in the same phase", they are unequivocally characterized chemically and spectroscopically, and go through a detectable catalyst cycle; they permit unequivocal reaction kinetics, etc., cf. [5]). Nevertheless, the catalyst (not the *catalysis* which remains homogeneous!) is *heterogeneous* in relation to the reactants (a catalyst solution is brought into contact with immiscible liquid or gaseous reactants), the participants in the reaction are *heterogeneous* in relation to one another, namely gaseous and liquid, and the reaction product is also present in a different liquid phase from the catalyst, i.e. in *heterogeneous* form. Thus, in the final analysis, the reaction is a three-phase reaction which is described (by tacit agreement!) as "biphasic" only because a gas phase is not counted separately in catalytic reactions.

The term "biphasic catalysis" has established itself only slowly, especially since the first commercial applications (by Shell in the SHOP process, [18]) named the process thus only tentatively (and in obviously rarely read patent applications).

Aqueous biphasic catalysis represents the most important special case of biphasic processes. Virtually in parallel to the propagation of biphasic catalysis, publications by Joó and his research group describing homogeneously catalyzed hydrogenations (and selective hydrogenations) in water appeared [19]. As catalytically active complexes, use was made of Rh compounds which had been modified by means of the monosulfonated derivative of triphenylphosphine (known as TPPMS, see below). The first experiments aimed at developing a hydroformylation process based on biphasic catalysis go back to Kuntz, then at Rhône-Poulenc [20]. The important factor, which finally brought success, in Kuntz's work was his clever and ingenious combination of the following features:

(1) the choice of rhodium as central atom of the oxo catalyst. This choice corresponded to the spirit of the times after, in 1975/76, the first oxo plants of Celanese [21] and Union Carbide [22] had commenced operation on the basis of Rh catalysts of the Wilkinson type [23] and their selectivity advantages had become evident,

(2) modification of the ligands of the homogeneous oxo catalyst by means of phosphines, which was likewise known from the Shell oxo process (albeit using cobalt as central atom [24]) and the example of Wilkinson [25], and

(3) the decisive choice of water as reaction medium and "mobile" or "liquid" support and thus as second phase in the process.

The experimental work of Kuntz struck on the ligand TPPTS (3,3,′ 3″[phosphinidyne]benzenesulfonic acid, trisodium salt, the triply sulfonated homologue of TPPMS), today the standard ligand in aqueous homogeneous applications (Fig. 3), and thus the opportunity of modifying the oxo-active $HRh(CO)_4$ by means of this readily water-soluble ligand (solubility: about 1100 g l^{-1}) so as to make the entire metal complex catalyst $HRh(CO)L_3$ water soluble and, even more importantly, to leave it in the aqueous phase without leaching.

The studies by Joó and Kuntz had a peculiar fate. It may well have been due to the remoteness from practice and the lack of incentive for new variants of hydrogenation processes that Joó's work almost became lost in the academic world. Likewise, the importance of the preliminary work of Kuntz was recognized neither by his own company nor by the academic community, which may well be attributed to the fact that the results were recorded in patents and academic teachers and researchers are reluctant to take notice of such apocryphal literature. There was therefore a need for an industrial impetus and the readiness to carry out process development work based on these studies. This occurred after 1982 in collaborative work between Ruhrchemie AG and Rhône-Poulenc (RCH/RP) [26]. The combination of a basic idea from Rhône-Poulenc with the competence of a team at the then Ruhrchemie AG with ex-

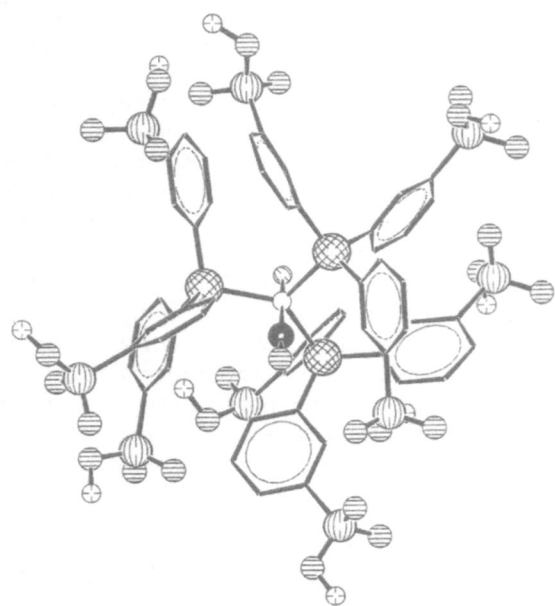

Fig. 3. TPPTS and the corresponding oxo catalyst

perience in the field of the oxo process and the translation of laboratory results into industrial practice led within two years to successful pilot plant development, the construction of the first production plant and its startup in July 1984 [27–29]. This very rapid progress was made possible only by decisive improvements in the basic idea, by processing improvements, the scale-up factor employed (1:24,000) and the systematic use of "simultaneous engineering". Looking back after 15 years, a very positive picture can be presented [27e].

Kuntz's recourse to water as solvent for a two-phase process was extraordinarily fortunate from both a chemical and process engineering point of view. Water displays particular advantages as solvent in processes during the course of which reactants and reaction products of different polarities participate or are formed [1–4]. A look at the properties of water which are important in this respect (cf. Table 2) shows this very clearly.

Water as a solvent has several anomalous features (e.g. anomalous density, the only nontoxic and liquid "hydride" of the nonmetals, melting point varying with pressure, dielectric constant) and with its two- or even three-dimensional structure has still not been fully researched.

Of direct importance for the aqueous biphase processes are the physiological (entries 2, 4 of Table 2), economic (1, 3, 6), ecological/safety-related (2, 4), process engineering (1, 6, 7, 9, 11, 12, 13), and chemical and physical properties (1, 5, 6, 8, 10, 12, 14) of water. The different properties interact and complement each other. Thus water, whose high Hildebrand parameter [31, 32] and high polarity advantageously influence organic chemical reactions (such as hydro-

Table 2. Properties of water as a liquid support of aqueous two-phase catalysis [29, 30]

Entry

1 polar and easy to separate from unpolar solvents or products; polarity may influence (improve) reactivity
2 inflammable, incombustible
3 ubiquitous and with suitable quality available
4 odour- and colourless, making contamination easy recognizable
5 formation of hexagonal two-dimensional surface structure and a tetrahedral three-dimensional molecular network, which influence the mutual (in)solubility significantly; chaotropic compounds lower the order by H-bonding breaking
6 high *Hildebrand* parameter as unit of solubility of non-electrolytes in organic solvents
7 density of 1 g cm^{-3} provides a sufficient difference to most organic substances
8 very high dielectric constant
9 high thermal conductivity, high specific heat capacity and high evaporation enthalpy
10 low refractive index
11 high solubility of many gases, especially CO_2
12 formation of hydrates and solvates
13 highly dispersable and high tendency of micelle formation; stabilization by additives
14 amphoteric behaviour in meanings of Brønsted

formylation), has sufficiently high polarity and density differences compared to organic (reaction) products to enable separation of the phases after the homogeneously catalyzed reaction.

On the other hand the high solvent capability for many compounds and gases, in some cases boosted by solvate or hydrate formation or by hydrogen bonding, facilitates reactions in the two-phase system. The chaotropic properties of many chemical compounds prevent the H_2O cage structures necessary for the formation of solvates and thus facilitate the transfer of nonpolar molecules from nonaqueous and aqueous phases.

Water is non-combustible and nonflammable, odorless and colorless, and is universally available: important prerequisites for the solvent of choice in catalytic processes. The dielectric constant or the refractive index can be important in particular reactions and in analyzing them. The favorable thermal properties of water make it highly suitable for its simultaneous double function as a mobile support and heat transfer fluid, a feature that is utilized in the RCH/RP process (see below).

Thus water as a solvent not only has the advantage of being a mobile support and simultaneously actually "immobilizing" the catalyst while retaining the homogeneous mode of reaction but, in particular, has positive effects on the environmental aspects of hydroformylation [33].

If the reactants and reaction products have polarities which are *very* different from that of water, special measures have to be taken in order to be able to continue to employ a two-phase process using water as solvent. This is the case in the hydroformylation of higher olefins ($> C_5$). The fall in the solubility in water associated with an increasing number of carbon atoms in the olefins used (and also in the resulting aldehydes) leads to a pronounced decrease in the reactivity (cf. Fig. 4).

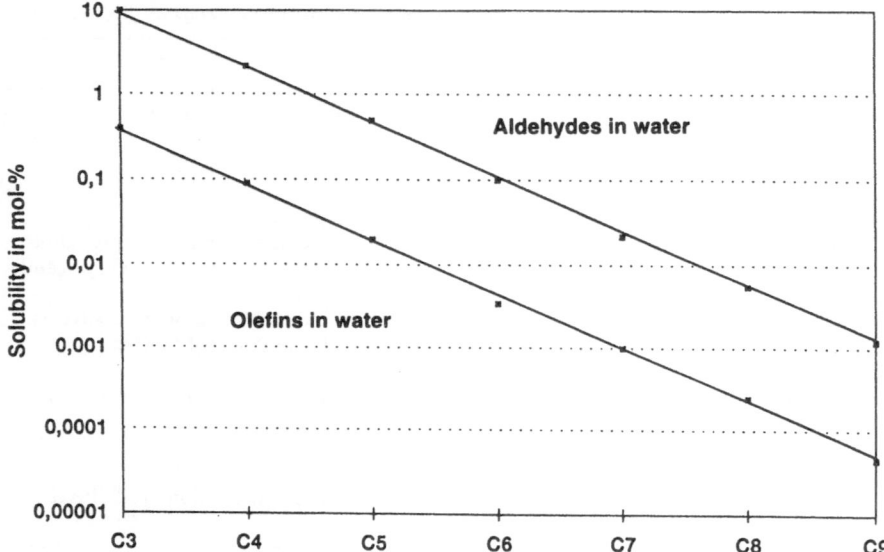

Fig. 4. Dependence of the solubility in water on the number of carbon atoms in the olefins used and in the resulting aldehydes [34]

There is still a dispute among experts as to the place in which the biphasic aqueous reaction actually takes place, although it is very probably not the bulk of the liquid but the interfacial layer between the aqueous and organic phases. In the case of aqueous biphasic hydroformylation, this question has been decided by methods of reaction modeling and comparison with experimentally proven facts, thus leading to scale-up rules and appropriate kinetic models as a basis for optimal reactor design [34].

2
State of the Art

2.1
Hydroformylation of Lower Olefins

The most important and oldest application of aqueous biphasic, homogeneous catalysis is hydroformylation (oxo process, Roelen reaction). This process is used to produce *n*-butyraldehyde, the desired main product of the reaction of propylene, which is converted by aldolization into 2-ethylhexenal and this is finally hydrogenated to give 2-ethylhexanol (2-EH), the most economically important plasticizer alcohol (Scheme 1):

2–EH is esterified with phthalic acid to produce the standard plasticizer dioctyl phthalate (DOP). With a capacity of roughly 5 million metric tons per year, *n*-butyraldehyde is among the most important applications of homogeneous catalysis [9, 35].

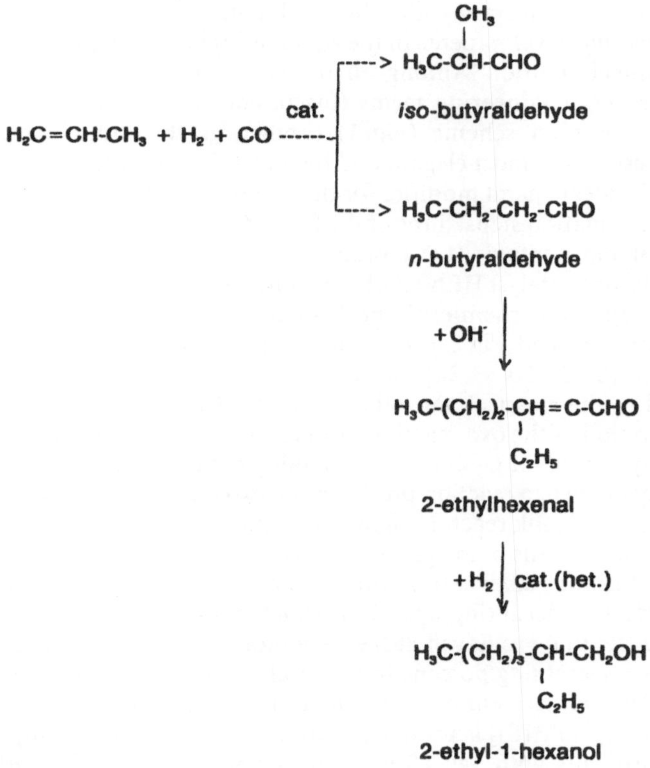

$$\text{H}_2\text{C=CH-CH}_3 + \text{H}_2 + \text{CO} \xrightarrow{\text{cat.}}$$

$$\begin{array}{c} \text{CH}_3 \\ | \\ \text{----> H}_3\text{C-CH-CHO} \\ \textit{iso-}\text{butyraldehyde} \end{array}$$

$$\text{----> H}_3\text{C-CH}_2\text{-CH}_2\text{-CHO}$$

$\textit{n-}$butyraldehyde

$$+\text{OH}^- \downarrow$$

$$\text{H}_3\text{C-(CH}_2)_2\text{-CH=C-CHO} \atop | \atop \text{C}_2\text{H}_5$$

2-ethylhexenal

$$+\text{H}_2 \Big| \text{cat.(het.)} \downarrow$$

$$\text{H}_3\text{C-(CH}_2)_3\text{-CH-CH}_2\text{OH} \atop | \atop \text{C}_2\text{H}_5$$

Scheme 1   2-ethyl-1-hexanol

The hydroformylation of propylene was an obvious choice as the first application of the new aqueous biphasic catalysis technology for several reasons:

(1) There is a balanced ratio of nonpolar reactants (such as propylene) and polar reaction products (butyraldehydes) which is particularly favorable for a two-phase application.
(2) The inherent problem of homogeneous catalysis, viz. the unsatisfactory separation of catalyst and reaction product, had not been optimally solved even not in the newer variants (e.g. Union Carbide's LPO process [36]).
(3) Other process engineering aspects of the processes also no longer met modern requirements after what was then over ten years of use. This applied particularly to the thermal balance of the LPO process and, after several energy crises, to its waste of energy.
(4) The more recent results of organometallic research with their possibilities for tailoring the oxo catalyst (especially in the area of very active and even more selective catalysts) could not be optimally applied to the LPO process, mainly because in this process the catalyst was still subjected to high thermal stresses during recycling.

It is therefore not surprising that the combination of modern *chemical* and *process engineering* developments in the aqueous biphasic oxo process led to the optimal technical solution. Among all the conceivable ways of designing a process in process engineering terms (discussion in [37]), the configuration based on the simplest scheme (Fig. 1) (known as the Ruhrchemie/Rhône-Poulenc process) is the most elegant and the most inexpensive.

TPPTS is the ideal ligand modifier for the oxo-active $HRh(CO)_4$. Without any expensive preformation steps, three of the four CO ligands can be substituted by the readily soluble, nontoxic (LD_{50}, oral: > 5000 mg kg^{-1}) TPPTS, which yields the hydrophilic oxo catalyst $HRh(CO)[P(m\text{-sulfophenyl-Na})_3]_3$ (Fig. 5). The fundamentals of the Ruhrchemie/Rhône-Poulenc oxo process and of the new TPPTS-modified Rh hydrido carbonyl complex have been described in a series of publications [25, 27, 30, 33, 34], cf. Fig. 6.

Compared to the classical and other variants of the oxo process, including thermal separation of the oxo reaction products from the catalyst, the procedure is considerably simplified. Owing to the solubility of the Rh complex in water and its insolubility in the oxo reaction products, the oxo unit is essentially reduced to a continuous stirred tank reactor followed by a phase separator (decanter) and a stripping column. Details of the process are available in the literature [27e – g].

The catalyst is not sensitive to sulfur or other oxo poisons. Together with the simple but effective decanting operation which allows organic impurities and other byproducts to be removed at the very moment of separation, accumulation of activity-decreasing poisons in the catalyst solution is prevented. Therefore, no special pretreatment or even purification steps are necessary. The oxo catalyst $HRh(CO)(TPPTS)_3$ is produced within the oxo reactor simply by reacting suitable Rh salts with TPPTS without any additional preformation step. Typical reaction conditions and compositions of crude products from the RCH/RP process based on a 14-year average are compiled in Table 3.

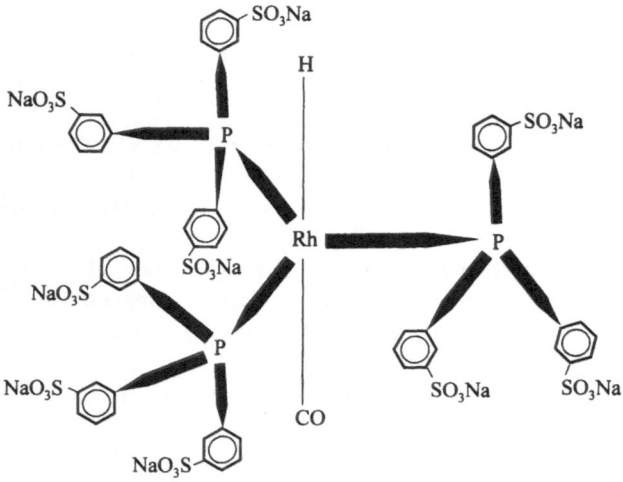

Fig. 5. The active catalyst of Ruhrchemie/Rhône-Poulenc's new oxo process

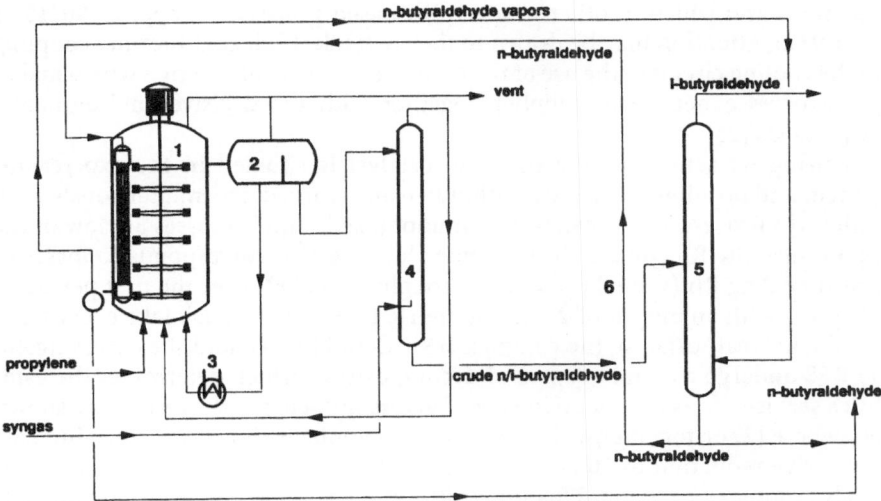

Fig. 6. Flow diagram of the RCH/RP process [27 e]

Table 3. Reaction conditions and results of RCH/RP process (15-year, average)

Parameter	Typical values
Temperature (°C)	120
Total pressure (Mpa)	5
CO/H$_2$ ratio	1.01
Ratio water/organic phase	6
Energy efficiency[a]	100
Conversion (%)	95
Results	
iso-Butanal (%)	4.5
n-Butanal (%)	94.5
Butanols (%)	< 0.5
Butyl formates (%)	traces
Heavy ends[b] (%)	0.4
n/iso ratio	95:5

[a] Exclusive of radiation and convection losses.
[b] Mainly 2-ethyl-3-hydroxyhexan-1-al.

The high selectivity of propene hydroformylation toward both the very highly valued C_4 *aldehydes* and the *sum of* C_4 *products* is a decisive factor. Both phenomena are the result of modifying the catalyst as compared to the classical oxo processes (see above). The high selectivity toward *the sum of* C_4 *products* is a specific result of biphasic operation and of the free availability of water during hydroformylation. The *n/iso* ratio, which is important for hydroformylation reactions and which in other modern Rh-based processes is approx. 90:10, is over 95:5. Attention may be drawn to the particularly elegant thermal coupling of the cooling circuit to the use of the (exothermic) heat of reaction which makes the process a net steam supplier, together with the decisive environmental advantages [25].

During its active life, the rhodium catalyst is situated in the oxo reactor system and no aliquot parts are withdrawn and worked up simultaneously as in other oxo processes. The catalyst is "immobilized" and Rh losses are low in the ppb range; the Rh content of the crude aldehyde also corresponds to losses of less than 2 kg Rh (valued presently at roughly $ 40,000) over the first period of 10 years with an output of 2 million metric tons of n-butanal. Like every technical and "real" catalyst, the complex $HRh(CO)(TPPTS)_3$ and the excess ligand TPPTS undergo a certain degree of decomposition, which determines the catalyst's service life as measured in years. Factors influencing the results are shown elsewhere [27e], together with comparative manufacturing costs which indicate a decisive reduction of 10 % compared to other oxo processes with their relatively complex operation. This underlines the favorable nature of the aqueous two-phase reaction system.

Starting in 1999, the Ruhrchemie/Rhône-Poulenc process will be operated in plants having a capacity of about 600,000 metric tons per year, which corresponds to over 10 % of the annual world production of C_4 products; the first licensed plant is operating in Korea (Hanwha Chem. Corp.). On the Ruhrchemie site, a butene hydroformylation plant [to produce n-valeraldehyde (pentanal)] is also operating without problems.

2.2
Hydroformylation of Higher Olefins

Compared to the conversion of propylene into n-butyraldehyde, the hydroformylation of higher olefins is, at a world production of about 1 million metric tons per year, of significantly less importance [9] and is concentrated (apart from small amounts of C_7-aldehyde from hexenes) on the reaction of C_8 and higher olefins. As shown in Fig. 4, the solubility in water of such olefins and the aldehydes containing one more carbon atom is about three powers of ten lower than that of propylene. This makes mass transfer more difficult, a fact which is held responsible for the reduced activity of the higher starting olefins in the aqueous two-phase process [41]. Attempts were initially made to deal with this problem by means of solubilizers [38], measures such as ultrasound which increase mixing [39], or specially tailored, e.g. surfactant, ligands [40]. Without going into the underlying dispute between experts as to the location of the aqueous two-phase catalytic reaction (cf. [34]), it may be mentioned that many other specific measures have been proposed, of which an arbitrary and incomplete selection is shown in Table 4.

The great number of very different proposals which address widely differing points in the oxo process does prove that there is currently no single ideal way of reacting higher olefins in aqueous biphasic hydroformylations. In addition, the oxo products, which are under a great deal of cost pressure, tolerate no cost-

Table 4. Examples of proposed measures for improving the hydroformylation of higher olefins

Measure	Reference
Substitution of the standard ligand by a tailor-made one	42
Addition of salts	43
Variation of the cationic part of the ligand/amphiphilic ligands	44
Other central atoms	45
Application of thermoregulating or polymer ligands	46
Addition of co- or promoter ligands	47
Exploitation of the effect of molecular recognition	48
Application of supported aqueous phase catalysts	49
Biphasic operation with heterogenized or heterogeneous catalysts	50
Changes of the reaction engineering concept	51

increasing process modifications either in terms of material costs or capital costs. For this reason, additives such as solubilizers or surfactants (which have to be changed according to the starting olefin and recovered subject to losses) or fundamental alterations in the process employed in the oxo plants are relatively improbable from an economic point of view. Furthermore, the campaign operation which is not uncommon for higher olefins prohibits, for cost reasons, a change of catalyst system or even only of ligand system with changing feed olefins. It may be pointed out that pH control of the hydroformylation [54, 55], which is suggested by process engineering considerations and by the effects on reactivity and product composition, is not sufficient in the presence of the great polarity and solubility differences of the reactants.

In the hydroformylation of higher olefins, even using the aqueous biphasic method, it has to be assumed that, owing to the reduction in the reaction rate caused by decreasing solubility of the olefins, the actual reaction should be single phase and the separation should be two phase. From today's point of view, this leaves the following routes to a solution:

(1) The use of "thermoregulating" ligands of the type shown in Fig. 7 [46a].

Oxocatalysts which are modified by means of such ligands take advantage of a temperature-dependent "cloud point" associated with the phosphorus-bonded poly(alkylene glycol ether). Thus, above the cloud point, the ligand (and thus the catalyst complex) loses its hydration shell, just as in the case of other compounds

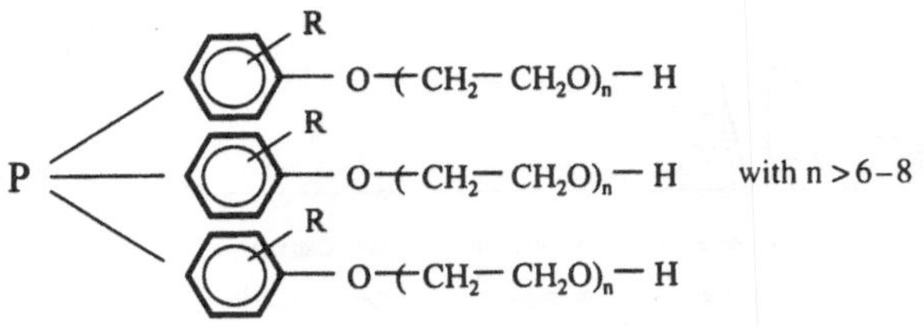

Fig. 7. Thermoregulating ligands as described by Jin and Fell [46 a]

of this type, causing the two-phase reaction mixture normally obtained when olefin is added to the catalyst solution to merge into a single phase, thereby initiating a rapid conversion that is no longer impeded by material transport problems. Subsequent lowering of the temperature causes the hydration shell to be reversibly restored, inducing the catalyst complex solution once again to separate out as an independent phase, this time from the reaction product, viz. the desired higher aldehydes (cf. Fig. 8 [52]).

Since the agent responsible for the merger and subsequent separation of the phases is the appropriately custom-designed ligand itself, there is no need to invest extra effort in the removal and recycling of an extraneous additive, and this must therefore be regarded as a promising avenue for further exploration on a commercially realistic scale. The means of solving the problem is thus the ligand tailored by chemical modification and the solution itself is the utilization of the physical effect of the cloud point.

(2) Another option is offered by the *chemical* solution. It is based on a problem-specific phase change between a two-phase and single-phase system by means of pH-controlled immobilization/re-immobilization [53]. The mode of action is demonstrated in Fig. 9.

In this procedure too, the actual oxo reaction is carried out without mass transfer problems in the monophase, but the actual separation is between two phases. This chemical variation of the process for reacting higher olefins can also include membrane steps [26], even if such additional process steps do involve the risk of drawbacks ranging from costs to material problems.

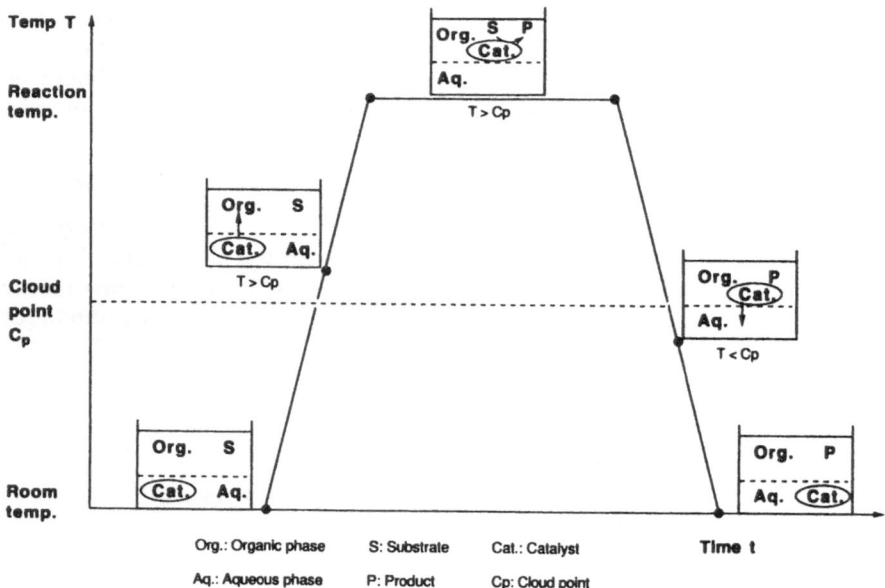

Fig. 8. Phase change in the case of thermoregulating ligands

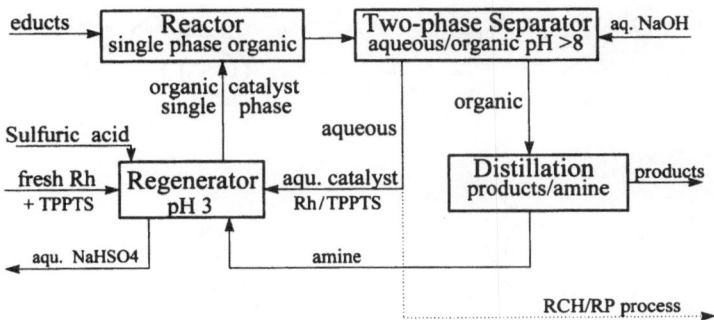

Fig. 9. Catalyst recycling by pH-induced phase separation

The development of thermoregulating ligands might well make a decisive contribution to solving the problems of aqueous, homogeneously catalyzed hydroformylations of higher olefins.

3
Water in Other Industrial Applications

Apart from the oxo process, a series of other reactions are carried out industrially, even if on a smaller scale. Kuraray carries out the hydrodimerization of butadiene and water to produce n-octanol (or 1,9-nonanediol) on a scale of about 5000 metric tons per year [55]. Applications which are significantly smaller up to now are, for example, the production of vitamin precursors by Rhône-Poulenc (cf. Scheme 2, [56]) and the production of substituted phenylacetic acids by carbonylation (Scheme 3) [57]) or of biaryls by Suzuki cross coupling (Scheme 4), both by Hoechst AG (now Clariant AG, [57, 58]).

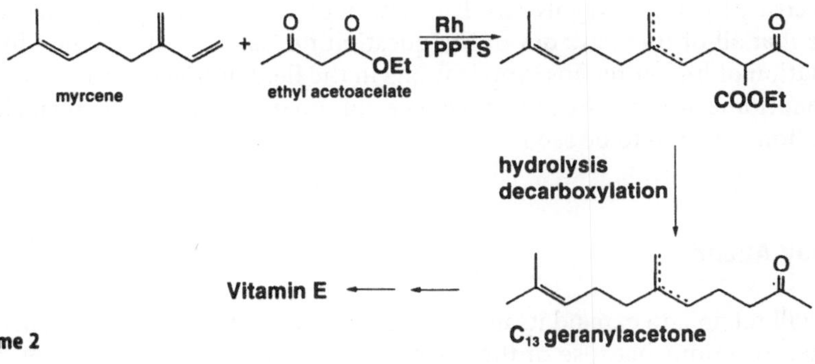

Scheme 2

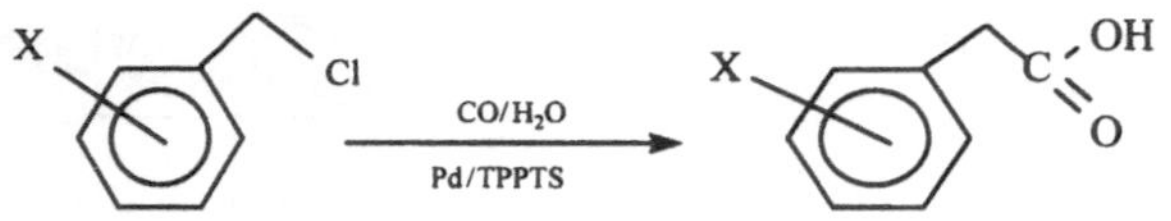

Scheme 3

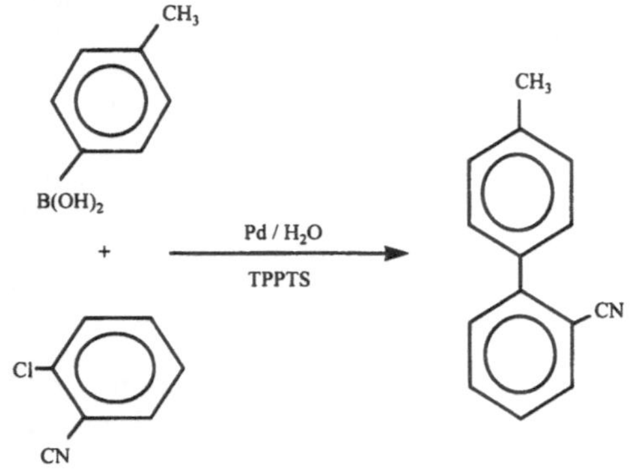

Scheme 4

4
Other "Modern" Solvents

The enormous interest in a search for solvents other than the "classical" solvents for homogeneous catalysis – aroused by the success of aqueous biphasic catalysis – has produced some initial possibilities, namely "fluorous phases" [59], "nonaqueous ionic liquids" ("NAIL", [60]) and supercritical CO_2 (scCO_2, [61]). A feature common to all three variants is that, as a result of the unusual properties of the solvents employed and the ligands used for modifying the catalyst complexes (partially fluorinated or perfluorinated solvents or phosphines and organochloroaluminates), they likewise make possible an (advantageous, cf. above) *mono*phase reaction and a *two*-phase separation of catalyst and reaction product. However, to date evidence of industrial suitability, ease of handling and competitiveness is lacking for all three possibilities, and all three are expensive due to costly ligands and solvents. Estimation of costs for all three variants indicates that all of them are out of the question for use in cost-oriented hydroformylation of higher olefins. Applications in the field of fine chemicals, especially enantiomerically pure intermediates for pharmaceutical or agricultural applications, remain to be seen.

5
The Way Ahead

Water will retain and expand its importance as a new solvent for homogeneously catalyzed reactions because of the great process engineering advantages. Four focal points will be at the forefront:

(1) Academic research has to fill the knowledge gap between applied and fundamental research as soon as possible and, building on this, make new advances.

(2) In organometallic research, it will be important to find new ligands and ligand systems and examine them for their properties, their potential for custom design and their compatibility.

(3) Fundamental studies regarding the site of the reaction and on modeling the course of the reaction will remain significant for the process engineering of aqueous biphasic catalysis.

(4) Biphasic catalysis, specifically the aqueous variant, will increasingly provide the opportunity to better exploit the potential of homogeneous catalysis compared to the heterogeneous modus operandi and to increase its share of the totality of catalytic processes.

In future the aqueous biphasic processes will grow in importance because of the great advantages of this homogeneously catalyzed variant. Dimerizations, telomerizations, hydrocyanations, hydrosilylations, aldolizations, Claisen condensations, and a wide variety of C–C coupling reactions will be objectives above and beyond the currently used syntheses. Only the future will tell what importance asymmetric/enantiomeric conversions, especially hydroformylations, will have here, although initial experience is encouraging.

6
References

1. Lubineau A (1996) Chemistry & Industry Feb 19, 123; Lubineau A, Augé J, Queneau Y (1994) Synthesis 8:714
2. Lubineau A (1998) In: Cornils B, Herrmann WA (eds) Aqueous-phase organometallic catalysis – concepts and applications. Wiley-VCH, Weinheim, chap 2.1
3. Li Ch-J (1993) Chem Rev 93:2023
4. Southern TG (1989) Polyhedron 8(4):407
5. Cornils B, Herrmann WA (eds) (1998) Aqueous-phase organometallic catalysis – concepts and applications, Wiley-VCH, Weinheim
6. Parshall GW, Ittel SD (1992) Homogeneous catalysis, 2nd edn.; Wiley-Interscience, New York
7. Dickson RS (1985) Homogeneous catalysis with compounds of rhodium and iridium. Reidel, Dordrecht, Boston
8. Cornils B (1980) Hydroformylation. In: Falbe J (ed) New syntheses with carbon monoxide, Springer, Berlin Heidelberg New York, p 177
9. Gick W (Celanese, Ruhrchemie works), private communication
10. cf. Cornils B (1980) Hydroformylation. In: Falbe J (ed) New syntheses with carbon monoxide, Springer, Berlin Heidelberg New York, p 158 ff
11. e.g. Abraham MH, Grellier PL, Nasehzadeh A, Walker RAC (1988) J Chem Soc Perkin Trans 2:1717
12. Nienburg H (1953) DE Patent 948 150 Chemische Verwertungsgesellschaft Oberhausen mbH; Falbe J (ed) (1980) New syntheses with carbon monoxide, Springer, Berlin Heidelberg New York, sect 1.6.2.1.2
13. Kolling H, Büchner K, Stiebling E (1965) DE Patent 949 737, (1952) GB Patent 736875 Ruhrchemie AG
14. Tummes H, Meis J (1965) US Patent 3 462 500 Ruhrchemie AG

15. Manassen J (1973) In: Basolo F, Burwell RL (eds) Catalysis in research, Plenum, London, p 183
16. cf. Panster P, Wieland St in [53], section 3.1.1.3; Artley TR (1985) Supported metal complexes. Reidel, Dordrecht; Ford WT (1984) CHEMTECH 7:436
17. Cornils B, preliminary results
18. Keim W (1984) Chem Ing Tech 56: 850; Bauer RS, Orinda HC, Glockner PW, Keim W (1972) US Patent 3 635 937 Shell Oil Comp
19. Joó F, Beck MT (1975) React Kin Catal Lett 2:257
20. Kuntz EG (1976) FR Patent 2 314 910, 2 349 562, 2 338 253, 2 366 237 Rhône-Poulenc
21. In Sep. 1971 Celanese Chem. Co. started commercial Rh-catalyzed propylene oxo unit at Bishop, TX – first one in the world; cf. Celanese Annual Report 1974, p 9; Hughes OR, Douglas ME (1971) DE-OS Patent 2 125 382 Celanese Corp
22. Fowler R, Connor H, Baehl RA (1976) CHEMTECH 772
23. Osborn JA, Wilkinson G, Young JF (1965) Chem Commun 17, 131; Osborne JA, Jardine FH, Young JF, Wilkinson G (1966) J Chem Soc A 1711
24. Slaugh, LH, Mullineaux RD (1966) US Patent 3 239. 566 and 3 239 571 Shell Oil Co
25. Cornils B, Kuntz EG (1995) J Organomet Chem 502:177
26. Cornils B (1998) Org Proc Res Dev 2:121
27. (a) Cornils B, Falbe J (1984) Proc 4th Int Symp on Homogeneous Catalysis, Leningrad, p 487; (b) Bach H, Gick W, Wiebus E, Cornils B (1984) Int Symp High-Pressure Chem Eng, Erlangen, Germany, Preprints, p 129; (c) Bach H, Gick W, Wiebus E, Cornils B, (1984) 8th ICC, Berlin, Preprints, vol V, p 417; Chem Abstr (1987) 106:198.051; cited also in Behr A, Röper M (1984) Erdgas und Kohle 37(11):485; (d) Bach H, Gick W, Wiebus E, Cornils B (1986) 1st IUPAC Symp Org Chem, Jerusalem, Abstracts, p 295; (e) Wiebus E, Cornils B (1994) Chem-Ing-Tech 66:916; (f) Cornils B, Wiebus E (1995) CHEMTECH 25:33; (g) Wiebus E, Cornils B (1996) Hydrocarb Proc March, 63
28. Cornils B (1998) Ind Catal News No 2
29. Cornils B (1998) J Mol Catal, A 143:10
30. Cornils B, Kohlpaintner CW, Wiebus E (1998) In: McKetta JJ (ed) Encyclopedia of chemical processing and design, Marcel Dekker, New York, Vol. 66, p 273
31. Hildebrand JH (1935) J Am Chem Soc 57:866
32. Duve G, Fuchs O, Overbeck H (1976) Lösemittel, 6th edn. Hoechst AG, Frankfurt/M, p 49; also cited in Reichardt C (1990) Solvents and solvent effects in organic chemistry, VCH, Weinheim, Fig 2-2
33. Cornils B, Wiebus E (1996) Recl Trav Chim Pays-Bas 115:211
34. Wachsen O, Himmler K, Cornils B (1998) Catalysis Today, 42:373
35. Beller M, Cornils B, Frohning CD Kohlpaintner CW (1995) J Mol Catal A 104:17
36. Cornils B (1980) Hydroformylation. In: Falbe J (ed) New syntheses with carbon monoxide, Springer, Berlin Heidelberg New York, p 174
37. Behr A, Keim W (1987) Erdöl Erdgas Kohle 103:126; Behr A (1998) Chem-Ing-Tech 70:685
38. Deshpande RM, Divekar SS, Bhanage BM, Chaudhari RV (1992) J Mol Catal A 75:L19; Purwanto P, Delmas H (1994) Symposium: Catalysis in Multiphase Reactors, Lyon, France, Abstracts C IV-2; Purwanto P, Delmas H (1995) Catal Today 24:135; Tinucci, L, Platona E (1989) EP Patent 0 380 154 Eniricerche SpA ; Hablot I, Jenck J, Casamatta G, Delmas H (1992) Chem Eng Sci 47:2689; Bahrmann H, Cornils B, Konkol W, Lipps W (1984) DE-OS Patent 3 412 335 Ruhrchemie AG; Bahrmann H, Lappe P (1992) EP Patent 0 602 463 Hoechst AG
39. Cornils B, Bahrmann H, Lipps W, Konkol W (1985) DE Patent 3 511 428 Ruhrchemie AG
40. Kanagasabapathy S, Xia Z, Papadogianakis G, Fell B (1995) J Prakt Chem/Chem Ztg 337:446; Fell B, Papadogianakis G (1991) J Mol Catal 66:143; Ding H, Hanson BE, Bartik T, Bartik B (1994) Organometallics 13:3761; Russell MJH, Murrer BA (1983) US Patent 4 399 312 Johnson Matthey plc; Cornils B, Konkol W, Bahrmann H, Bach HW, Wiebus E (1984) DE Patent 3 411 034 Ruhrchemie AG; Bahrmann H, Papadogianakis G, Fell B (1989) EP Patent 0 435 084 Hoechst AG
41. Bahrmann H, Cornils B, Konkol W, Lipps W (1984) DE Patent 3 420 491 Ruhrchemie AG

42. Andriollo, A et al (1997) J Mol Catal A 116:157; Bahrmann H et al (1996) J Organomet Chem 520:97; Smith RT, Ungar RK, Sanderson LJ, Baird MC (1983) Organometallics 2:1138; Smith RT, Baird MC (1982) Inorg Chim Acta 62:135; Haggin J (1995) Chem Eng News April 17, 25; Ding H, Kang J, Hanson BE, Kohlpaintner, CW (1997) J Mol Catal A 124:21; Amrani Y, Sinou D (1984) J Mol Catal 24:231; Herd O, Langhans KP, Stelzer O, Weferling N, Sheldrick WS (1993) Angew Chem 105:1097; Manassen J, Dror Y (1983) US Patent 4 415 500 Yeda Res and Dev Comp; Herrmann WA, Gooßen LJ, Köcher C, Artus GRJ (1996) Angew Chem 108:2980; (1995) 107:2602; Herrmann WA et al (1995) J Mol Catal A 97:65

43. Ding H, Hanson BE (1995) J Mol Catal A 99:131

44. Herrmann WA, Kellner J, Riepl H (1990) J Organomet Chem 389:103; Russell MJH, Murrer BA (1980) DE Patent 3 135 127 Johnson Matthey plc; Buhling A, Kamer PCJ, van Leeuwen PWNM, Elgersma JW (1995) J Mol Catal A 98, 69; Buhling A, Kamer PCJ, van Leeuwen PWNM, Elgersma JW (1997) J Mol Catal 116:297; Lavenot L, Roucoux A, Patin H (1997) J Mol Catal A 118:153; Bahrmann H, Cornils B, Konkol W, Lipps W, Wiebus E et al (1984) EP Patent 0 163 234, (1984) EP Patent 0 183 200, (1984) DE-OS Patent 3 447 030, (1985) EP Patents 0 216 315, (1987) EP Patent 0 302 375 Ruhrchemie AG/Hoechst AG

45. Ritter U, Winkhofer N, Schmidt H-G, Roesky HW (1996) Angew Chem Int Ed Engl 35:524; Herrmann WA, Kulpe J, Kellner J, Riepl H (1988) EP Patent 0 372 313 Hoechst AG

46. (a) Jin Z, Yan Y, Zuo H, Fell B (1996) J Prakt Chem/Chem Ztg 338:124; (1997) J Mol Catal A 116:55; (b) Chen J, Alper H (1996) In: Cornils B, Herrmann WA (eds) Applied homogeneous catalysis with organometallic complexes, vol 2. VCH, Weinheim, p 844; Bergbreiter DE (1987) CHEMTECH Nov 688; Bergbreiter DE, Zhang L, Mariagnanam (1993) J Am Chem Soc 115:9295; Cenini S, Ragaini F (1996) J Mol Catal A 105:145; Manassen J (1971) Platinium Metal Rev 15:142; Mercier F, Mathey F (1993) J Organomet Chem 462:103; Bayer E, Schurig V (1975) Angew Chem Int Ed Engl 14:493; (1976) CHEMTECH March, 212

47. Chaudhari RV, Bhanage BM, Deshpande RM, Delmas H (1995) Nature 373:501; Fell B, Papadogianakis G (1993) J Prakt Chem/Chem Ztg 335:75

48. Sawamura M, Kitayama K, Ito Y (1993) Tetrahedron Asymmetry 4:1829; Monflier E, Fremy, G, Castanet Y, Mortreux A (1995) Angew Chem 107:2450

49. Arhancet JP, Davis ME, Merola JS, Hanson BE (1989) Nature 339:454; (1990) J Catal 121:327; Dav ME (1992) CHEMTECH August, 498; Joó F, Beck MT (1984) J Mol Catal 24:135; Fremy G, Monflier E, Carpentier J-F, Castanet Y, Mortreux A (1995) Angew Chem 107:1608

50. Anderson JR et al (1997) J Mol Catal A 116:109; Yamashita K et al (1993) EP Patent 0 552 908 Asahi Kasei KK

51. Healy FJ, Livingston JR, Mozeleski EJ, Stevens JG (1994) US Patent 5 298 669 Exxon Chem Patent Inc; Gosser LW, Knoth WH, Parshall GW (1977) J Mol Catal 2:253; Bahrmann H, Haubs M, Kreuder W, Müller T (1988) EP 0 374 615 Hoechst AG; IT Horváth et al (1998) J Am Chem Soc 120:3133

52. Jin Z, Zheng X (1998) In: Cornils B, Herrmann WA (eds) Aqueous-phase organometallic catalysis – concepts and applications, Wiley-VCH, Weinheim, p 233

53. (a) Bahrmann H (1996) In: Cornils B, Herrmann WA (eds) Applied homomogeneous catalysis with organometallic complexes, vols 1–2, VCH, Weinheim; (b) vol 2, p 644; (c) Bahrmann H, Haubs M, Müller T, Schöpper N, Cornils B (1997) J Organomet Chem 545/546:139

54. Cornils B, Wiebus E et al (1984) DE-OS Patent 3 413 427 and EP Patent 0 158 246 Ruhrchemie AG

55. Deshpande RM, Purwanto P, Delmas H, Chaudhari RV (1997) J Mol Catal A 126:133

55. N. Yoshimura (1998) In: Cornils B, Herrmann WA (eds) Aqueous-phase organometallic catalysis – concepts and applications, Wiley-VCH, Weinheim, chap 6.7; Bahrmann H (1996) In: Cornils B, Herrmann WA (eds) Applied homomogeneous catalysis with organometallic complexes, vol 2, VCH, Weinheim, p 351

56. Mercier C, Chabardes P (1994) Pure Appl Chem 66:1509

57. Kohlpaintner CW, Beller M (1997) J Mol Catal 116:259
58. S Haber (1998) In: Cornils B, Herrmann WA (eds) Aqueous-phase organometallic cataly-
    sis – concepts and applications, Wiley-VCH, Weinheim, chap 6.10
59. Horváth IT, Rábai J (1994) Science 266:72
60. Chauvin Y, Olivier-Bourbigou H (1995) CHEMTECH Sept 26; Olivier H, (1998) In: Cornils
    B, Herrmann WA (eds) Aqueous-phase organometallic catalysis – concepts and applica-
    tions, Wiley-VCH, Weinheim, chap 6.10
61. Subramaniam B, McHugh MA (1986) Ind Eng Chem Proc Des Dev 25:1; Rathke JW, Kling-
    ler RJ, Krause TR (1991) Organometallics 10:1350; Kainz S, Koch D, Baumann W, Leitner
    W (1997) Angew Chem 109:1699; Angew Chem Int Ed Engl 36:1628

Solvent-free Reactions

André Loupy

Laboratoire des Réactions Sélectives sur Supports – CNRS UMR 8615, Paris-South University, Building 410, 91405 Orsay-Cédex, France, *E-mail: aloupy@icmo.u-psud.fr*

For reasons of economy and pollution, solvent-free methods are of great interest in order to modernize classical procedures making them more clean, safe and easy to perform. Reactions on solid mineral supports, reactions without any solvent/support or catalyst, and solid-liquid phase transfer catalysis can be thus employed with noticeable increases in reactivity and selectivity. A comprehensive review of these techniques is presented here. These methodologies can moreover be improved to take advantage of microwave activation as a beneficial alternative to conventional heating under safe and efficient conditions with large enhancements in yields and savings in time.

Keywords: Mineral solid supports, Phase transfer catalysis, Solid state reactions, Reactivity, Selectivity, Microwave activation.

Topics in Current Chemistry, Vol. 206
© Springer-Verlag Berlin Heidelberg 1999

1
Introduction: General Interest for Solvent-free Processes

Nowadays, one of the main duties assigned to the organic chemist is to organize research in such a way that it preserves the environment and to develop procedures that are both environmentally and economically acceptable. One major objective is therefore to simplify and accommodate in a modern way the classical procedures with the aim of keeping pollution effects to a minimum, together with a reduction in energy and raw materials consumption.

Among the most promising ways to reach this goal, solvent-free techniques hold a strategic position as solvents are very often toxic, expensive, problematic to use and to remove. It is the main reason for the development of such modern technologies. These approaches can also enable experiments to be run without strong mineral acids (i.e. HCl, H_2SO_4 for instance) that can in turn cause corrosion, safety, manipulation and pollution problems as wastes. These acids can be replaced advantageously by solid, recyclable acids such as clays.

1.1
Reactivity

An enhancement in kinetics can result from increasing concentrations in reactants when a diluting agent such as a solvent is avoided [Eq. (1)].

$$A + B \longrightarrow \text{product} \qquad v = k\,[A]\,[B] \tag{1}$$

As concentrations in reactive species are optimal, reactivity is increased and only *mild conditions* are required. In several cases, difficult reactions (even impossible) using solvents are easily achieved under solvent-free conditions. Another unquestionable advantage lies in the fact that *higher temperatures*, when compared to classical conditions, can be used without the limitation imposed by solvent boiling points.

The last, but not least, benefit is the possibility of using solvent-free techniques coupled with *microwave irradiation*. This new type of activation is now more frequently employed but often the presence of solvents prevents its use for safety reasons. This difficulty can be overcame by solvent-free processes.

1.2
Selectivity

The layout of reacting systems can be increased when high concentrations and/or aggregation of charged species are involved. It can lead to some modifications in mechanisms resulting in a decrease in molecular dynamics and induce subsequent *special selectivities* (stereo-, regio- or enantioselectivity). Weak interactions can, for instance, appear (such as π-stacking) which are usually masked by solvents, inducing further consequences on selectivity.

1.3
Simplification of Experimental Procedures

Firstly, complex apparatus is not needed and, for instance, reflux condensers are not required, which in turn allows the handling of a smaller quantity of material as there is no solvent. It also allows an operation to be carried out *with increased amounts of products in the same vessels.*

Washing and extraction steps are made easier or even suppressed. In the case of equilibrated reactions leading to light polar molecules (MeOH, EtOH or H_2O), equilibrium can be easily shifted by a simple heating just above the boiling points or under reduced pressure. With the usual procedure this operation is impeded by the presence of solvent necessitating an azeotropic distillation using a Dean–Stark apparatus [Eq. (2)].

$$\text{Ex: RCOOH + R'OH} \xrightleftharpoons{H^+} \text{RCOOR' + } H_2O \nearrow \qquad (2)$$

1.4
Overall Benefits

Solvent-free techniques represent a clean, economical, efficient and safe procedure which can lead to substantial savings in money, time and products. They can be efficiently coupled to non-classical methods of activation that include ultrasound and microwaves.

2
Solvent-free Techniques

Three types of experimental conditions without solvents can be considered.

2.1
Reactions on Solid Mineral Supports [1–4]

Reactants are first impregnated as neat liquids onto solid supports such as aluminas, silicas and clays or via their solutions in an adequate organic solvent and further solvent removal in the case of solids. Reaction in "dry media" is performed between individually impregnated reactants, followed by a possible heating. At the end of the reaction, organic products are simply removed by elution with diethyl ether or dichloromethane.

2.2
Reactions Without any Solvent, Support, or Catalyst

These heterogeneous reactions are performed between neat reactants in quasi-equivalent amounts without any adduct. In the case of solid-liquid mixtures, the reaction implies either solubilization of solid in the liquid phase or adsorption of liquid on the solid surface as an *interfacial* reaction.

2.3
Solid-Liquid Phase Transfer Catalysis (PTC) [7–9]

Reactions occur between neat reactants in quasi-equivalent amounts in the presence of a catalytic quantity of tetraalkylammonium salts or cation complexing agents. When performed in the absence of solvent, the liquid organic phase consists of the electrophilic reagent then possibly the reaction product (Scheme 1). Nucleophilic anionic species can be generated *in situ* by reacting their conjugated acids with solid bases of increased strength due to ion-pair exchange with $R_4N^+X^-$ [Eq. (3)].

$$K^+ OH^- + R_4N^+ Br^- \rightleftharpoons R_4N^+ OH^- + K^+ Br^-$$

$$R_4N^+ OH^- + NuH \rightleftharpoons R_4N^+ Nu^- + H_2O$$

(3)

solid phase	$M^+, Nu^- + R_4N^+, X^- \rightleftharpoons R_4N^+, Nu^- + M^+X^-$
liquid organic phase	$R\text{-}Nu + R_4N^+, X^- \longleftarrow R_4N^+, Nu^- + R\text{-}X$

Scheme 1

3
Reactions on Mineral Solid Supports

3.1
The Different Types of Supports

This technique was initially described by Keinan and Mazur [1]. Solvents are replaced by solid supports which can simply act as an inert phase towards reactants or as a catalyst according to their active sites.

3.1.1
Amorphous Supports: Aluminas, Silicas

Aluminas. Non-activated aluminas (chromatographic grade) are basic supports versus rather acidic carbonic acids (pKa ≤ 15), able to abstract hydrogen and to subsequently react. When calcinated up to 400–600°C, they behave as Lewis acids due to liberation of the surface hydroxyl groups.

Silicas. Silicas are rather acidic with a large number of silanol groups on their surfaces. Quartz and Fontainebleau sand are particular cases, very pure non-hydrated silicas.

3.1.2
Layered Supports: Clays

Clay minerals consist of a large family of fine-grained crystalline silicate sheets with arrangements of tetrahedral and octahedral layers. Interlamellar cations can be exchanged (for instance, with H^+, leading to K10 and KSF montmorillonites which are very strong solid acids).

3.1.3
Microporous Supports: Zeolites

Zeolites are crystalline, microporous aluminosilicates with molecular-sized intracrystalline channels and cages. Guest molecules with molecular diameters smaller than zeolites (from 3 to 15 Å) can enter the interior of zeolite crystals (*intercalation*) giving rise to shape and size selective sorption and, consequently, highly selective reactions.

3.1.4
Surface Areas

The surface of the supports are listed below in m^2/g:

Aluminas	200–500
Silicas	500–600
Clays	70–700
Fontainebleau sand	1.4
Activated carbons	300–900

3.2
Reactions on Alumina

3.2.1
Anionic Reactions

Anionic Activation. Due to physicosorption of water molecules, alumina can behave as a hydrated oxide (Al_2O_3, nH_2O) with amphoteric properties. It can, therefore, lead to specific interactions with ions acting both as a donor (base) versus cation and as an acceptor (Lewis or Brönsted acid) towards anion [Eq. (4)]:

(4)

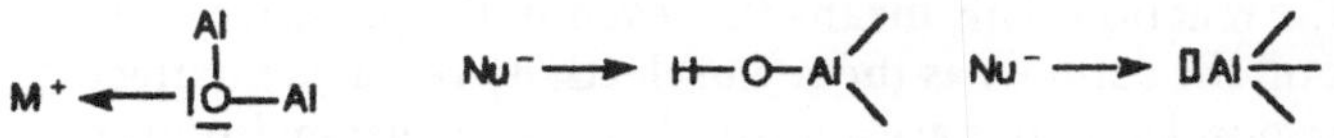

It acts therefore as a polar support with ionising and dissociating power in the same manner as protic solvents [Eq. (5)] [11]:

(5)

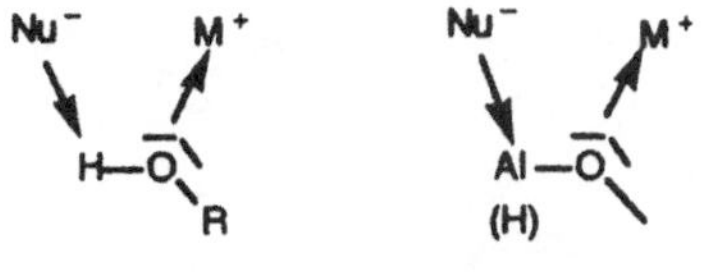

Aluminas are consequently very efficient supports for anionic reactions [Eq. (6)] [10]:

$$Na^+\,CN^- + ArX \xrightarrow{Al_2O_3} ArCN \quad 62\text{-}95\%$$
$$CH_3CO_2^-\,K^+ + RX \xrightarrow{Al_2O_3} CH_3CO_2R \quad 85\text{-}97\%$$
$$Na_2S + nOctBr \xrightarrow{Al_2O_3} nOct\text{-}S\text{-}nOct \quad 97\%$$
$$K^+\,SCN^- + nBuBr \xrightarrow{Al_2O_3} nBuSCN \quad 100\%$$

(6)

Reduction Reactions [12, 13]. The regioselectivity of the reduction of α-enones by $M^+BH_4^-$ was studied as a function of cation and medium effects, especially under "dry media" conditions onto alumina. The effect of added diethyl ether was evaluated [Table 1, Eq. (7)].

(7)

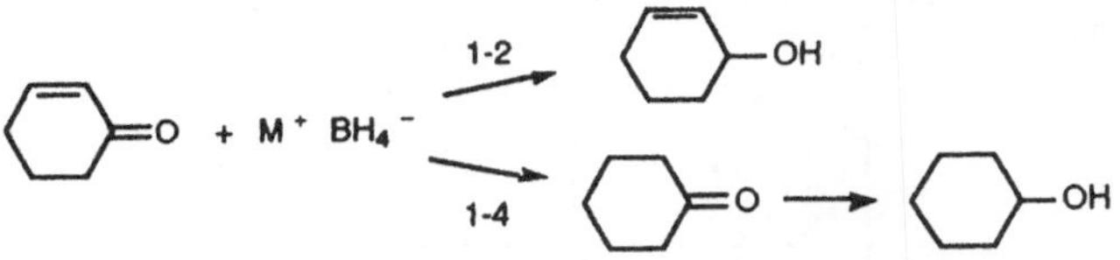

Table 1. Regioselectivity and half-time reaction for 2-cyclohexenone reduction

		Alumina "dry media"	Alumina + Et_2O	Et_2O solution	THF solution
Li^+	1-2/1-4	57/43	**59/41**	**97/3**	**52/48**
	$t_{1/2}$	30 min	20 h		
NBu_4^+	1-2/1-4	6/94	6/94	12/88	12/88
	$t_{1/2}$	5 min	45 min		

Results show undoubtedly that diethyl ether does not alter regioselectivity but only delays the reaction. When compared to reactions performed in ethereal solutions, regioselectivity is clearly different, thus excluding the intervention of solvent in the reaction. This means that even in the presence of a solvent, reaction occurs on the support as the regioselectivity remains unaffected.

Finally, regioselectivities are almost identical in tetrahydrofuran (THF) solution and onto alumina. Consequently, the proposed reactive species can be deducted by analogy (Scheme 2), with assumed rather identical donor-ability (basicity) of oxygen atoms for THF and Alumina.

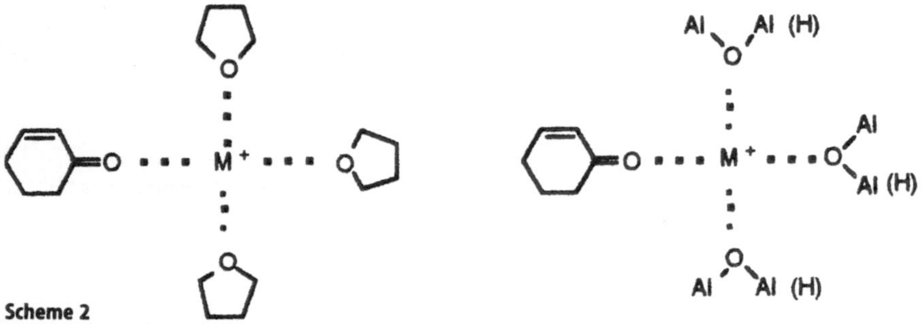

Scheme 2

3.2.2
Non-Nucleophilic Polar Medium

Bromine Addition to Alkenes. Alumina can advantageously replace protic solvents thus avoiding secondary reactions due to their nucleophilicity. This situation is evidenced in the bromation of alkenes [14]. When performed in methanol, bromine addition leads to a mixture of a *trans*-dibromo adduct and a *trans*-bromo ether compound. The latter results from competitive attack by protic solvent on the bromonium ion intermediate. This byproduct can be suppressed using Br_2/alumina, as the support behaves as a non-nucleophilic polar medium (Scheme 3).

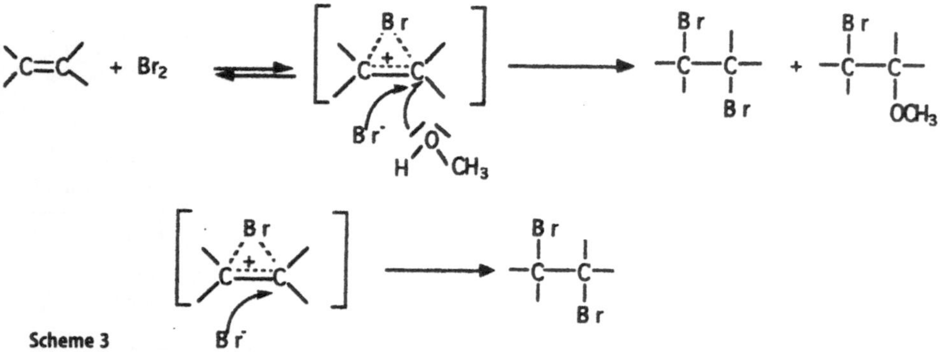

Scheme 3

Hydrohalogenation. Hydrohalogenation has also become more convenient with the use of alumina (or silica). As a typical example, 1-octene reacts with HBr only very slowly in solution and very quickly onto alumina without competitive radical addition [Eq. (8)] [15].

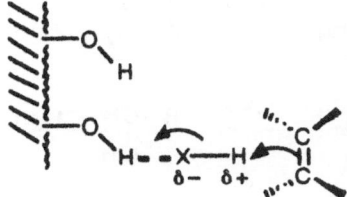

$$(8)$$

According to a radical mechanism **2** is the main product whereas **1** is obtained by ionic addition (Table 2).

Table 2. Hydrobromination of 1-hexene

Conditions	*1*	*2*
CH_2Cl_2	12	88
Alumina	96	–
Silica	93	–

The support is responsible for activation due to the hydroxyl groups on its surface. A mechanism for addition induced by the surface is thus proposed. (Scheme 4)

Scheme 4

Alternative to High Dilution Techniques. The macrocyclization of terminal dibromoalkanes with sodium sulfide was performed onto alumina (Na_2S/Al_2O_3). In this case, the use of a solid support for intramolecular cyclization represents a viable alternative procedure to the more traditional high dilution technique in solution [Eq. (9)] [16, 17].

$$Br(CH_2)_nBr + Na_2S/Al_2O_3 \longrightarrow (CH_2)_nS + (CH_2)_n \overset{S}{\underset{S}{\big|}} \qquad (9)$$

(unimolecular)   (bimolecular)

The same procedure was then extended to α,α'-dibromo-*m*-xylene leading to dithia[3.3]metacyclophanes [18] (Scheme 5) with satisfactory yields (62–65%) after 1–2 hours.

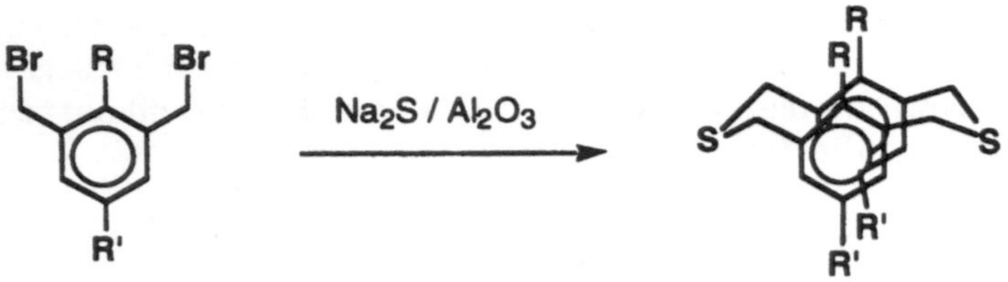

Scheme 5

3.2.3
Alumina Acting Both as Base and Support

Non-activated γ grade alumina for chromatography is sufficiently basic to promote hydrogen abstraction to rather acidic carbon acids (pKa ≤ 15) [19]. In such a way, numerous anionic condensations are described, for example, an aldolization [Eq. (10)] [20] leading to aurone, a basic compound in flower pigmentation, a Knoevenhagel reaction [Eq. (11)] [21], or a case of a Henry reaction [Eq. (12)] [22].

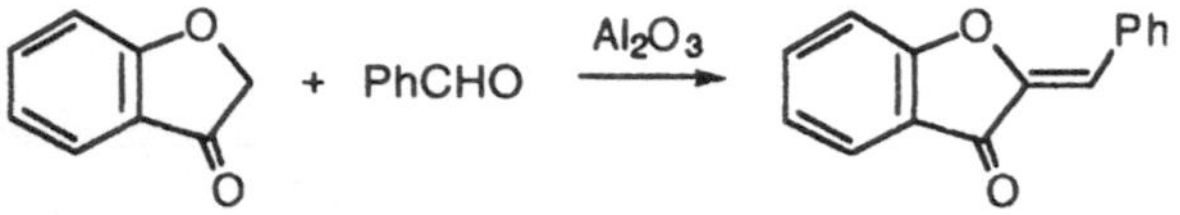

$$\text{(10)}$$

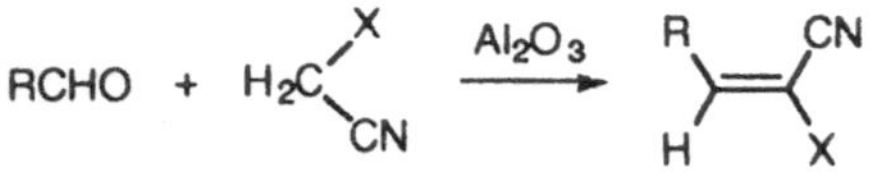

$$\text{(11)}$$

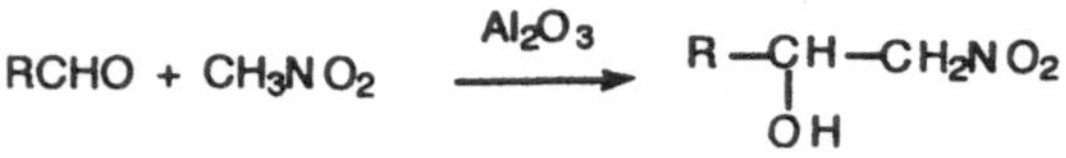

$$\text{RCHO} + \text{CH}_3\text{NO}_2 \xrightarrow{\text{Al}_2\text{O}_3} \text{R}-\underset{\underset{\text{O}\,\text{H}}{|}}{\text{CH}}-\text{CH}_2\text{NO}_2 \qquad \text{(12)}$$

In a typical work, the Michael reaction of several 1,3-dicarbonyl compounds, nitroalkanes and thiols as donors with various α,β-unsaturated carbonyl acceptors on the surface of alumina in dry media have been described [23]. It was concluded that a "dramatic improvement" was obtained using this process when compared to the existing methods. The important features of this methodology are: (a) no need for base, (b) no undesirable side reaction, (c) extremely fast addition, (d) mild reaction conditions, (e) easy set-up and work-up, (f) no toxic and expensive materials involved, and (g) high yields.

However, due to the amphoteric character of alumina, acid catalysis due to OH surface groups or to Al atoms can take place in a bifunctional catalysis mechanism [Eq. (13)] [24].

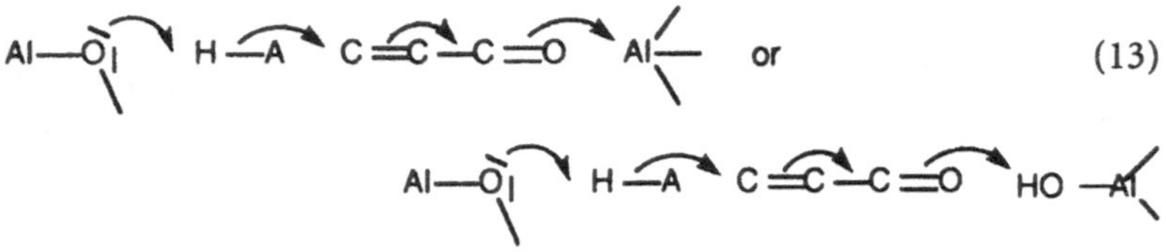

$$\text{(13)}$$

3.2.4
Impregnated Bases on Alumina KF/Al₂O₃ [25, 26]

Whereas KF can be considered as a weak base, when impregnated on alumina, it becomes very strong and is able to ionize extremely weak carbon acids (up to pKa ≈ 30) [27]. This enhancement in basicity results from ionic dissociation $K^+F^- \rightleftharpoons K^+ + F^-$ on the surface of alumina due to its amphoteric character (Scheme 6).

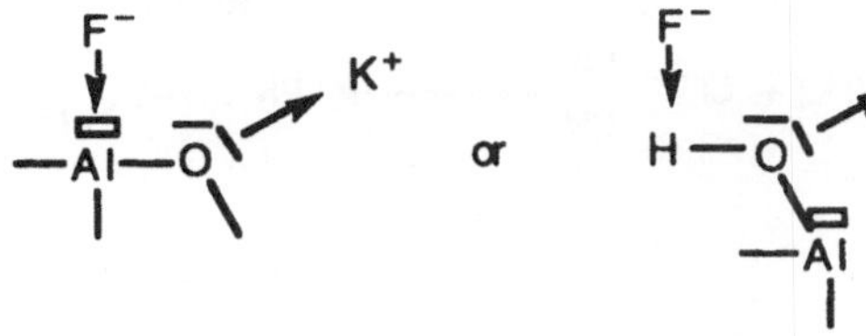

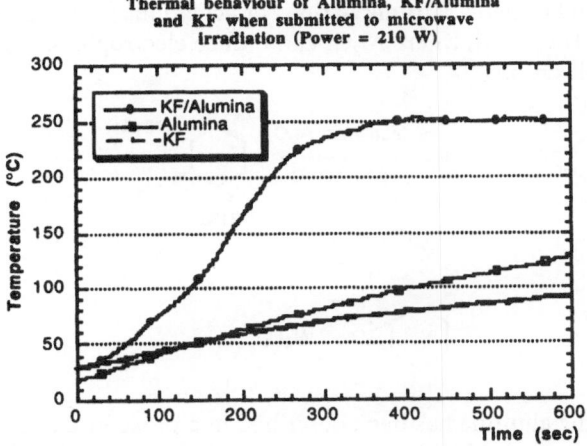

Scheme 6

This dissociation has been proved by considering the effects of microwaves on KF and KF/alumina. The rise in temperature in the latter case is characteristic of a more polar species and then of a strong increase in polarity (ionic dissociation) [28] (Scheme 7).

Thermal behaviour of Alumina, KF/Alumina and KF when submitted to microwave irradiation (Power = 210 W)

Scheme 7

KF/Alumina has therefore very often been used with better results when compared to analogous reactions carried out in solutions (reaction time, temperature, ease of work-up, yield, selectivity, etc.) [4]. However, in this case too, one can expect KF-Al₂O₃ to act as a bifunctional catalyst, F⁻ to act as the base and alumina as the acid [Eq. 14] [24]:

$$F^- \frown H\overset{|}{\underset{/}{C}} \quad \overset{\frown}{\underset{}{C=O \cdots Al}} \quad \text{or} \quad F^- \frown H\overset{|}{\underset{/}{C}} \quad \overset{\frown}{\underset{}{C=O \cdots H \, O-Al}} \qquad (14)$$

Among some typical recent works in this field, one can consider several nucleophilic substitutions [29], eliminations [30], Michael reactions [30], and Knoevenhagel condensations [30] [Eq. (15)].

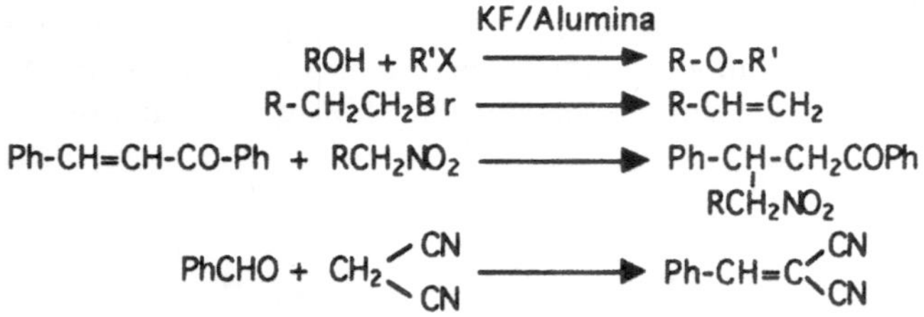

KF/Alumina
ROH + R'X
R-O-R'
R-CH₂CH₂Br
R-CH=CH₂
Ph-CH=CH-CO-Ph + RCH₂NO₂
Ph-CH-CH₂COPh
RCH₂NO₂
PhCHO + CH₂
CN
CN
Ph-CH=C
CN
CN

(15)

3.2.5
Activated Alumina: Lewis Acid and Support

High activation temperatures (calcination) result in elimination of physicosorbed H_2O onto alumina. At 400 °C, for instance, about 50% of the hydroxyl groups are lost; at 600°C, 80% are lost and at 800°C almost 100% are removed [4]. Consequently, activated alumina behaves as a strong Lewis acid due to liberation of aluminum sites. This support can be therefore used as a typical Lewis acid catalyst, e.g. in Friedel–Crafts alkylations or acylations [31] and in epoxide ring opening [32] [Eq. (16)], where Al_2O_3 can induce electrophilic assistance due to oxygen complexation:

activated Al₂O₃
24h, 100°C
O
+ PhNH₂
NHPh
OH
91%
PhNH₂
Al₂O₃

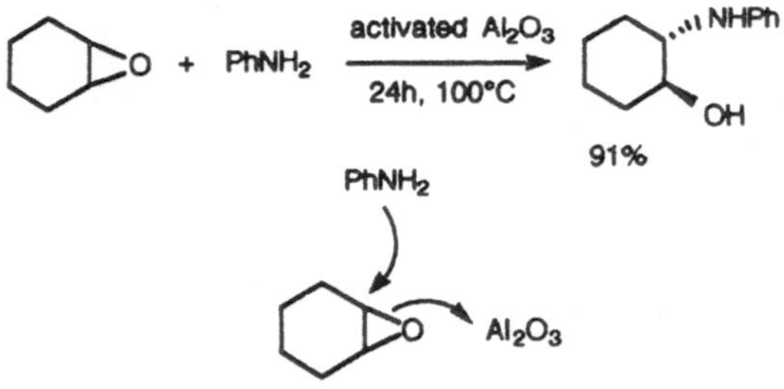

(16)

Activated acidic alumina has been described in a procedure to protect alcohols with methoxymethyl chloride using ultrasound as a non-classical way of activation [Eq. (17)] [33].

OH
+ ClCH₂OCH₃
activated alumina
0-5°C 6h
US
OCH₂OCH₃
83%

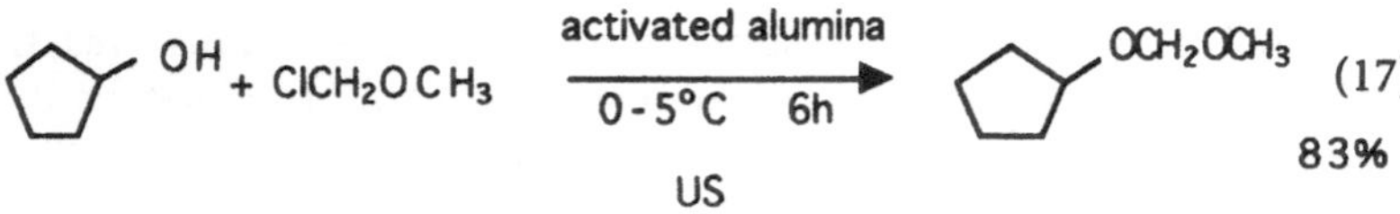

(17)

Aldehydes undergo efficient *E*-stereoselective Wittig olefination with alkylidene triphenyl phosphoranes in the presence of activated alumina (pre-treated at 200°C) under mild conditions [34] with high yields. The same reactions without alumina are less selective and require high temperatures (benzene, reflux) or long stirring times at room temperature [Eq. (18)].

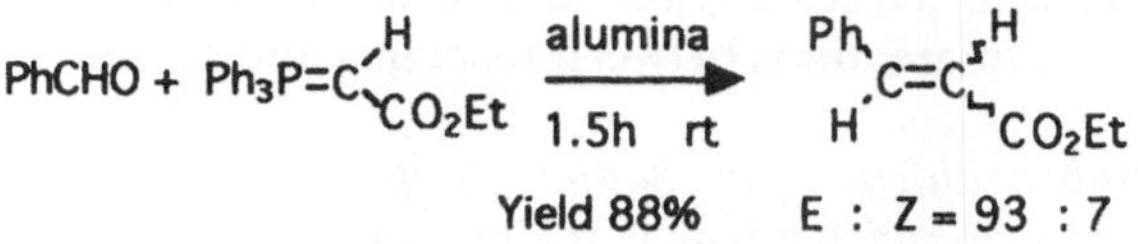
PhCHO + Ph3P=C
H
CO2Et
alumina
1.5h rt
Ph
H
C=C
H
CO2Et
Yield 88%
E : Z = 93 : 7

(18)

The high *E*-stereoselectivity and reactivity could be rationalized if we take into account the activation of carbonyl group by complexation with Al_2O_3 (Scheme 8).

Scheme 8

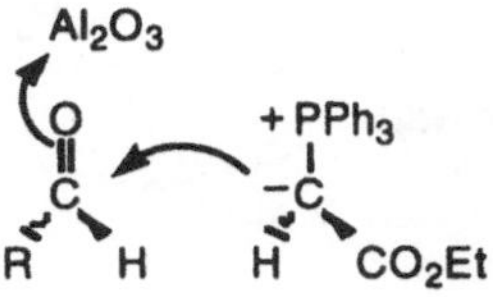
Al2O3
O
C
R
H
+PPh3
C
H
CO2Et

Pagni, Kabalka et al. have shown that alumina activity deeply affects the diastereoselectivities of the heterogeneous Diels–Alder reactions of cyclopentadiene with acrylate esters [Eq. (19), Table 3] [35].

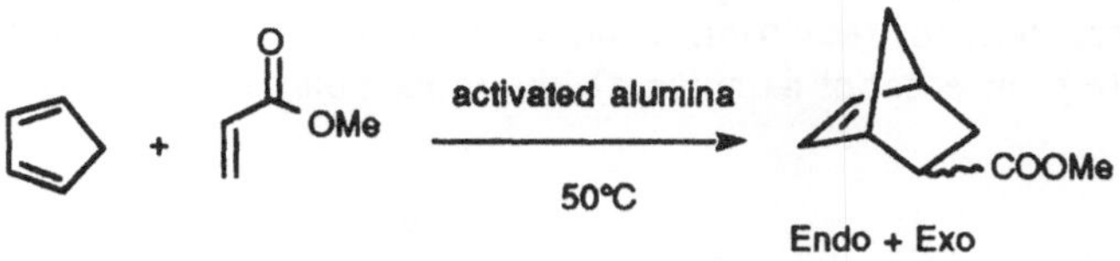
O
OMe
activated alumina
50°C
COOMe
Endo + Exo

(19)

Table 3. Diastereoselectivities for Diels–Alder reaction of cyclopentadiene with methyl acrylate on alumina

Alumina activity	*endo*/*exo*
unactivated	5.8
200 °C	7.0
300 °C	10.3
400 °C	52

The diastereomeric excess increases steadily as the activity of the alumina increases, i.e. when its Lewis character increases. Results obtained for menthyl acrylate cycloaddition with cyclopentadiene were rather similar but not as important [36] (*endo*/*exo* increase from 2.4 to 8.1 changing non-activated alumina for a pre-heated one at 200 °C).

3.2.6
Examples of Modified Selectivities

In several cases special induced selectivities due to alumina can be proved. They may result from specific interactions between reactants and supports.

Reduction of p-Nitrobenzaldehyde by Sodium Sulphide [37]. The reduction by Na_2S, following a radical mechanism, is regiospecific on the nitro moiety in ethanol as solvent whereas only the aldehyde group is reduced on alumina (Scheme 9).

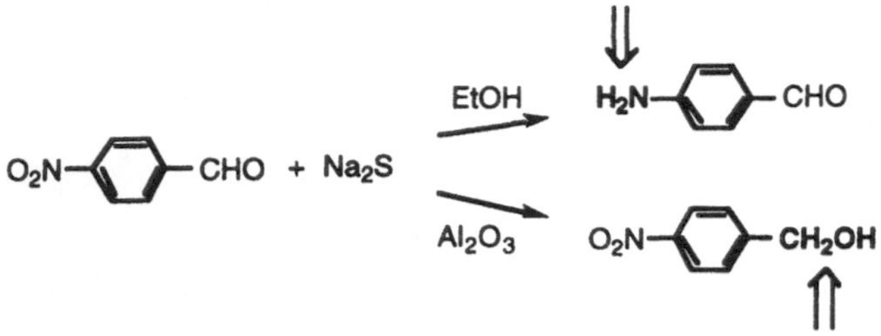

Scheme 9

This noticeable behaviour can be explained by the specific adsorption of the nitro function (the most polar one) on the surface of alumina by interaction with hydroxyl groups in the support. In such a case, the aldehyde group is free and easily accessible for reduction (Scheme 10). In the absence of such a strong interaction (e.g. in ethanol as solvent), the most polar function is selectively reduced.

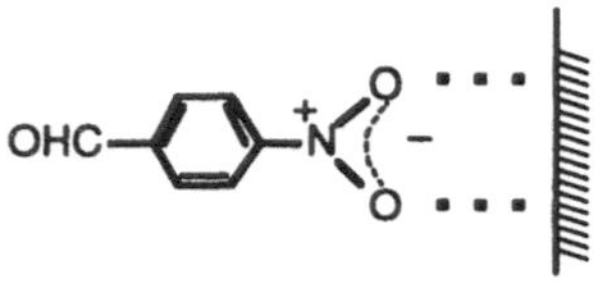

Scheme 10

Stereoselective Additions of Phenols to DMAD. Phenols are prone to add on activated acetylenic compounds such as dimethylacetylene dicarboxylate (DMAD) [Eq. (20)].

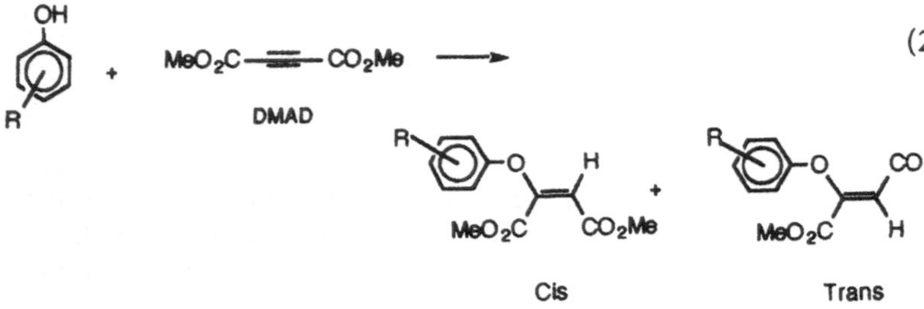

(20)

Whereas in carbon tetrachloride as the solvent, the thermodynamic ratio *cis/trans* = 37/63 was obtained, reaction onto alumina in dry media gave specifically the *cis*-adduct (kinetic product).

This significant change in selectivity can be explained by a specific adsorption of DMAD on the support including two binding polar sites (ester moieties) with hydroxyl groups of alumina, thus leading only to *cis*-addition (Scheme 11).

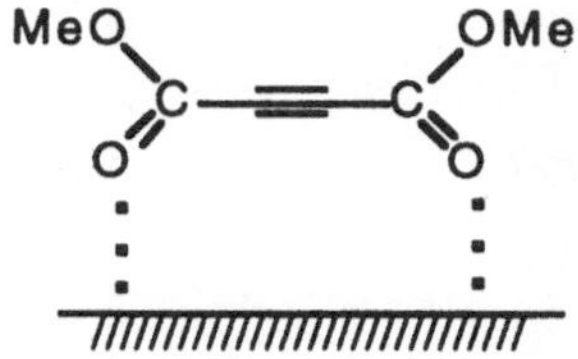

Scheme 11

Acylation of Aromatic Ethers. A simple and improved procedure for regioselective acylation of aromatic ethers with carboxylic acids on alumina in the presence of trifluoroacetic anhydride has been described by Ranu et al. [Eq. (21)] [39]:

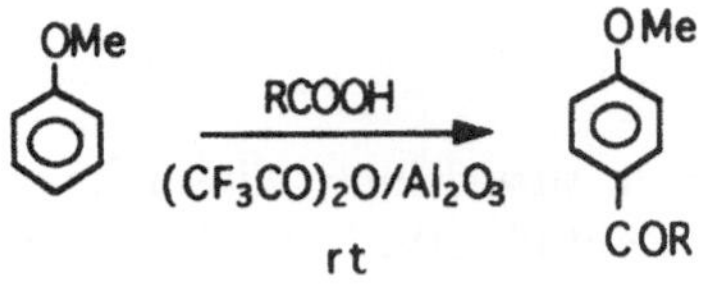

$$(21)$$

Acetylation of Unsymmetrical Diols [40]. The effect of the presence of chromatograsulfic grade alumina on the acetylation of a series of unsymmetrical 1,5-diols has been investigated. For diols containing both a primary and a secondary hydroxyl group, it was observed that higher yields of the more hindered secondary acetates were formed in the presence of alumina than the corresponding reactions in solution (Scheme 12).

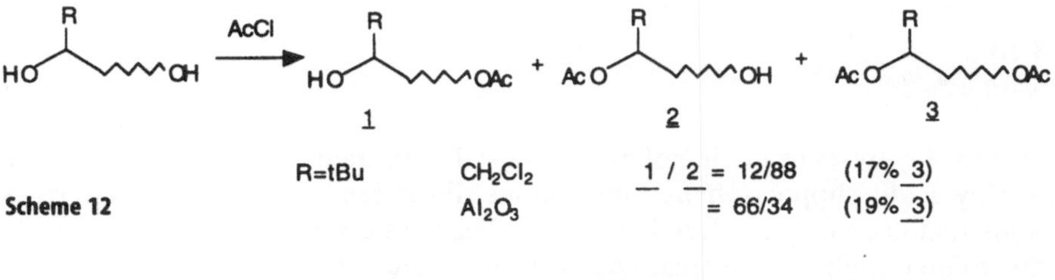

		1 / 2 = 12/88	(17% 3)
R=tBu	CH_2Cl_2		
Scheme 12	Al_2O_3	= 66/34	(19% 3)

These results are consistent with a model for the reaction in which adsorption of an unsymmetrical diol to the surface of Al_2O_3 occurs primarily via the least hindered hydroxyl group. Interaction of the adsorbed hydroxyl group with the surface effectively shields that site leaving only the non-adsorbed group available for reaction with the added acetylating agent.

Photochemical Reactions. The addition of allene to a steroidal enone on Al_2O_3 revealed a reversal of stereochemistry to that observed in solution [41] (Scheme 13).

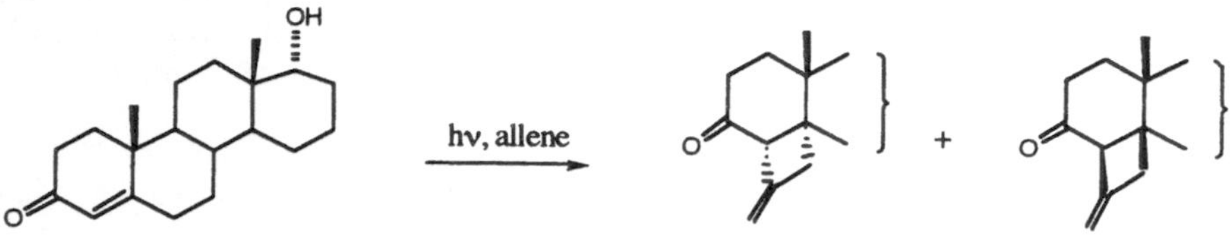

	Methanol (- 70°C)		91.5	:	8.5
Scheme 13	Alumina (20°C)		6	:	94

Photolysis of *trans*-stilbene adsorbed on alumina afforded a [2 + 2] dimer, no isomerization to *cis*-stilbene occurred. This behaviour is unlike that observed in the gas and solution phases where isomerization is dominant [Eq. (22)] [42].

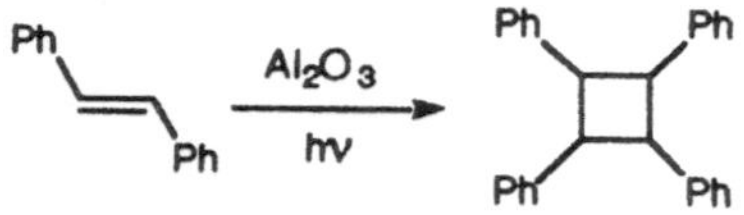

$$\text{(22)}$$

Photolysis of benzyl alcohol on alumina, surprisingly, allows the production of dibenzyl ether [11]; this behaviour is not observed in solution. A plausible mechanism for the formation of the ether involves the photogeneration of a benzyl cation which subsequently reacts with benzyl alcohol [Eq. (23)].

$$PhCH_2OH \xrightarrow[hv]{Al_2O_3} PhCH_2OCH_2Ph + H_2O \qquad (23)$$

3.3
Reactions on Silica

3.3.1
Silica Gels

Due to the presence of silanol groups (Si-OH) on their surfaces, silica gels are weakly acidic supports; hence amorphous silicas can be used to catalyze reactions that are easily catalyzed with acid. They are essentially used as supports due to their high surface areas and large pore volumes.

The field of applications is therefore very similar to the one with alumina, reactants being impregnated on silica gels prior to reactions. As typical examples, they are applied in reduction reactions [43] especially in silica-gel supported zinc borohydride [44], oxidations including mainly $KMnO_4/SiO_2$ [45] and several cases of anionic activations in dry media [46].

Very recently, it has been shown that the ring opening of epoxides can be efficiently promoted on the surface of silica with impregnated lithium halides. The reactivity of salts was shown to follow the order LiI > LiBr ≫ LiCl, and the reactivity was strongly increased by adding one equivalent of water to this system [Eq. (24)] [47].

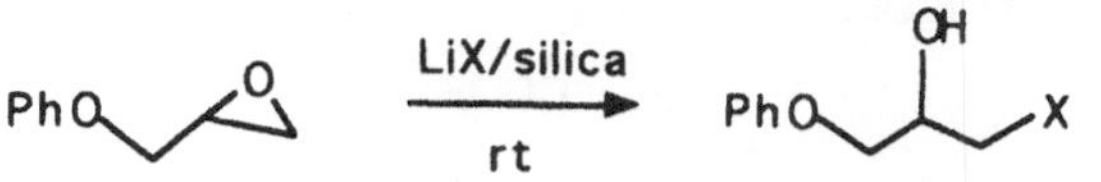

$$\text{PhO} \diagdown \diagdown^{O} \quad \xrightarrow[\text{rt}]{\text{LiX/silica}} \quad \text{PhO} \diagdown \overset{\overset{\displaystyle OH}{|}}{\diagdown} X \qquad (24)$$

LiCl	22 days	50%
LiCl + H$_2$O	6 days	91%
LiI	15 min	96%

The proposed mechanism involved electrophilic assistance by silanol groups as well as by lithium cation (Scheme 14).

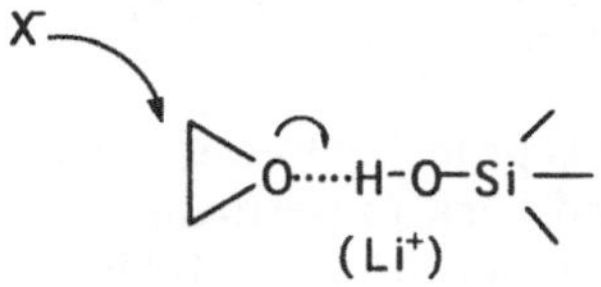

Scheme 14

Alone, silica can behave as a weak acid able to promote some Diels–Alder reactions with good selectivity [Eq. (25)] [48], or Wittig reactions in high yields and purities [49].

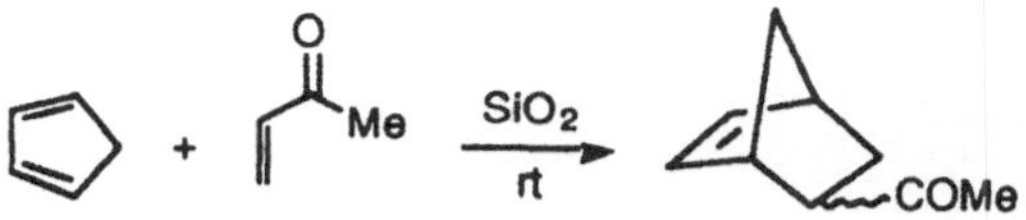

Yield : 90 %
endo : 96 %   (25)

3.3.2
Fontainebleau Sand

Fontainebleau sand is a non-hydroxylated microcrystalline silica (purity > 99.9%) with a very low specific area (1.4 m^2/g) and, consequently, a weak adsorbent power. It is a very convenient medium that acts as a *dispersant* which has in addition a large ability to adsorb thermal effects.

It is used as a dispersion agent to prevent polymerization or product decomposition taking place when an uncontrolled temperature rise occurs in the course of an exothermic reaction. In turn, it also plays the role of a diluent as solvent but with unquestionable benefits in cost and safety; for example, reduction of carbonyl compounds [Eq. (26)] [50], indole acylation [Eq. 27)] [51], and thiophenol alkylation [Eq. (28)] [52].

$$\text{cyclohexanone} \xrightarrow[\text{5mn, 60°C}]{\text{NaBH}_4} \text{cyclohexanol-OH} + \text{cyclohexenol-OH}$$

$$(1:1)$$

without sand 60 %
with sand 97%

(26)

$$\text{indole} \xrightarrow[\text{2) ClCOOEt}]{\text{1) KOtBu}} \text{N-COOEt indole}$$

without sand 20 %
with sand 82 %

(27)

$$2 \text{ PhSH} + \underset{Br}{\overset{Br}{>}}\text{-COOEt} \xrightarrow{\text{KOH}} \underset{PhS}{\overset{PhS}{>}}\text{-COOEt}$$

without sand 40 %
with sand 75 %

(28)

The discovery of the catalytic effect of sand on the activity of $KMnO_4/NaIO_4$ or $KMnO_4/NaClO$ oxidant systems provides a convenient method for double-bond cleavage under mild conditions [Eq. (29)] [53].

$$PhCH=CH_2 \xrightarrow[\text{18h rt}]{KMnO_4 + NaIO_4} PhCOOH$$

without sand 0%
with sand 75%

(29)

The role played by sand here is probably to catalyze the $NaIO_4$ oxidation of the low valent manganese formed back to permanganate and/or to prevent permanganate from further reduction beyond $Mn(V)$ so that it could easily be oxidized by sodium metaperiodate to regenerate permanganate [53].

3.4
Reactions on Clays [3, 54, 55]

3.4.1
Structure of Clay Minerals

Clays are the most universal of all the minerals occurring at the surface of the earth and consequently allow both cheap and environmentally friendly organic chemistry [56, 58]. Natural clays were among the earliest solid acid catalysts used to promote cracking and isomerization reactions in the oil industry [59].

The mineral clays are *hydrous aluminosilicates* (*montmorillonites*). Their ability to accommodate a broad range of guest molecules is brought about by the extensive expansion of its lamellar structures as indicated in Scheme 15. The lamellar system consists of aluminosilicate sheets incorporating alternatively $[SiO_4]^{4-}$ tetrahedra and $[AlO_4(OH)_2]^{7-}$ octahedra [60].

In the natural clays the main interlayer cations (present to maintain the neutrality of total charges) are sodium and calcium. These cations can be exchanged by treatment with solutions of other ions such as, for instance, H^+ leading to K10 and KSF montmorillonites [Eq. (30)].

$$\text{Montmorillonite - Ca}^{2+} + H_2SO_4 \longrightarrow \text{Montmorillonite - H}^+ + CaSO_4 \quad (30)$$

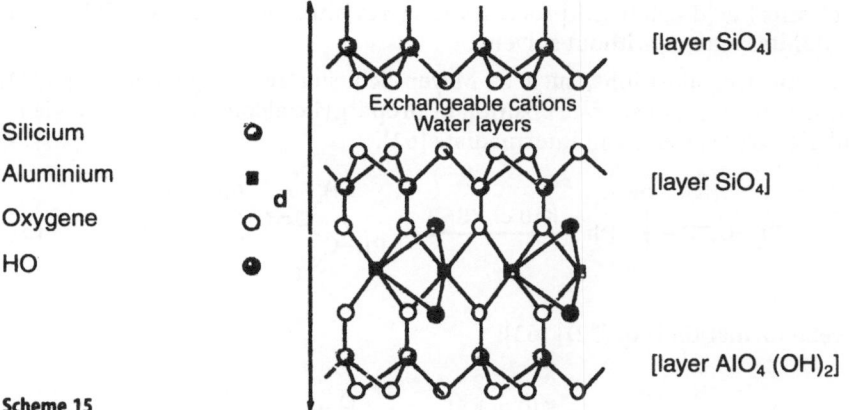

Silicium   ○

Aluminium   ■

Oxygene   ○

HO   ◉

Scheme 15

The increasing sequence of acidity according to exchanged cations is $H_3O^+ > Al^{3+} > Ca^{2+} > Na^+$.

3.4.2
Acidity of Clays [61]

Acidities of montmorillonites can be evaluated according to their Hammett functions (H_0) (Table 4). From these values, the acidity of clay minerals can be compared to that of HNO_3 (−5) or H_2SO_4 (−12).

Table 4. Hammett functions (H_0) of several clays

Natural montmorillonite	+ 1.5 to	− 3.0
Hydrogen montmorillonite	− 5.6 to	− 8.2
Natural kaolinite	− 3.0 to	− 5.6
Hydrogen kaolinite	− 5.6 to	− 8.2

These supports are, therefore, very strong solid acids with considerable benefits due to their ease of handling, low cost, recyclability and non-polluting character. They can very favourably replace mineral acids such as HNO_3 and H_2SO_4 (corrosive, difficult to use and leading to polluting wastes) in all acid-catalyzed processes, resulting in a very simple and safe procedure.

3.4.3
Acid-Catalyzed Reactions on Clays

Montmorillonite clays have been widely used as acidic catalysts in a number of reactions and there are clear indications that they are more efficient and selective in certain processes than the commonly used Brönsted and Lewis acids. All

the classical acid-catalyzed processes were revisited using K10 or KSF montmorillonites with or without solvent:

- cationic transpositions, such as Meyer–Schuster rearrangement [Eq. (31)]
  which consists of the transposition of propargylic alcohol in α-enone via an
  allylic carbocation as an intermediate [62].

$$\text{Ph-C}\equiv\text{C}\overset{\text{Ph}}{\underset{\text{OH}}{\overset{|}{\underset{|}{\text{C}}}}}\text{-Ph} \xrightarrow{\text{K10 or KSF}} \text{Ph-C}\overset{}{\underset{O}{}}\overset{H}{\underset{Ph}{}}\text{C}=\text{C}\overset{Ph}{\underset{Ph}{}} \qquad 95\% \qquad (31)$$

- acetal formation [Eq. (32)] [63]:

$$\overset{}{\underset{}{}}\text{C}=\text{O} + \overset{\text{HO}}{\underset{\text{HO}}{}}\Big] \xrightarrow[20°C]{\text{K10 or KSF}} \overset{}{\underset{}{}}\text{C}\overset{O}{\underset{O}{}}\Big\rangle \qquad > 90\% \qquad (32)$$

- addition of alcohols to double bonds,
- epoxide ring opening,
- Diels–Alder cycloaddition,
- ene reactions,
- β-eliminations,
- glycosidations [56–58].

Generally, the new procedure brings enhanced results and upgrades the classical methods so that better efficiency and safety are obtained together with a
lower consumption of raw materials and energy.

As a typical improvement, let us consider anthraquinone (AQ) synthesis
[Eq. (33)]. This important product is classically and industrially obtained by
cyclodehydration of o-benzoylbenzoic acid in boiling concentrated sulfuric acid
for several hours ($\geq$ 8 h at 170 °C). These conditions lead to many problems, be it
handling, treatment, AQ purification or generation of polluting wastes.

$$\text{[o-benzoylbenzoic acid]} \xrightarrow{\text{Acid}} \text{[anthraquinone]} \qquad (33)$$

Among all the acidic supports tested, bentonites or montmorillonites were
shown to be especially efficient since 30 minutes in a metallic bath at 350 °C are
now sufficient in "dry media" when a 1:2 (w/w) mixture of o-benzoylbenzoic
acid with clay is used. The yield is rather similar for both methods but the new
process enables a safe and simplified manipulation and treatment (AQ is
obtained pure directly by sublimation) [64]. However, some loss in catalytic
activity is observed after several reuses of the same clay. It is thought this limitation can be overcome by using microwaves as activation procedure in place
of the traditional heating (see below) [65] since within 5 minutes yield is

maintained ($\geq 90\%$), now with the possibility of reusing the catalyst more than 50 times.

3.4.4
Clay-Supported Inorganic Reagents

Several clay-supported reagents have been prepared by treating K10 montmorillonite with acetone solvate of metal salts and subsequent removal of the solvent under reduced pressure.

Among the most common, and now commercially available, reagents we can mention:

- clayfen, clay-supported iron(III) nitrate;
- claycop, clay-supported copper(II) nitrate.

These two reagents act as efficient oxidant or nitrating species in very mild and efficient conditions [3, 54].

- clayzic, clay-supported zinc chloride, which behaves as one of the most efficient catalysts for Friedel–Crafts alkylations [Eq. (34)] [66] or acylations [Eq. (35)] [67].

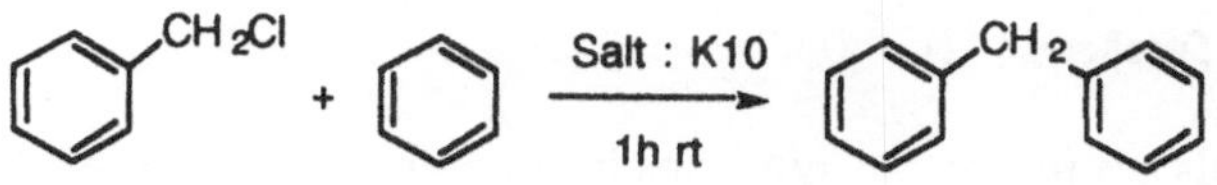

$$\text{(34)}$$

Salt / K10	/	CuCl$_2$	NiCl$_2$	ZnCl$_2$	ZnI$_2$	Zn(NO$_3$)$_2$	ZnSO$_4$
Yield (%)	12	42	67	80	80	49	24

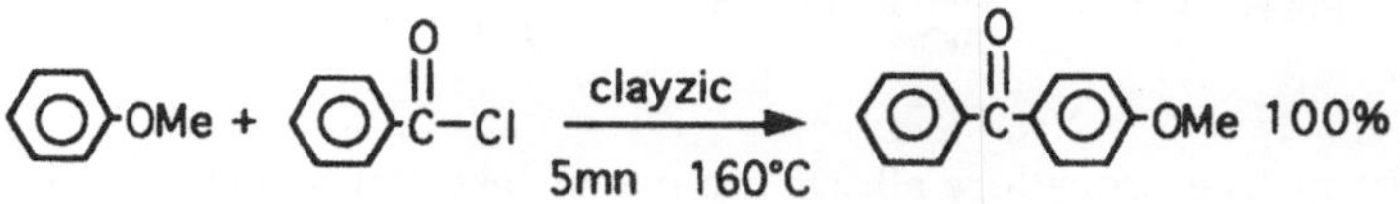

$$\text{(35)}$$

3.5
Limitations and Perspectives

The benefits brought by the supported reagent chemistry are considerable: efficiency, low cost, possibility of reusing the supports, implementation of nontoxic and cheap materials, ease of set-up and work-up, minimization of pollution, possibility to work in solvent-free conditions.

However, some relative limitations can be observed. They are essentially related to the heating mode of activation. The supports involved are generally rather poor heating conductors (isolating species) and consequently generate significant gradients in temperature inside the reaction vessels. As the temperature rise is slow and non-homogeneous, reactions can be slow. On the other hand, when submitted to microwave exposure, they behave as good adsorbents

of electromagnetic waves. As a consequence, the temperature is then homogeneous throughout and reaches a high value very quickly. It is the reason why
later (Sect. 6) we consider the coupling of microwaves and dry media supported
organic synthesis [68].

4
Non-Catalyzed Solid State Reactions

Organic reactions were found to take place efficiently in the solid state or at
interfaces between liquids and solids in a solvent-free medium (it must be
emphasized that the presence of solvent in such cases is detrimental as it induces
dilution and destructuring effects). Thus, the simple mixture of neat reactants in
quasi-equivalent amounts is sometimes sufficient to induce reation and, in these
cases, we can take advantage of very efficient procedures, outstanding yields and
selectivities [5, 6].

4.1
Alkylation of Sulfur Anions

4.1.1
Dithioacetal Synthesis [52]

Dithioacetals are important synthons in organic synthesis as acyl-masked reactants. The synthesis carried out by Corey and Seebach [69] is a two-step procedure alkylation with a reactive iodo electrophile followed by monomethyla-
tion of disulfide [Eq. (36)].

$$PhSH \ + \ CH_2I_2 \ \xrightarrow[EtOH]{NaOH} \ PhSCH_2SPh \ \xrightarrow[CH_3I]{nBuLi} \ H_3C-CH\begin{smallmatrix}SPh\\ \\SPh\end{smallmatrix} \quad 54\% \qquad (36)$$

The solvent-free procedure allows the use of less reactive chloro electrophiles
(here 1,1-dichloroethane) when compared to the technique using solvent. The
reaction can then be performed in a one-step procedure leading to a better yield
and an interesting simplified process as the tedious separation of mono- and
dimethylated products necessary in the previous procedure is now avoided.
[Eq. (37)]

$$2\ PhSH \ + \ 2\ KOH \ + \ H_3C-CH\begin{smallmatrix}Cl\\ \\Cl\end{smallmatrix} \ \xrightarrow[\text{no solvent}]{24h,\ 60°C} \ H_3C-CH\begin{smallmatrix}SPh\\ \\SPh\end{smallmatrix} \quad 81\% \qquad (37)$$

4.1.2
Dibenzylsulfone Synthesis [70]

This product, used in various industrial applications, is obtained by alkylation
of sodium formaldehyde sulfoxylate in dimethylformamide (DMF) at 100°C.

Because of its instability at this temperature, this salt has to be added incrementally during the reaction. The final yield is limited to around 25 % [Eq. (38)] [71].

$$2 \text{ PhCH}_2\text{Br} + \underset{\text{(excess)}}{\text{HOCH}_2\text{SO}_2{}^- \text{Na}^+} \xrightarrow[\text{20h 100°C}]{\text{DMF}} \text{PhCH}_2\text{SO}_2\text{CH}_2\text{Ph} \quad \textbf{25\%} \quad (38)$$

The solvent-free procedure allows operation at a lower temperature (50 °C) at which the salt remains stable. The yield can be elevated up to 75 % under simplified and mild conditions [Eq. (39)] [70].

$$\underset{\text{(2 éq.)}}{2 \text{ PhCH}_2\text{Br}} + \underset{\text{(1 éq.)}}{\text{HOCH}_2\text{SO}_2{}^- \text{Na}^+} \xrightarrow[\text{20h 50°C} \atop \text{no solvent}]{\text{K}_2\text{CO}_3 \ (1,5 \ \text{éq.})} \text{PhCH}_2\text{SO}_2\text{CH}_2\text{Ph} \quad \textbf{75\%}$$

$$(39)$$

4.2
Solid State Organic Reactions [5, 6]

Solid state organic reactions are usually carried out by keeping a mixture of finely powdered reactant and reagent at room temperature. In some cases, solid state reactions are accelerated by heating, shaking, irradiating with ultrasound or microwaves, or by grinding the reaction mixture with a mortar and pestle.

Classical organic reactions were thus performed:

- benzylic acid and pinacol rearrangement,
- Baeyer–Villiger oxidation [Eq. (40)] [72],

$$(40)$$

R=R'=Ph	CHCl₃ : 13%	No solvent : 85%
R=Ph R'= CH₂Ph	: 46%	97%

- Grignard, Reformatsky and Luche reactions,
- reduction of ketones with NaBH₄,
- Wittig reaction, and
- aldol condensation [Eq. (41)] [73].

Some aldol condensations proceeded more efficiently and stereoselectively in the absence of solvent than in solution:

R = H aq. EtOH 50 % : 11 %	No solvent : 97 %	
R = Me 3 %	99 %	(41)

- dehydration, rearrangement and etherification of alcohols.

In this field, a new simplified method for esterification of secondary and tertiary alcohols has recently been described by Le Bigot et al. [Eq. (42)] [74].

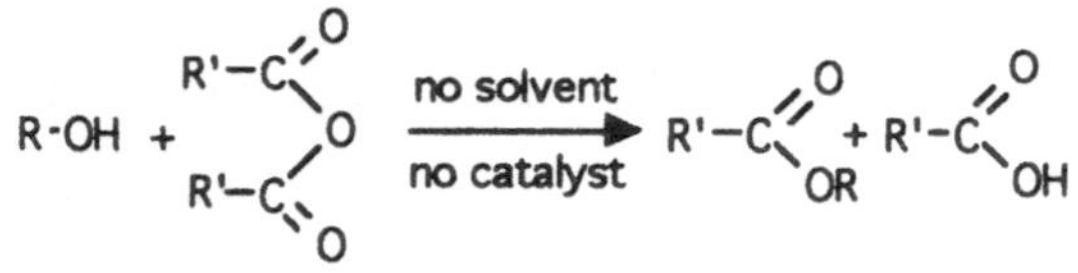

$$R\text{-}OH + \begin{array}{c} R'-C \overset{O}{\diagdown} \\ \diagup \\ O \\ R'-C \overset{O}{\diagup} \\ \diagdown \\ O \end{array} \xrightarrow[\text{no catalyst}]{\text{no solvent}} R'-C \overset{O}{\underset{OR}{\diagdown}} + R'-C \overset{O}{\underset{OH}{\diagdown}} \tag{42}$$

ex	R= iPr	R' = Me	3h	90°C	94%
	iPr	Et	"		95%
	tBu	Et	7h	105°C	75%

- *N*-glycosylation reactions of glycopyranosyl halides and silylated uracil or thymine in the presence of silver trifluoroacetate gives exclusively one anomer (in solid-solid conditions), while the fusion method lead to an anomeric mixture [Eq. (43)] [75].

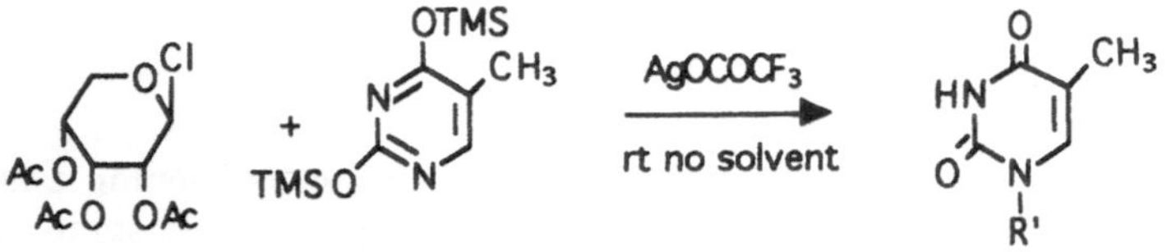

$$\tag{43}$$

solid-solid   β anomer only
fusion   α : β = 1 : 2.4

R' = pyranosyl moiety

4.3
Enantioselective Solid-Solid Reactions

When solid-solid reactions are carried out in an inclusion crystal with a chiral host, the reactions can be monitored to proceed enantioselectively. As a typical example, the Wittig reaction according to Eq. (44) was studied as a 1:1 inclusion crystal of ketone and a chiral diol as host [76].

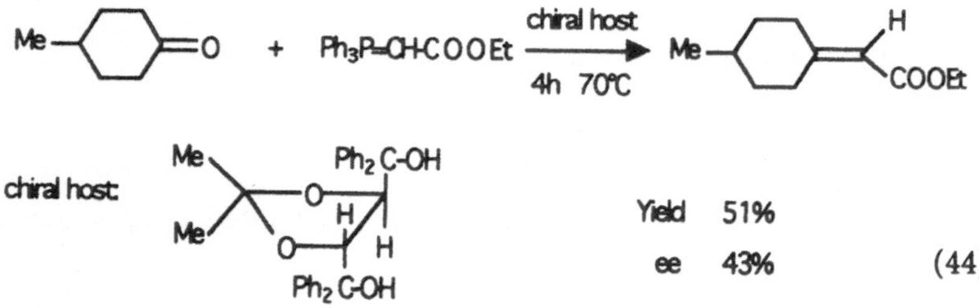

chiral host:

Yield   51%

ee   43%

$$\tag{44}$$

Asymmetric synthesis of spirodione **A** was obtained by an enantioselective cyclization of **B** resulting from annelation of 2-formylcyclohexanone with methyl vinyl ketone [Eq. (45)].

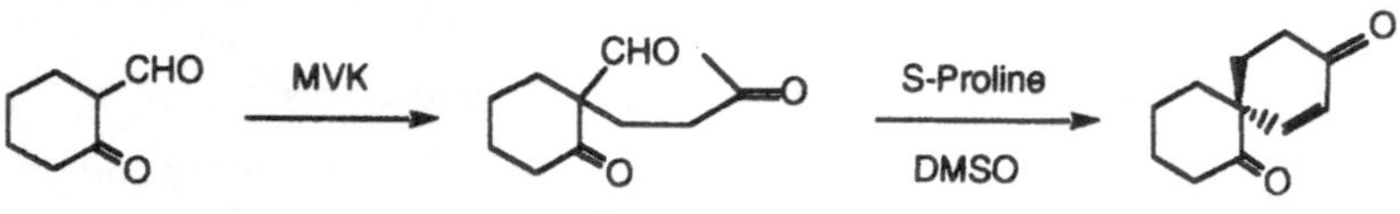

$$\tag{45}$$

B   A

The cyclization carried out in DMSO as solvent using catalytic amounts of (S)-proline gave A in a yield of 70% and 22.4% enantiomeric excess (ee). These yields were improved by carrying out the reaction in the absence of solvent [77] and the ee was then 42.6%, representing a twofold improvement over the solvent procedure.

It can be concluded that, in many cases, solid state reactions proceed much faster and with increased selectivity than the solution reactions, probably because they bring into play a very high concentration of reactants.

5
Solvent-Free Phase Transfer Catalysis (PTC)

In this case, which is specific of anionic reactions, no solvent or support is involved but reactions are induced by addition of a catalytic amount of a phase transfer agent (tetraalkylammonium salts, crown ethers, etc.).

This technique, that involves a higher concentration of reactants as the electrophile acts both as reagent and as organic phase, allows some reactions to be achieved that are almost impossible when performed in solvents. Yields are often better and, further, obtained using milder conditions (time, temperature) [7, 9]. This has been exploited in pharmaceutical and biological chemistry [9], polymer chemistry [78] and liquid crystal chemistry [79].

5.1
Some Comparative Examples

5.1.1
Etherifications

Long-chain alkylating agents, poorly reactive, are almost inert under classical PTC conditions in the presence of solvents. Yet, they present a significant interest when reacted with some other molecules (e.g. detergents, liquid crystals, organic conductors, etc.) or as such for their lipophilicity. Under solvent-free PTC, they can react with good yields under rather mild conditions.

As an example, phenols react with long-chain bromoalkanes under harsh conditions (refluxing DMF). In the case of p-hydroxybenzaldehyde, under liquid-liquid PTC, only a Cannizaro reaction occurred after one week. On the other hand, solvent-free PTC resulted in a quantitative yield in long chain ether within 5 h [Eq. (46)] [80].

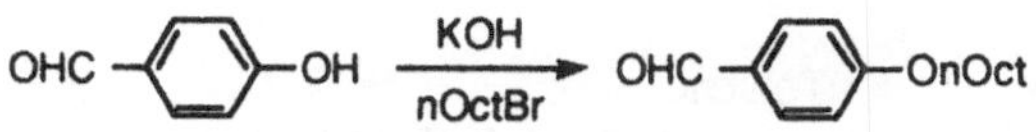

$$OHC-\langle\bigcirc\rangle-OH \xrightarrow[nOctBr]{KOH} OHC-\langle\bigcirc\rangle-OnOct \qquad (46)$$

Solvent-free PTC	**Aliquat 2%**	**5 h**	**85 °C**	**97 %**
Liquid-liquid PTC	aq. KOH-C_6H_6	8 d	80 °C	0 %
DMF		6 h	153 °C	76 %

5.1.2
Esterifications

The alkylation of potassium benzoate with n-octyl bromide under classical solid-liquid PTC (crown ether, chloroform) leads to only 58% of n-octyl ester after 40 h at 85°C. Under solvent-free conditions (2% of Aliquat), yield is quasi-quantitative (95%) within 2 h at the same temperature [Eq. (47)] [81].

$$\text{PhCOO}^-\text{K}^+ \;+\; n\text{Oct Br} \;\xrightarrow[\substack{\text{no solvent}}]{\text{Aliquat 2\%}}\; \text{PhCOOnOct} \quad 95\% \qquad (47)$$

$$\qquad\qquad 1\;:\;1 \qquad\qquad 2h \quad 85°C$$

Carboxylate anions can be generated in situ from their carboxylic acid precursors and subsequently alkylated [Eq. (48)] [82].

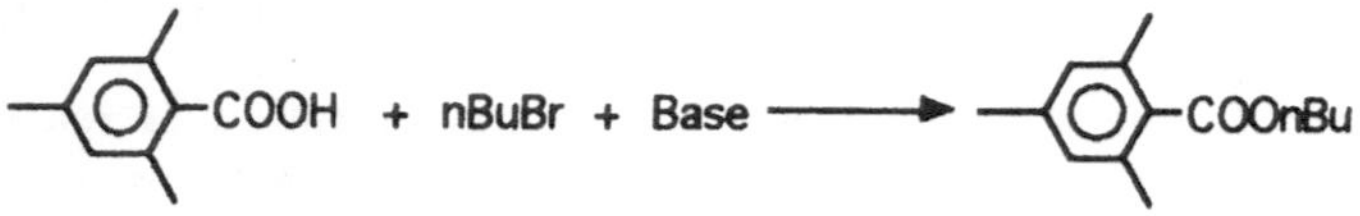

$$\qquad\qquad \text{COOH} + n\text{BuBr} + \text{Base} \longrightarrow \qquad \text{COOnBu} \quad (48)$$

solvent-free	PTC	K$_2$CO$_3$	Aliquat 2%	2h	85°C	98%
resin grafted on polystyrene			toluene-water	92h	75°C	90%

With very long-chain halides, the only working procedure is the solvent-free one [Eq. (49)].

$$\text{(49)}$$

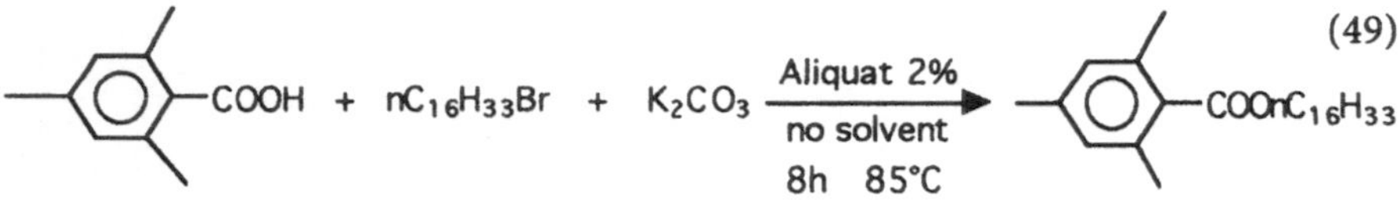

$$\qquad\qquad \text{COOH} + n\text{C}_{16}\text{H}_{33}\text{Br} + \text{K}_2\text{CO}_3 \xrightarrow[\substack{\text{no solvent}\\8h\quad 85°C}]{\text{Aliquat 2\%}} \qquad \text{COOnC}_{16}\text{H}_{33}$$

$$\qquad\qquad\qquad\qquad\qquad\qquad\qquad\qquad\qquad\qquad\qquad\qquad\qquad 88\%$$

5.1.3
Saponification of Hindered Esters [82]

Mesitoic ester saponification is extremely difficult under classical conditions. It can be conveniently performed within 5 h at 85°C with improved yield using the solvent-free technique that, in addition, does not require expensive catalysts [Eq. (50)].

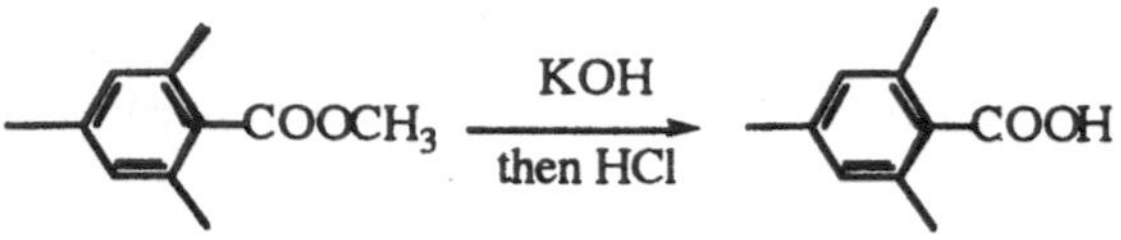

$$\qquad\qquad \text{COOCH}_3 \xrightarrow[\substack{\text{then HCl}}]{\text{KOH}} \qquad \text{COOH} \qquad (50)$$

Classical conditions:	Toluene	30h	75°C	0%
	+ crown ether (1 eq.)	30h	75°C	58%
	+ cryptand [2.2.2] (1 eq.)	12h	25°C	70%
Solvent-free PTC Aliquat 2%		**5h**	**85°C**	**93%**

5.2
Base-Catalyzed Isomerizations [83]

The different solvent-free techniques have been compared during safrole → isosafrole isomerization (Table 5). Solvent-free PTC requires by far the least drastic conditions and only a stoichiometric equivalent amount of base. The use of a very strong base (KOtBu) means that the catalyst can be removed and an identical result can be achieved by increasing the time from 5 min to 3 h at 80 °C.

Table 5. Comparative methods for base-catalyzed isomerization of safrole

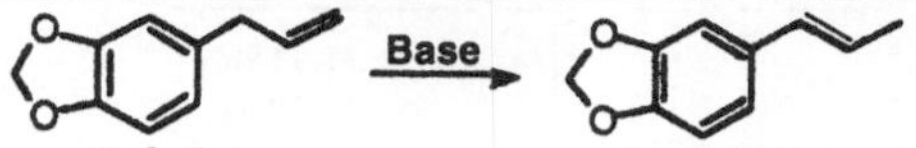

	Safrole		Isosafrole	
KF-Al$_2$O$_3$ (20 eq.)	ethylene glycol	20 min, 150 °C	75%	
	dry media	20 min, 150 °C	91%	
KOH (1.1 eq.), Aliquat 5%	no solvent	5 min, 80 °C	96%	
KOtBu (1.1 eq.), Aliquat 5%	no solvent	5 min, 80 °C	99%	
KOtBu (1.1 eq.)	no solvent	3 h, 80 °C	96%	

5.3
β-Elimination of Bromo Acetals [84–86]

In presence of a base, α-bromo acetals can be converted into ketene acetals, products difficult to obtain by the classical processes. The reaction was studied in the presence of KOH in solvent-free conditions. The effects of a phase transfer agent as well as that of ultrasound, a non-classical method of activation especially efficient in heterogeneous solid-liquid systems due to cavitation phenomenon, were studied (Table 6) [87].

Table 6. β-Elimination from bromo acetal under solvent-free conditions

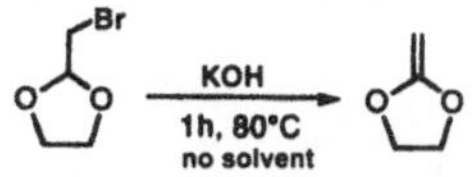

nBu$_4$NBr	U.S.	Yield
–	–	37%
+	–	68%
–	+	65%
+	+	81%

Whereas yields are not changed when PTC or ultrasound is used alone, coupling these two techniques revealed a synergy able to induce an improvement in yield. This effect is even stronger when a larger cycle is involved (Table 7).

Table 7. β-Elimination from bromo acetal under solvent-free conditions

nBu$_4$NBr	U.S.	Yield
+	–	41%
–	+	22%
	+	70%

More recently, it has been shown that these reactions can be performed under even better conditions using microwaves as activation method (Table 8) [86].

Table 8. Comparison of different modes of activation for β-elimination from bromo acetal

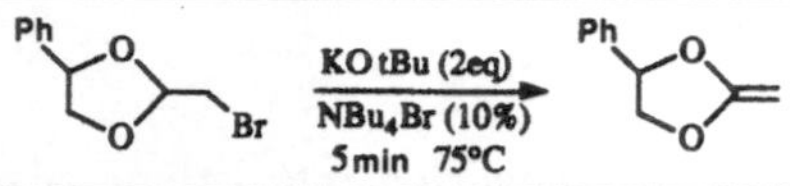

Microwaves	87%
Ultrasound	55%
Conventional heating	36%

5.4
Chlorine-Bromine Exchange [88, 89]

This example illustrates a striking case of a reaction impossible to carry out in a solvent [Eq. (51)].

$$R\text{-}Cl + M^+Br^- \underset{2}{\overset{1}{\rightleftharpoons}} R\text{-}Br + M^+Cl^- \tag{51}$$

This transformation is potentially highly valuable as it allows the formation of more electrophilic R-Br molecules that are more expensive and less commercially available than the corresponding chlorides.

Unfortunately, this reaction is classically limited by its high reversibility in the presence of solvent as two favourable phenomena are involved to promote reversion: nucleophilic strength of anions in dipolar aprotic media is $Cl^- > Br^-$ (harder site), and the electrophilic strength of alkyl halides is RBr>RCl (least bonding energy). Equilibrium is therefore naturally shifted to the left (90%).

To favour reaction (1) and to prevent reaction (2), reaction of Cl⁻ must be avoided. The solution is, of course, to work in solvent-free conditions.

To occur, these halide displacements require anionic activation which can be achieved by PTC coupled with the addition of a transfer agent (NR_4^+, X^-) to provoke ion-pair exchange.

In the best situation, lattice energies for LiBr and LiCl are 788 and 834 kJ/mol, respectively. Consequently, ion-pair exchange occurs preferentially with LiBr of a lower lattice energy. LiBr can thus be selectively activated in the presence of LiCl [Eq. (52)].

$$\text{Li}^+ \text{Br}^- + \text{Li}^+ \text{Cl}^- + \text{R}_4\text{N}^+ \text{X}^- \rightleftharpoons \text{R}_4\text{N}^+ \text{Br}^- + \text{Li}^+ \text{Cl}^- + \text{Li}^+ \text{X}^-$$

$$(52)$$

The exchange R-Cl → R-Br is thus possible using a slight excess of LiBr (1.2 eq.) in the presence of a catalytic amount of Aliquat 336 as phase transfer agent [Eq. (53)].

$$\text{PhCH}_2\text{Cl} + \text{LiBr} \xrightarrow[\substack{\text{2h, 60°C} \\ \text{no solvent}}]{\text{Aliquat (2\%)}} \text{PhCH}_2\text{Br} \quad 94\%$$
$$(1,2 \text{ éq.})$$

$$(53)$$

$$\text{nOctCl} + \text{LiBr} \xrightarrow[\text{1h, 98°C}]{} \text{nOctCl} \quad 94\%$$

5.5
Selective Alkylations of β-Naphthol [90, 91]

It is well established since the work by Kornblum [92] that the regioselectivity of alkylation of an ambident anion as β-naphthol anion is essentially dependent on the reaction medium. Whereas O-alkylation is selectively obtained under anionic activation conditions (dipolar aprotic solvents or PTC), selective C-alkylation remains an unsolved problem.

C-Alkylation of enolates are orbital-controlled reactions. Therefore, a good selectivity implies minimization of charge-controlled processes and, consequently, operating under conditions where ionic associations are optimal. This is the case when solid lithiated bases are used in the absence of solvent, ion-pairing interaction between oxyanion and Li⁺ (two hard sites) being the highest in solvent-free conditions. As two products may result from C-alkylations (mono- and di-C), the selectivity is closely related to base strength (LiOH or LiOtBu) and/or relative amounts of reagents [Eq. (54)].

$$(54)$$

Thus, by a judicious choice of the different heterogeneous solvent-free conditions, each one of the four possible products can be selectively obtained with excellent yields under mild conditions, the reactions being extremely easy to perform, results which cannot be achieved in organic solvent (→ mixtures) (Scheme 16).

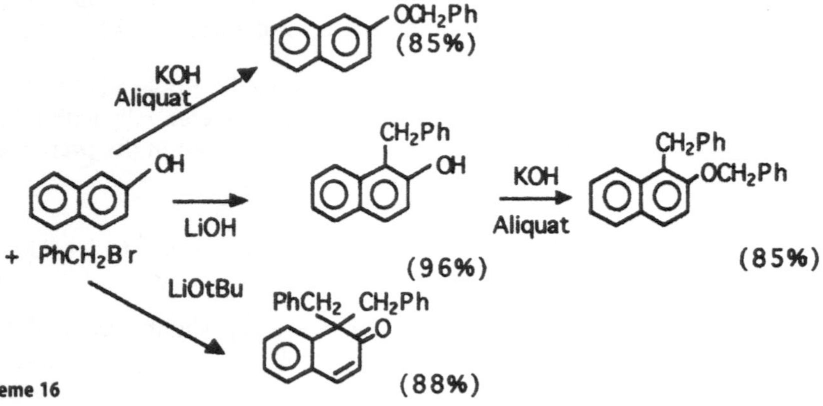

Scheme 16

5.6
Michael Addition

These processes are more frequently limited by the reversibility of the addition on base effect. It can be minimized by using weak bases and dilute solutions.

In the case of the addition of diethyl ethylmalonate anion, it is thus convenient to use Et_3N as a weak base in a highly diluted solution of CH_2Cl_2 and consequently to work under high pressure (10 kbar) [93].

Under solvent-free PTC conditions, catalytic amounts of base (K_2CO_3 or KOH) and of phase transfer agent (Aliquat) are sufficient, simulating high dilution conditions. Retro-Michael reaction can be consequently limited [Eq. (55)] [94].

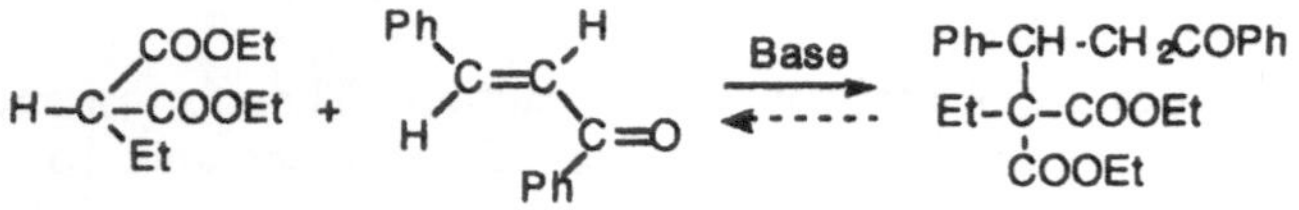

NEt_3-CH_2Cl_2 high dilution	10 kbar	20h	70°C	81%
K_2CO_3 (6%) Aliquat 6%	no solvent	2h	20°C	82%

More interesting is the Michael addition of diethyl acetamido malonate as an amino acid precursor [Eq. (56)] [94, 95].

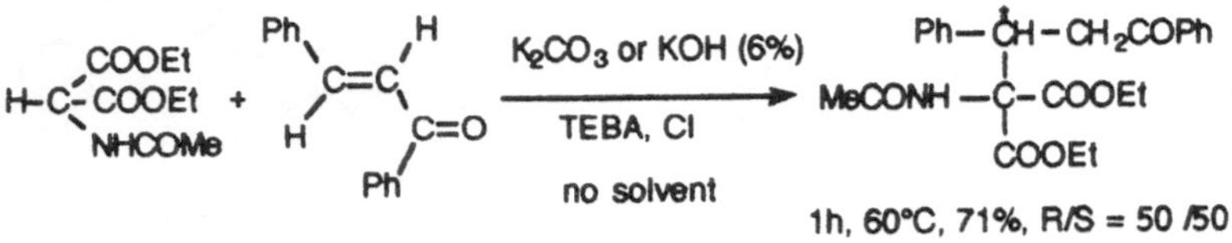

Under ultrasonic activation, the yield can be increased up to 96% [95].

5.7
Asymmetric Michael Addition

The reaction shown in Eq. (56) can be studied using optically active catalysts to envisage asymmetric induction in this Michael addition. For this purpose, chiral tetraalkylammonium salts derived from β-amino alcohols are considered (Scheme 17) either from N-methylephedrine 1 or cinchonine 2.

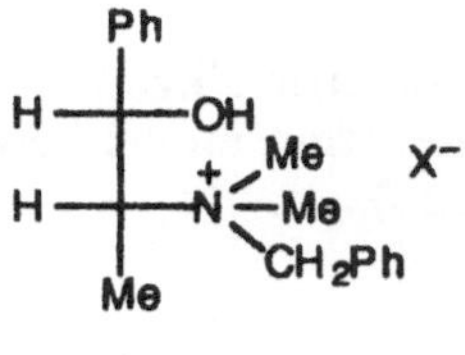

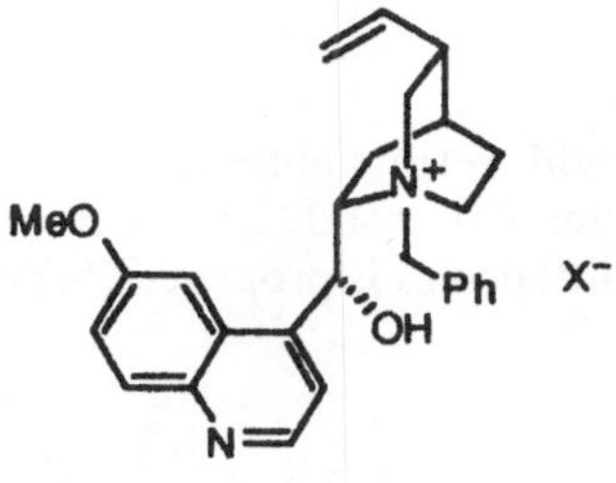

Scheme 17

According to the literature data, enantiomeric excesses are usually limited under classical PTC conditions with solvent. On the other hand, the absence of solvent can bring about an increased rigidity of the reacting system. This can result in a decrease in molecular dynamics and consequently an improvement in enantioselectivity.

The results we obtained [96] led to the best ee which increased in the sequence: CCl_4 > toluene > no solvent (Table 9). By introducing substituents on the aromatic ring of catalysts, changes in ee were observed. Electron-donating groups are favourable to asymmetric induction as shown on Hammett plot correlation (Scheme 18) [97].

Table 9. Effect of a chiral catalyst on the asymmetric Michael addition of diethyl acetamido malonate on chalcone

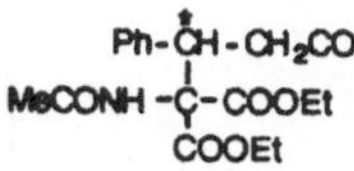

Catalyst	1 (–)	1 (–)	1 (–)	1 (–)	1 (+)
Solvent	/	Toluene	Mesitylene	CCl_4	/
S/R	80:20*	64:36	62:38	56:44	19:81
ee	60	28	24	12	62

* 90:10 with

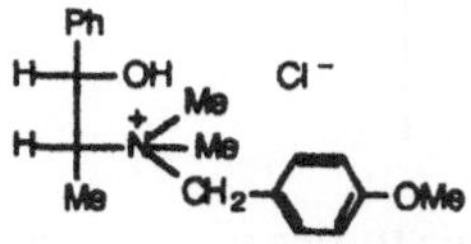

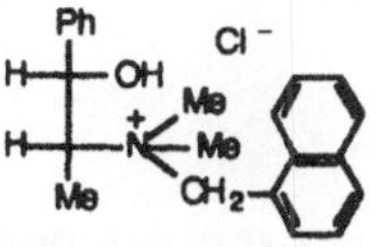

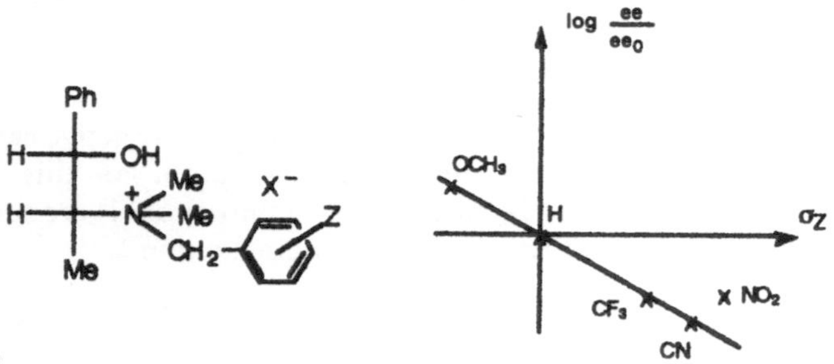

Scheme 18

By analogy with results obtained from asymmetric alkylations of indanone derivatives under PTC conditions, a π–π interaction model between catalyst and electrophilic species is proposed (Scheme 19).

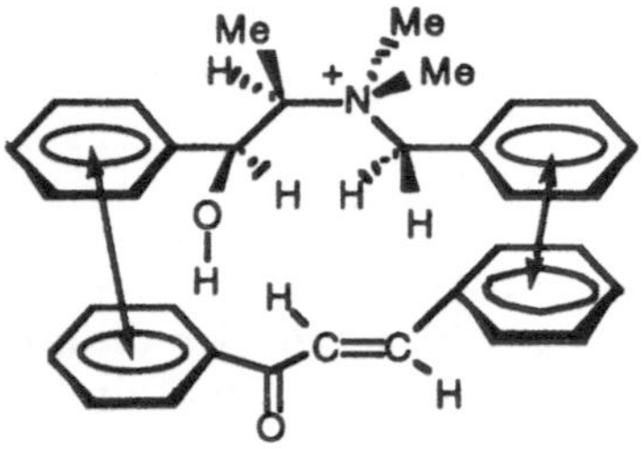

Scheme 19

Such π–π interactions, which occur at approximately 3.5 Å over an aromatic ring [98], are beneficial in selective synthesis. Specifically, asymmetric induction achieved with chiral auxiliaries and chiral catalysts can be enhanced by π-stacking effects [98]. These are optimized under solvent-free conditions as there are no interactions able to interfere in such a system when compared to the solvent procedure.

On the contrary, Conn et al. [99] observed and explained an opposite behaviour by postulating π–π interactions between enolate and catalyst during Michael addition of 6,7-dichloro-5-methoxy-1-indanone to methyl vinyl ketone (Scheme 20).

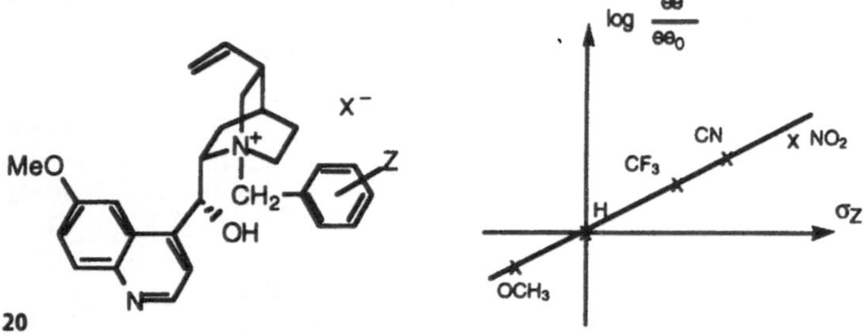

Scheme 20

Similar considerations were extended to the studies of regio- and diastereoselectivities in the Michael addition of 2-phenylcyclohexanone [Eq. (57), Table 10] [100–102].

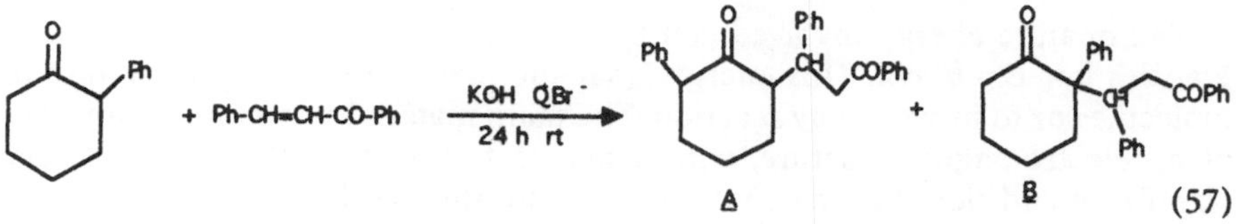

$$(57)$$

Table 10. Effect of Q^+ on the regioselectivity of Michael addition of 2-phenylcyclohexanone anion on chalcone

PTC	Q^+	Yield (%)	A/B
liquid-liquid	TBAB	20	60/40
	ephedrinium	27	48/52
solid-liquid	TBAB	46	**91/9**
	ephedrinium	44	70/30
no solvent	TBAB	42	40/60
	ephedrinium	48	**27/73**

The 2,2-regioisomer was favoured using the ephedrinium salt instead of TBAB. It is formed from the thermodynamic enolate while the 2,6-regioisomer came from the kinetic enolate. The results obtained show the preference for the thermodynamic enolate using the ephedrinium salt as catalyst, probably as a result of the stabilization of the enolate through a $\pi-\pi$ interaction between the catalyst and the enolate.

6
Microwave (MW) Activation in Solvent-Free Reactions

New strategies have recently been developed aimed at working without solvent. Furthermore, it is also possible to activate processes by physical means such as ultrasound, pressure or microwaves.

Among these new non-conventional methods in organic synthesis [103], microwave irradiation takes a particular place as it induces interactions between materials and waves of an electromagnetic nature assimilated to dielectric heating. This original procedure involves heating the materials which then become reactive in situations where traditional treatments failed to give any reaction at all.

6.1
Generalities

Microwaves range from 1 cm to 1 m in wavelength in the electromagnetic spectrum and are situated between the infrared and radio frequencies. The frequency band allowed by worldwide legislation, be it for industrial, scientific, medicinal or domestic purposes, corresponds to $v = 2450$ Mhz (i.e. $\lambda = 12.2$ cm under vacuum).

The quantic energy involved can be evaluated to 0.3 cal/mol according to Planck's law $E = h \cdot c/\lambda$. This energy is far too low to induce any excitation of molecules or to provoke any reaction. The main material-wave interactions are of an electromagnetic nature, with a penetration depth into materials very significant and close to a few centimeters, as for the wavelength.

6.1.1
Microwave Heating

Heating of products submitted to microwave exposure can only result from material-wave interactions. It is brought about by the transformation into heat of a part of the energy contained in the electromagnetic wave.

Polar molecules display the property that they can be oriented along an electric field (*dipolar polarization* phenomenon). In the absence of this phenomenon, dipoles are orientated at random and molecules submitted to Brownian movement only. In the presence of a continuous electric current, all the dipoles are lined up together in the same direction. If submitted to an alternating current, the electric field is inversed at each alternance with a subsequent tendancy for dipoles to move together to follow the field. Such a characteristic induces stirring and friction of molecules which dissipates as internal homogeneous heating (Scheme 21).

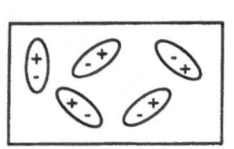 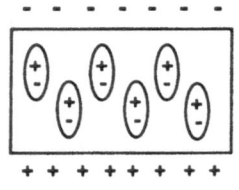 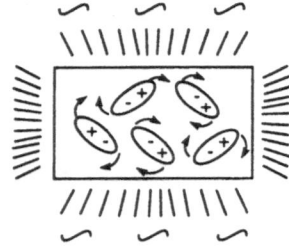

without constraint   submitted to **continous**   **alternative** electric current
                            electric current

Scheme 21. Influence of electric field on a dielectric product

From this heat dissipation inside the materials, the result is a final repartition in temperature much more homogeneous when compared to classical heating. Heating by microwaves is therefore an original procedure bringing the following advantages: speed, no inertia, heat affects the product only, ease of use, quick energy transfer in the whole mass without any superficial overheating.

6.1.2
Advantages of Microwave Exposure [104]

From the interactions between materials and electromagnetic waves heat is produced according to an original process characterized by a heating taking place in the core of the materials without superficial overheating, with a subsequent very homogeneous temperature. The profiles of gradients in temperature are inverted when going from classical heating (Δ) to microwave (MW) irradiation (Scheme 22).

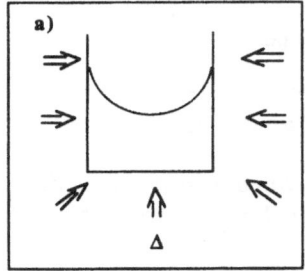 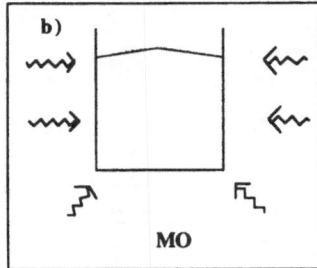

Scheme 22. Gradient in temperature in solid submitted to (a) traditional heating by conduction (b) microwave exposure

Other main benefits of microwave heating are:

- the selective heating of polar molecules (due to dipolar polarization),
  e.g. 50 ml liquid, 1 min, 600 W,
  DMF 140 °C ($\mu = 10.8$ Debye); H_2O, 80 °C ($\mu = 5.9$ Debye); CCl_4, 25 °C
  ($\mu = 0$ Debye)
- very fast heating

The rise in temperature can be up to 10 °C/s with, consequently, a system especially efficient in the case of poor heating conductors.

6.1.3
Specific Effects of Microwaves (Purely Non-thermal)

Microwaves can be used to promote many chemical syntheses [105]. The materials-wave interactions produce heating of the reaction medium by polar molecules (solvents, reagents or complexes, solid supports). To these purely thermal effects can be added specific effects due to MW radiation.

To determine these effects a strict comparison is required between MW and conventional heating (Δ), all other conditions being identical (time, temperature, pressure, same profile of elevation in temperature). If the results obtained are different, the origin of these specific effects could be due to:

- a better homogeneity and speed of heating,
- the intervention of hot spots with high localized microscopic temperatures [106, 107],
- variations in activation parameters $\Delta G^{\neq} = \Delta H^{\neq} - T \cdot \Delta S^{\neq}$ [108, 109].

Due to previous organization of the polar system under microwaves (dipolar polarization), activation parameters, and essentially $\Delta S^{\neq}$, can be modified. This was experimentally proved by Lewis et al. [108] during imidization of polyamic acid either by conventional heating or under MW activation [Eq. (58)].

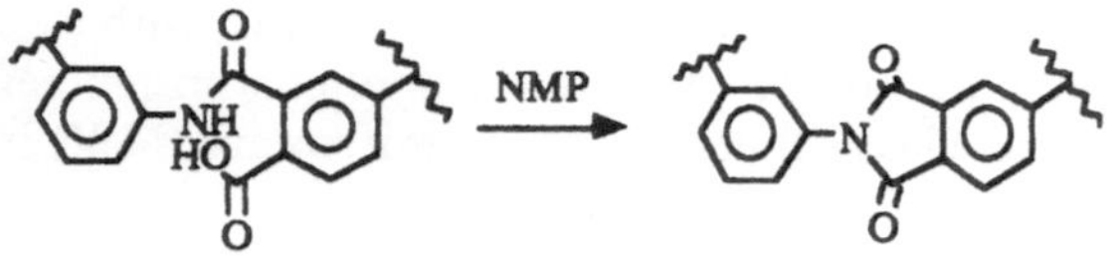

$$\Delta G^{\ddagger} \qquad\qquad \ln A \qquad\qquad (58)$$
$$\text{kJ/mol}$$

	$\Delta G^{\ddagger}$ kJ/mol	$\ln A$
Microwave	57 ± 5	13 ± 1
Thermal	105 ± 14	25 ± 4

The energy of activation is largely reduced with a corresponding decrease in lnA, (preexponential Arrhenius factor), a property linked to the entropic effects.

6.2
Organic Synthesis Under Microwaves

The applications that can implement such a technique are derived from two main types of reactions:

- thermal reactions, which need high temperatures for long reaction times. Microwaves will bring acceleration of reactions, low decomposition of products and consequently enhanced yields.
- equilibrated reactions, with displacement of equilibrium by vaporization of small polar molecules [Eq. (59)].

$$RCOOH + R'OH \underset{\xleftarrow{\qquad}}{\xrightarrow{H^+ \text{ or enzyme}}} RCOOR' + H_2O \nearrow$$
$$RCOOMe + R'OH \underset{\xleftarrow{\qquad}}{\xrightarrow{H^+ \text{ or } B^-}} RCOOR' + MeOH \nearrow \qquad (59)$$

This allows the *in situ* preformation of nucleophilic salts followed by anionic reaction (PTC), both steps being favoured independently by microwaves (Scheme 23).

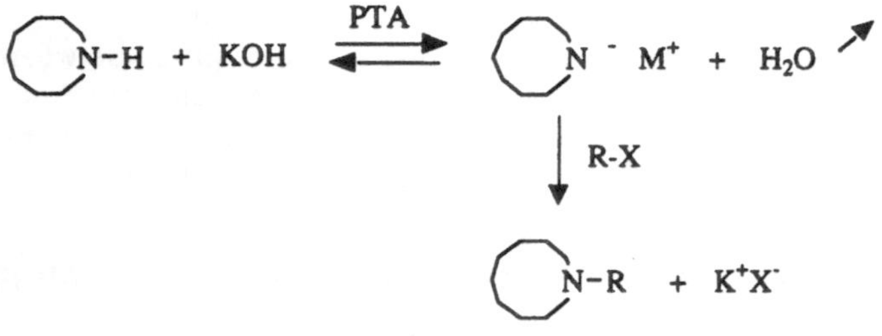

Scheme 23

6.2.1
Benefits

The pioneering works are due to Gedye, Giguere et al. [110, 111] who advocated the use of domestic ovens and solvents for their experiments. To date, more than

500 publications have appeared, testifying to the exceptional interest in the method. The most common benefits described are:

- very rapid reactions, frequently a few minutes, brought about by high and homogeneous temperatures and combined with pressure effects (if conducted in closed vessels),
- higher degree of purity achieved due to short residence time at high temperatures, no local overheating, minor decomposition and minor occurrence of secondary reactions, and
- yields often better, obtained within shorter times and with purer products.

6.2.2
Limitations

The boiling points of solvents are reached rapidly, often posing safety problems (e.g. explosions). To solve these problems, the operation has to be carried out in closed vessels (generally made of Teflon, a material transparent to MW and resistant up to 250 °C and 80 psi) and using only small amounts of products (roughly 1/10 of the total volume). This of course constitutes a serious limitation (e.g. reduction in MW efficiency as the penetration depth is far below λ, scaling up, etc.).

Another main limitation is the absence of measurement and control of temperature. However, these limitations can be overcome by the following two approaches:

(1) to use solvent-free techniques,
(2) to operate with a monomode reactor with permanent control of temperature (see below).

6.2.3
Equipment [112]

Two types of reactors can be used in the laboratory:

- multimode systems

These domestic ovens (with limited power at 800 – 1000 W) are characterized by a non-homogeneous distribution of electric field due to several reflections on the metallic walls of the oven. Their use for synthetic purposes requires a previous cartography to determine the hot spots of high energy using a filter paper sheet impregnated with a solution of cobalt chloride [113] (Scheme 24).

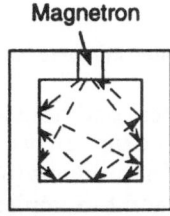

Scheme 24. Dispersion energy in a multimode oven

Two other drawbacks follow from a construction aspect as there is no modification of power as the oven only operates by sequential irradiation between the maximum and zero and no possibility of *in situ* temperature check.

– monomode reactors (e.g. Synthewave Prolabo)

These drawbacks led to the development of monomode applicators that focus the electromagnetic waves in an accurately dimensioned wave-guide. This allows a homogeneous distribution of the electric field and can be used with a low emitted power with a high energetic yield (Scheme 25).

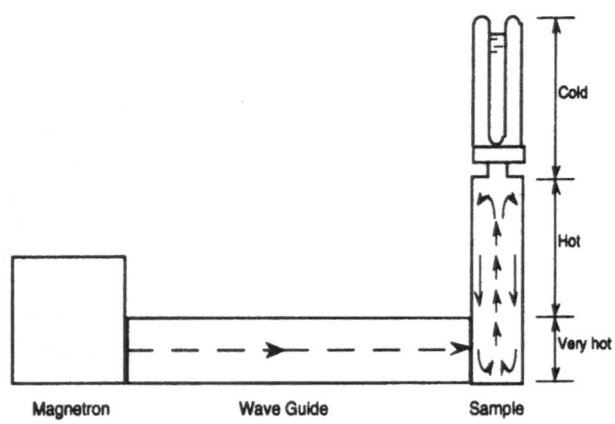

Scheme 25. Dispersion energy in a monomode reactor

The Synthewave 402 reactor from Prolabo presents a number of benefits: temperature measurement by infrared detection [114] on the surface of the product, temperature control using power modulation from 15 to 300 W, monitoring of the reaction by a computer to program power or temperature, the use of open vessels allowing reactions to be run under normal or reduced pressure or controlled atmospheres and under stirring with an easy addition of reagents. Monomode reactors offer increased efficiency and reliability. They lead to considerable improvements in yields of organic synthesis by preserving thermal stabilities of products with real low emitted power and good homogeneity in temperature.

6.3
Solvent-free Organic Synthesis Under Microwaves

It has been shown that solvent-free conditions are especially adaptable to microwave activation as reactions can be run safely under atmospheric pressure in the presence of significant amounts of products [62, 68, 112, 115].

In this section, typical and recent examples of the coupling of solvent-free reactions with microwave activation will be described. Comparisons with conventional heating methods realized either in an oil or sand bath previously heated to the same temperature observed in microwave experiments will be given where available.

6.3.1
Reaction on Solid Supports

Mineral oxides are often very poor conductors of heat but behave as very efficient microwave adsorbents, this resulting in turn in a very rapid and homogeneous heating. Consequently, they display very strong specific microwave effects with significant improvements in temperature homogeneity and heating rates enabling faster reactions and less degradation of final products when compared to classical heating.

6.3.1.1
Acidic Supports (Clays)

Many examples of the use of acidic supports include:

- Meyer-Schuster rearrangement [Eq. (60)] [62]:

$$Ph\text{-}C\equiv C\text{-}\overset{\overset{\displaystyle Ph}{|}}{\underset{\underset{\displaystyle OH}{|}}{C}}\text{-}Ph \xrightarrow{KSF} Ph\text{-}\overset{\overset{}{}}{\underset{\underset{\displaystyle O}{\|}}{C}}\text{-}CH=C\overset{\displaystyle Ph}{\underset{\displaystyle Ph}{}} \qquad (60)$$

5 min   170°C   MW   98 % $\Delta < 2$ %

- Fries rearrangement [Eq. (61)] [116]:

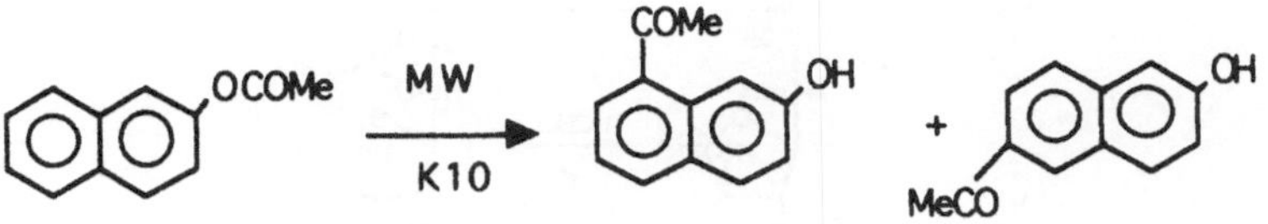

10 mn   Yield 70%   (9 : 1)   (61)

- 4-phenylcoumarin synthesis [Eq. 62)] [117]:

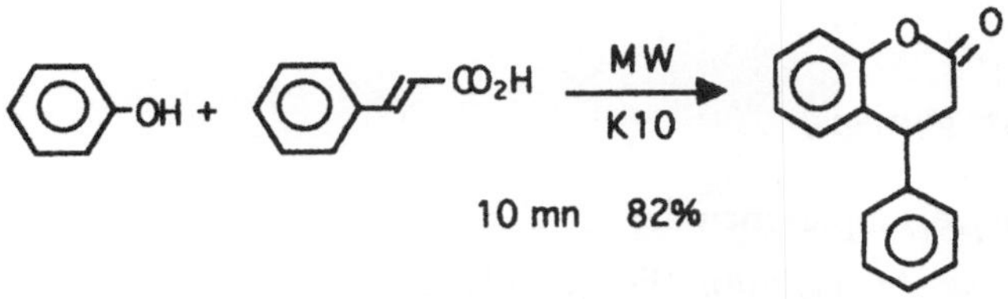

10 mn   82%   (62)

– deacetylation of 5-nitrofurfural diacetate [Eq. (63)] [118]:

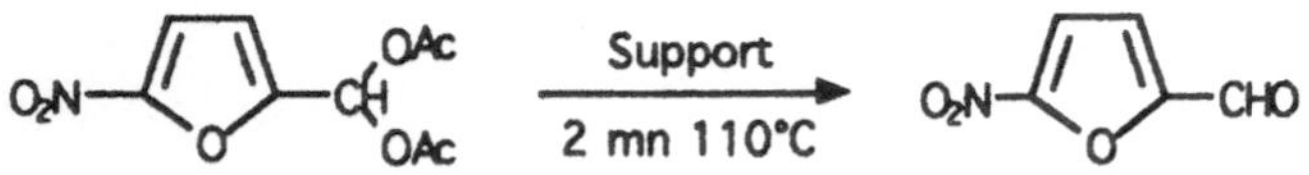

$$\tag{63}$$

		Alumina	MW	5%
K10 clay	dispersion	M W	99%
K10 clay	dispersion	Δ	< 2%
K10 clay	impreg.	MW	80%

– synthesis of fused heterocyclic quinones [Eq. 64)] [65, 119]:

$$\tag{64}$$

Classical method	PCl$_5$ (1.5 eq)	AlCl$_3$ (1.5 eq)	Nitrobenzene	4 h	140°C	70%
Dry media	**K10**	**MW**		**3 min**	**320°C**	**92 %**
		Δ		1 h	320°C	41 %

– acetalization of 1-galactono-1,4-lactone (Table 11) [120]:

This lactonic diol, byproduct of sugar beet, can be converted to acetals with crystal-liquid properties when treated with long-chain halides.

Table 11. Acetalization of 1-galactono-1,4-lactone with dodecanol [R = CH$_3$(CH$_2$)$_{10}$]

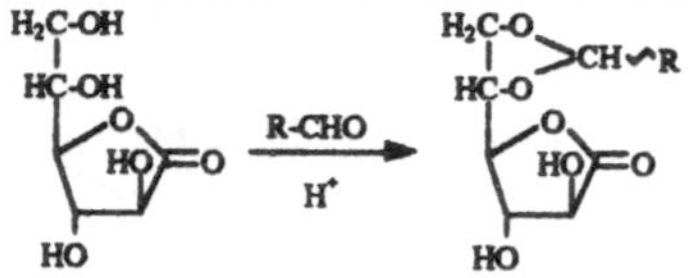

Classical conditions: DMF, H$_2$SO$_4$, anhydrous CuSO$_4$	24 h	40°C	25%
Dry media + MW KSF	10 min	155°C	66%
K10	10 min	155°C	89%

6.3.1.2
Neutral Supports (Alumina)

In this category, examples include:

– potassium acetate alkylation [Eq. (65)] [115, 121]:

$$CH_3COO^-K^+ \ + \ nC_8H_{17}X \ \xrightarrow{\ Al_2O_3\ } \ CH_3COOnC_8H_{17} \tag{65}$$

2 min   186°C   MW   98 %   Δ < 2 %

- imine synthesis [Eq. (66)] [122]:

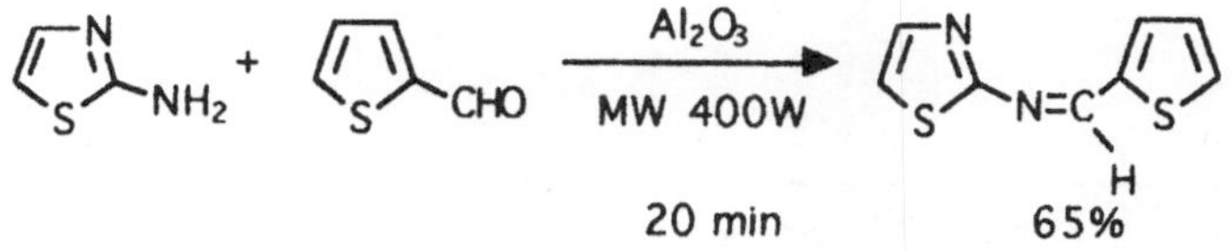

$$\tag{66}$$

MW 400W

20 min   65%

- Michael additions [Eq. (67)]:

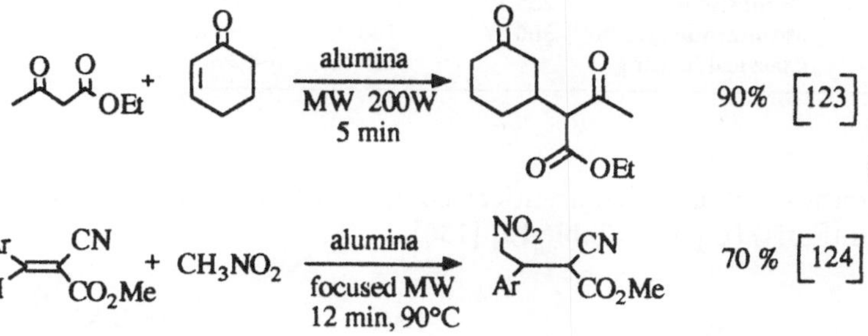

alumina
MW 200W
5 min

90% [123]

$$\tag{67}$$

Ar CN + CH_3NO_2 alumina / focused MW 12 min, 90°C NO_2 CN / Ar CO_2Me 70 % [124]

- deprotections by thermolysis on alumina: benzylic esters [Eq. (68)] [125], silylated ethers [Eq. (69)] [126] and α,α-diacetates [Eq. (70)] [128]

$$Ph\,CO_2CH_2Ph \ \xrightarrow[\text{7 min 800W}]{\text{Alumina}} \ PhCO_2H \qquad 92\% \tag{68}$$

$$PhCH_2O-\underset{\underset{Me}{|}}{\overset{\overset{Me}{|}}{Si}}-tBu \ \xrightarrow[\text{11min 800W}]{\text{alumina}} \ Ph\,CH_2OH \qquad 82\% \tag{69}$$

$$Ph\text{-}CH\overset{OAc}{\underset{OAc}{<}} \ \xrightarrow[\text{40 s 800W}]{\text{Alumina}} \ Ph\,CHO \qquad 98\% \tag{70}$$

- oxidation over $KMnO_4$/alumina: selective oxidations of arenes [Eq. (71)] [128], oxidation of β,β-disubstituted enamines (Table 12) [129]:

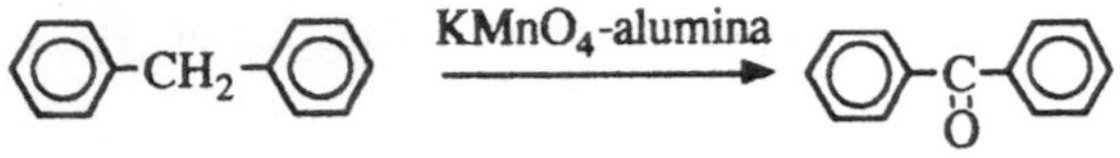

$$\text{Ph–CH}_2\text{–Ph} \xrightarrow{\text{KMnO}_4\text{-alumina}} \text{Ph–}\underset{\underset{O}{\parallel}}{C}\text{–Ph} \tag{71}$$

	ClCH$_2$CH$_2$Cl (83)	282 h	r.t.	91%
MW	**Dry media**	**30 min**	**110°C**	**96%**
Δ	Dry media	30 min	110°C	70%

Table 12

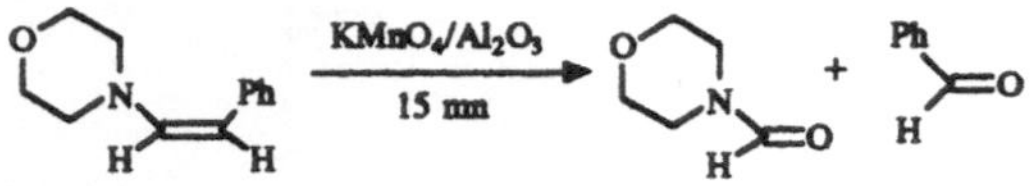

MW	Domestic oven	255 W		73%
	Monomode reactor	**300 W**	**140°C**	**83%**
Δ	Classical heating		140°C	<2%

– synthesis of a pharmaceutical compound: 1,4-dihydropyridine (Hantzsch synthesis) [Eq. (72), Table 13] [130].

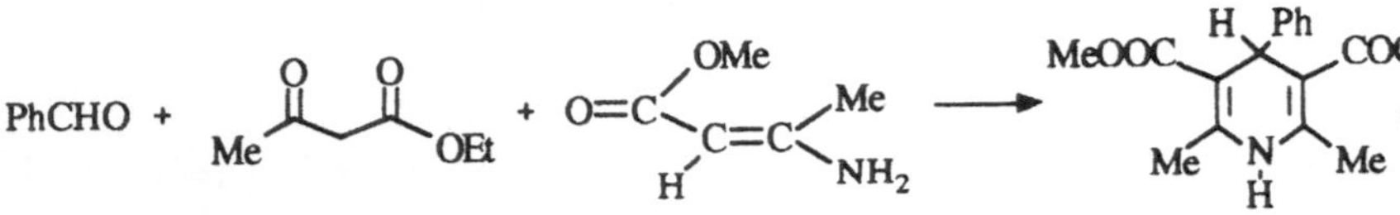

$$\tag{72}$$

Table 13. Synthesis of 1,4-DHP

Activation	Method	Conditions	Yield (%)
Δ	classical method	EtOH, reflux, 8 h	55
MW	alumina "dry media"	80–85°C, 6 min	40
MW	**alumina + ε DMF**	**12°C, 6 min**	**>95**
Δ	alumina + ε DMF	120°C, 6 min	5
Δ	alumina + ε DMF	120°C, 1 h	30

In this case, the temperature reached by alumina/materials-wave interactions is not high enough (80–85°C) to promote an efficient reaction. A small amount of a polar molecule (few drops of DMF) allows a rise in temperature up to 120°C and for a quantitative yield for the microwave-promoted reaction to be obtained.

The benefits obtained with microwave-assisted dry chemistry was recently advocated for combinatorial chemistry and applied to the high output, automated, one-step, parallel synthesis of various substituted pyridines using this procedure [131].

6.3.1.3
Basic Supports (KF/Alumina)

Among the varied examples of the use of basic supports are:

– Knoevenhagel condensations [132].

Many cyclic compounds with an acidic methylene group can be condensed with aldehydes by adsorption on KF/alumina and subsequent MW irradiation. For instance, rhodamine gives 5-alkylidene products of biological interest [Eq. (73)].

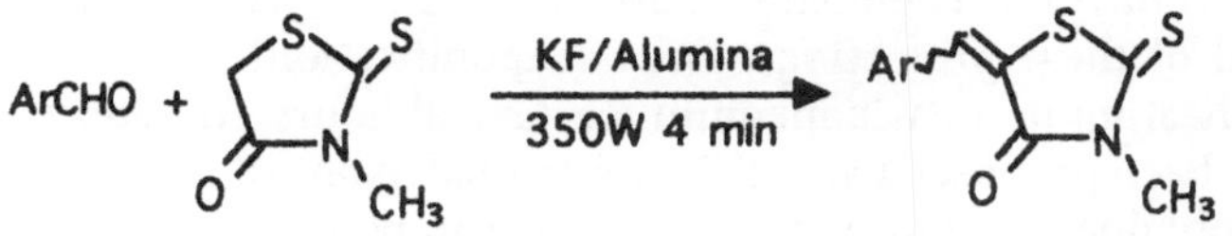

$$
\text{ArCHO} + \underset{\substack{O \\ N-CH_3 \\ S \quad S}}{} \xrightarrow[\text{350W 4 min}]{\text{KF/Alumina}} \text{Ar} \underset{\substack{O \\ N-CH_3 \\ S \quad S}}{} \qquad (73)
$$

– synthesis of dithioacetals [133].

The preparation of dithioacetals by reacting active methylenes and 5-methyl methanesulfonothioate on KF/Al$_2$O$_3$ has been described [Eq. (74)]. Potential antiviral phosphonates are thus prepared.

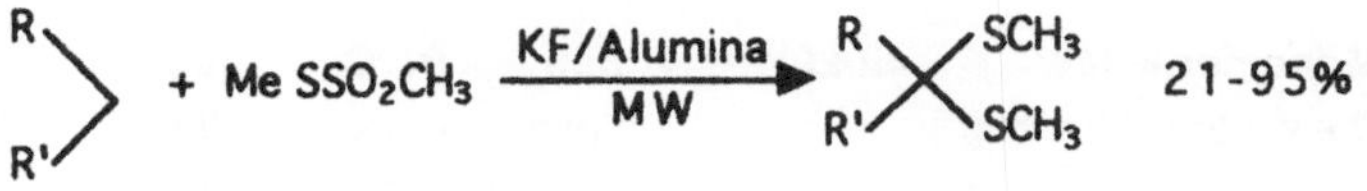

$$
\underset{R'}{\overset{R}{>}} + \text{Me SSO}_2\text{CH}_3 \xrightarrow[\text{MW}]{\text{KF/Alumina}} \underset{R'}{\overset{R}{\times}} \overset{\text{SCH}_3}{\underset{\text{SCH}_3}{}} \qquad 21\text{-}95\% \qquad (74)
$$

$$
\text{R, R' = CO}_2\text{R, CN, Ph, P(O)(OEt)}_2
$$

– 1,3-dipolar cycloaddition of diphenylnitrilimine (DPNI) [134].

Prior treament of hydrazonoyl chloride with a base is necessary to generate the reactive species DPNI, which subsequently adds to electrophilic double bonds. The overall reactions were performed by simultaneous impregnation of hydrazonoyl chloride and dipolarophile on KF/alumina as basic system. DPNI is formed *in situ* and reacts in a one-pot procedure (Scheme 26).

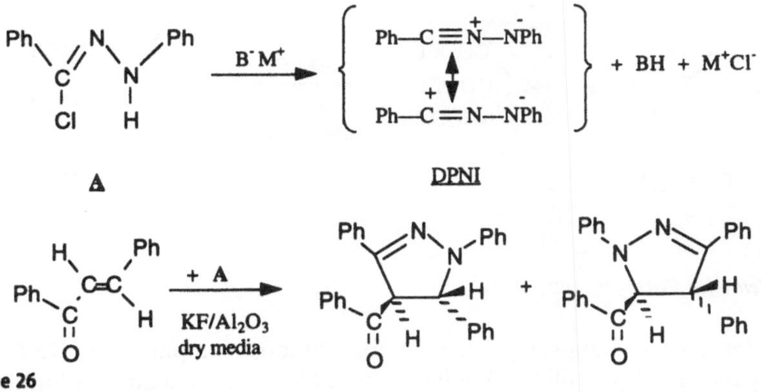

$$
\underset{\substack{\text{Cl} \quad \text{H} \\ \mathbf{A}}}{\overset{\text{Ph}}{\underset{\text{C}}{}}\!\!=\!\!N\!-\!\overset{\text{Ph}}{N}} \xrightarrow{\text{B}^-\text{M}^+} \left\{ \begin{array}{c} \text{Ph-C} \equiv \overset{+}{N}\!-\!\overset{-}{N}\text{Ph} \\ \updownarrow \\ \text{Ph-C} \overset{+}{=} N\!-\!\overset{-}{N}\text{Ph} \end{array} \right\} + \text{BH} + \text{M}^+\text{Cl}^-
$$

$$
\underset{\text{DPNI}}{}
$$

$$
\underset{\substack{O}}{\overset{H}{\underset{Ph}{}}\!C\!=\!C\!\overset{Ph}{\underset{H}{}}} \xrightarrow[\substack{\text{KF/Al}_2\text{O}_3 \\ \text{dry media}}]{+ \text{A}} \quad \underset{O}{\overset{\text{Ph}}{\underset{\text{Ph}}{}}} + \quad \underset{O}{\overset{\text{Ph}}{\underset{\text{Ph}}{}}}
$$

Scheme 26

MW	Domestic oven	700W	12 mn	123°C	90%
	Monomode reactor	30W	6 mn	170°C	90%
Δ	Oil bath		6 mn	170°C	<2%
	Oil bath		24 h	170°C	90%

Scheme 26 (continued)

– saponification of peracetylated glycosides [135].

Acetylation is one of the most popular methods chosen to protect hydroxyl groups in carbohydrate chemistry. Therefore an unavoidable subsequent step is the removal of these protecting groups (saponification). This is usually performed by heating in a basic medium for several hours. An alternative to this method has been proposed using KOH impregnated onto alumina as a base. The result was obtained within 2 min under microwaves, the improvement being attributed essentially to a strong specific microwave effect (Scheme 27).

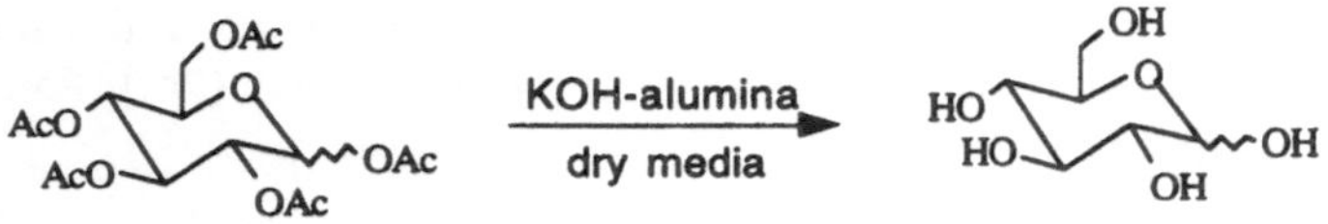

Monomode MW	40-20W	2 min	97°C	98%
Conventional heating		2 min	100°C	0%
Conventional heating		3 h	100°C	a)

a) incomplete reaction + mixtures of partially deacetylated products

Scheme 27

– ring opening of a fatty epoxide [136].

Tetradecyl oxirane is reacted with diethyl acetamidomalonate in basic medium. The presence of LiCl, co-impregnated with KF on alumina, is necessary here to insure the electrophilic assistance to ring opening. The main product is a lactone, formed after epoxide ring opening and subsequent cyclization [Eq. (75)].

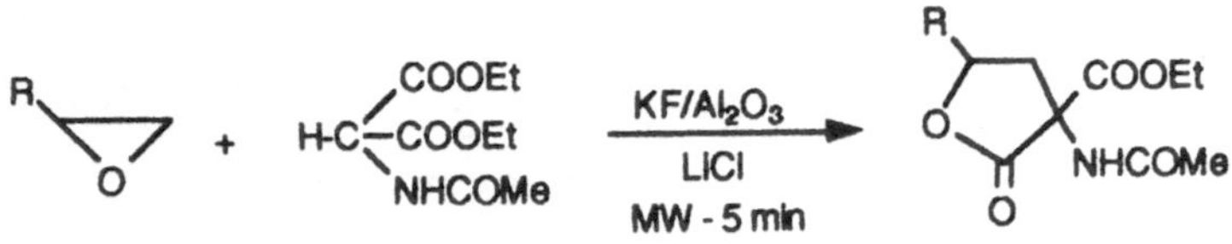

$$R = nC_{14}H_{29} \qquad\qquad 90\% \qquad (75)$$

6.3.2
Phase Transfer Catalysis [87, 137]

Due to ion-pair exchange, there is formation of loose ion pairs Nu^-, NR_4^+ which are very reactive, lipophilic and polar species. They are consequently highly sensitive to MW exposure [138] producing, in turn, an important rise in temperature. This solvent-free procedure is therefore very prone to microwave coupling.

- fatty ester syntheses (Table 14) [115, 138, 139].

Table 14. Fatty ester synthesis

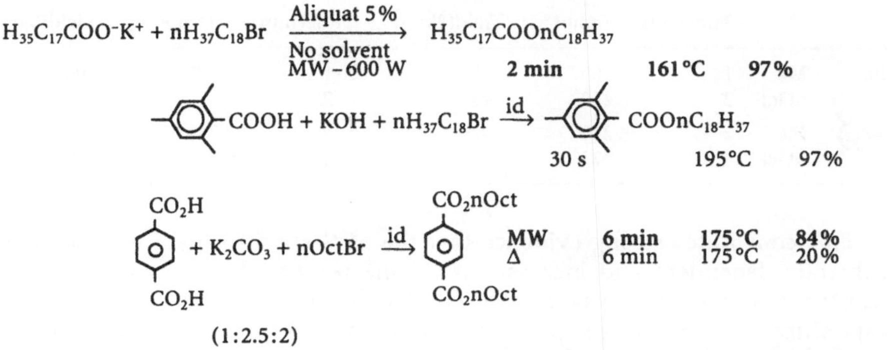

$H_{35}C_{17}COO^-K^+$ + $nH_{37}C_{18}Br$
Aliquat 5%
No solvent
MW - 600 W
$H_{35}C_{17}COOnC_{18}H_{37}$
2 min
161 °C
97%
COOH + KOH + $nH_{37}C_{18}Br$
id
$COOnC_{18}H_{37}$
30 s
195°C
97%
CO_2H
+ K_2CO_3 + nOctBr
id
CO_2nOct
MW
Δ
6 min
6 min
175 °C
175 °C
84%
20%
CO_2H
CO_2nOct
(1:2.5:2)

- jasminaldehyde synthesis [140].

This is a typical example of aldol condensation leading to a very important product in perfume chemistry with some impurities due to self-condensation of *n*-heptanal (Table 15). The best result is achieved using the last system described, i.e. with MW within one minute and giving only 18% of self-condensation.

Table 15. Jasminaldehyde synthesis

PhCHO + $CH_3(CH_2)_6CHO$
Base
Aliquat (5%)
Ph
CHO
H
C_5H_{11}
+ $CH_3(CH_2)_5$–CH
CHO
C_5H_{11}

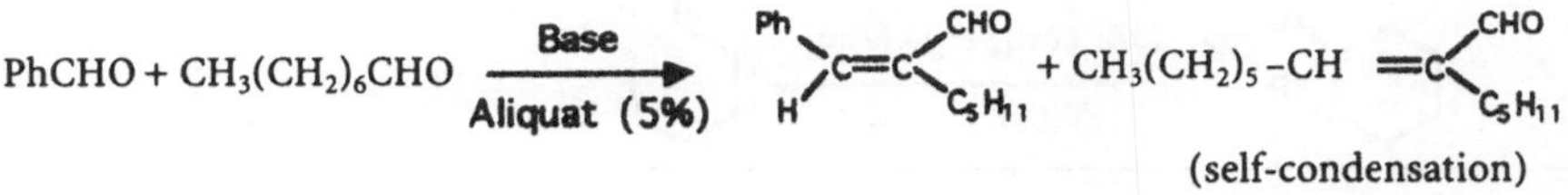

					(self-condensation)
K_2CO_3	Δ	60 h	rt	75%	25%
	MW (350 W)	4 min	141°C	75%	15%
KOH	Δ	24 h	rt	70%	30% ↘
	MW (350 W)	1 min	118°C	82%	18%

- ester saponifications [141]

The study here was performed with different substituents R and aromatic esters. [Eq. (76)]:

$$\text{Ar COOR} \xrightarrow[\text{then HCl}]{\text{KOH(2eq) - Aliquat (10\%)}} \text{Ar COOH} \qquad (76)$$

Results were obtained with monomode MW at a power of 90 W and compared with conventional heating under strictly the same conditions (Table 16).

Table 16. Saponifications of aromatic esters under solvent-free PTC conditions

Ar	R	MW irradiation			Conventional heating		
		Time(min)	Temp(°C)	Yield(%)	Time(min)	Temp(°C)	Yield(%)
Ph	Me	1	205	96	1	205	90
	nOct	2	210	94	2	210	72
(ring)	Me	2	240	87	2	240	38
	nOct	4	223	82	4	223	0

This constitutes strong evidence that the MW specific effect is noticeably substrate dependent and increases when the reaction becomes increasingly difficult, as pointed out by Lewis et al. [142]. In particular, the almost impossible saponification of hindered and long-chain esters can be achieved easily under solvent-free PTC when coupled with MW.

– dealkoxycarbonylation of activated esters – Krapcho reaction [143]

Usually, this reaction is performed in DMSO at high temperatures with the necessary addition of alkaline salts. In order to avoid the use of DMSO at reflux and the tedious work-up, a new procedure, consisting of a salt (LiBr) and a phase transfer agent (NBu_4Br) without solvent coupled with MW irradiation, has been developed (Table 17).

Table 17. Krapcho reaction of β-cyclic keto esters

	Ex : R = C_2H_5				
Δ	Classical method: DMSO-CaCl$_2$ (5eq)	3 h	160°C	20%
MW	Solvent-free PTC	30 W	15 min	160°C	94%
Δ	Solvent-free PTC	oil bath	15 min	160°C	0%
	Solvent-free PTC	oil bath	3 h	160°C	60%

Clearly, the excellent results obtained under MW are not only due to thermal effects. When compared with conventional heating, two main benefits appear: a large reduction in time with simplified experimental conditions and prevention of the degradation of product at high temperature.

– N-alkylation of azaheterocycles [144, 145].

Under MW irradiation, a number of azaheterocycles (pyrrole, imidazole, pyrazole, indole, carbazole, phthalimide, etc.) react remarkably fast with alkyl halides to give exclusively N-alkyl derivatives (Table 18).

Table 18. *N*-Alkylation of phthalimide using K_2CO_3, TBAB

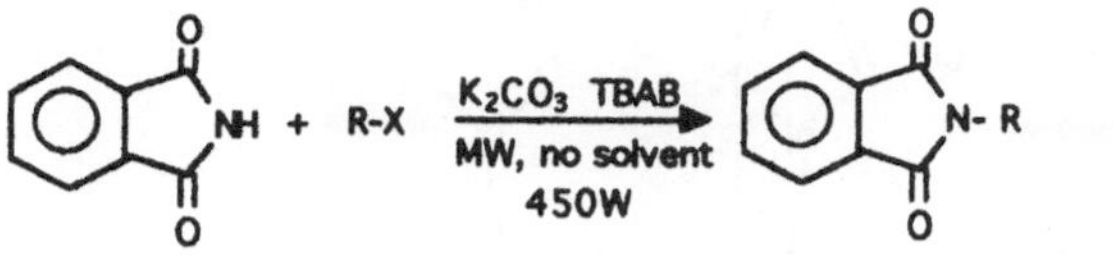

PhCH$_2$Cl	4 min	93%
nC$_4$H$_9$Br	4 min	73%
nC$_{10}$H$_{21}$Cl	4 min	51%

– selective dealkylations of aromatic esters [146].

Ethyl isoeugenol and ethoxy anisole can be selectively dealkylated into **2** (demethylation) or **3** (deethylation) using a base (KOtBu or NaOH) in the presence of 18–6 crown ether (10%). By addition of ethylene glycol (EG), the selectivity is entirely inverted from deethylation to demethylation. In both cases, strong accelerations were observed under MW which are absolutely necessary to achieve demethylation (Table 19).

Table 19. Selective dealkylations of ethoxy anisole

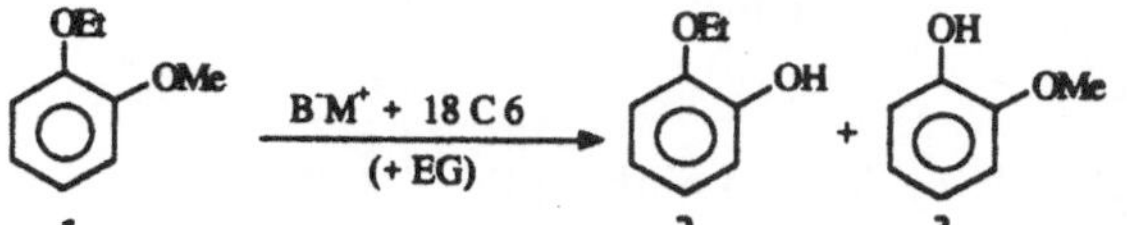

B⁻M⁺	EG (ml)	Exp conditions			% 1	% 2	% 3
KOtBu	0	MW (60 W)	20 min	120 °C	7	–	90
	0	Δ	20 min	120 °C	48	–	50
	0	Δ	2 h	120 °C	28	–	60
KOtBu	2	MW (60 W)	75 min	180 °C	–	72	23
	2	Δ	75 min	180 °C	98	–	–
	2		20 h	180 °C	63	26	–
NaOH	5	MW (60 W)	2 h	205 °C	5	77	10
	5	Δ	2 h	205 °C	94	–	–

6.3.3
Neat Reaction Without Support or Catalyst

Reactions include:

– *N*-alkylation of 1,2,4-triazole [147].

In this case, it was shown that MW irradiation produces specific effects both on reactivity and selectivity (Scheme 28). Due to strong acceleration of the first

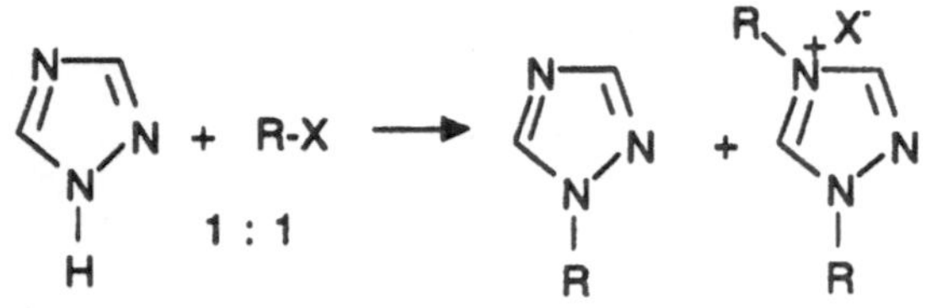

RX = PhCH$_2$Br

| MW | 5 mn | 165°C | 100 | 0 | (yield 80%) |
| Δ | 5 mn | 165°C | 0 | 100 | (yield 14%) |

Scheme 28

alkylation, N_1-alkylated product is selectively obtained under MW in good yields whereas only the dialkylated product is obtained under conventional heating and that in poor yield.

- synthesis of 1-arylpiperazines [Eq. (77)] [148]:

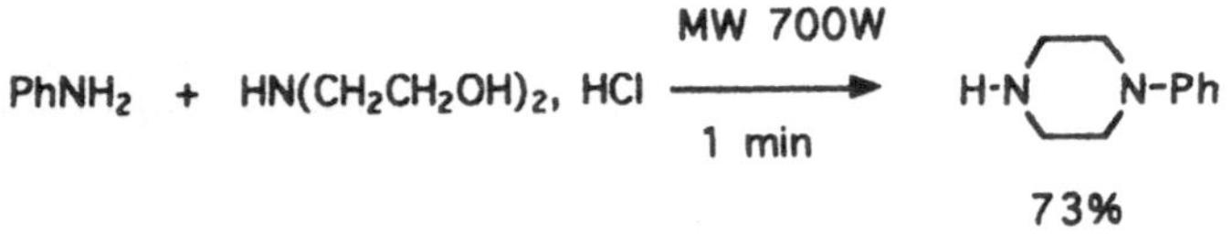

$$PhNH_2 \ + \ HN(CH_2CH_2OH)_2, \ HCl \ \xrightarrow[\text{1 min}]{\text{MW 700W}} \ H\text{-}N\diagdown N\text{-}Ph \quad (77)$$

73%

- synthesis of N-carboxyalkyl maleimides and phthalimides [149]

Maleic and phthalic anhydrides condense with amino acids under MW to afford the desired products in excellent yields [Eq. (78)]:

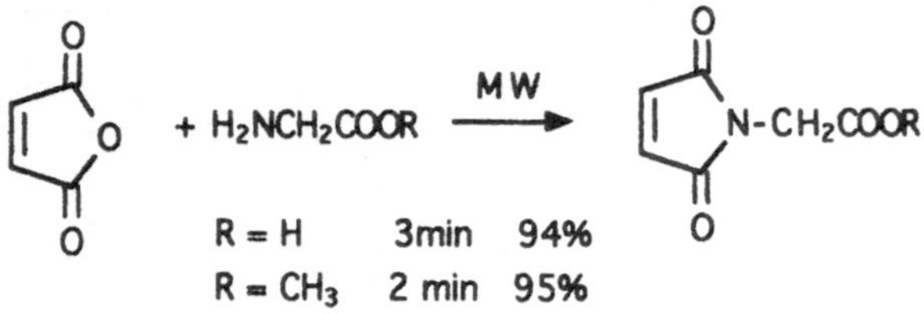

$$+ \ H_2NCH_2COOR \ \xrightarrow{MW} \quad N\text{-}CH_2COOR \quad (78)$$

R = H   3min   94%
R = CH$_3$ 2 min 95%

- 1,3-dipolar cycloaddition of nitrones.

[2 + 3]-dipolar cycloaddition of C-phenyl-N-methyl nitrone to fluorinated dipolarophiles lead to isoxazolidines of biological interest. In classical experiments, reactions were performed in refluxing toluene to give limited yields after long reaction times. Yields and experimental conditions were improved first in solvent-free conditions and then further under microwave irradiation (Scheme 29) [150].

It is worthy of note that, for the second example, yields are equivalent, be it with or without solvent; however, reaction time, without solvent, is dramatically reduced: 3 min vs 24 h when toluene is used and at very close operating temperature.

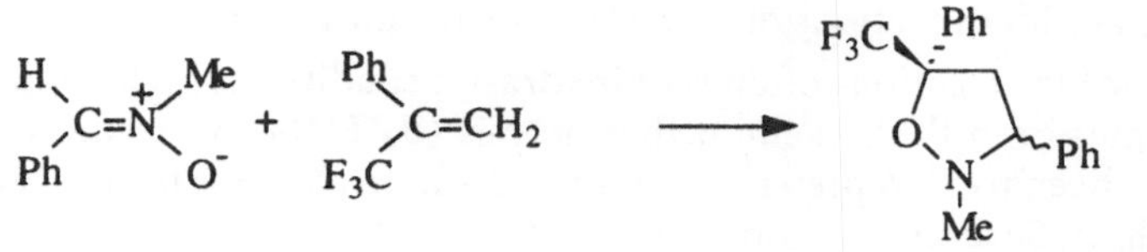

			50 : 50	
MW	**no solvent**	**3 mn**	**170°C**	**98%**
Δ	no solvent	3 mn	170°C	55%
Δ	toluene	48 h	110°C	65%

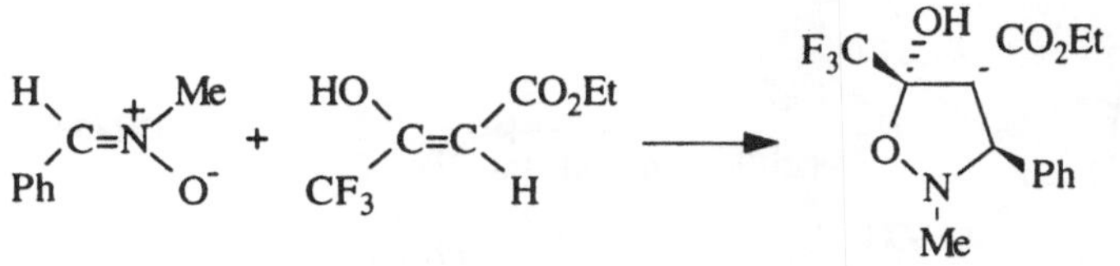

MW	**no solvent**	**3 mn**	**119°C**	**88%**
Δ	no solvent	3 mn	119°C	64%
Δ	toluene	24 h	110°C	65%

Scheme 29

More recently, the same procedure has been applied to unreactive nitrone cycloaddition to alkenes and produced high yields with an interesting comparison with classical heating (Δ) and ultrasonic (US) activation showing the MW procedure to be far better [151] (Scheme 30).

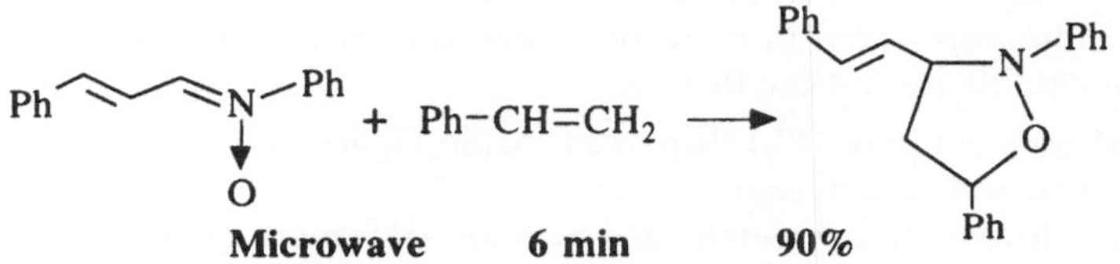

Microwave	**6 min**	**90%**
Thermal	34 h	80%
Ultrasound	1 h	87%

Scheme 30

– Diels–Alder cycloaddition of vinylpyrazoles [152].

Vinylpyrazoles undergo Diels-Alder cycloadditions within 6–30 min to give acceptable yields of easily purified products. This methodology eliminates the most important drawbacks of the classical conditions and allows the reaction to be extended to poorly reactive dienophiles, such as ethyl phenyl propiolate, not accessible by classical heating [Eq. (79)].

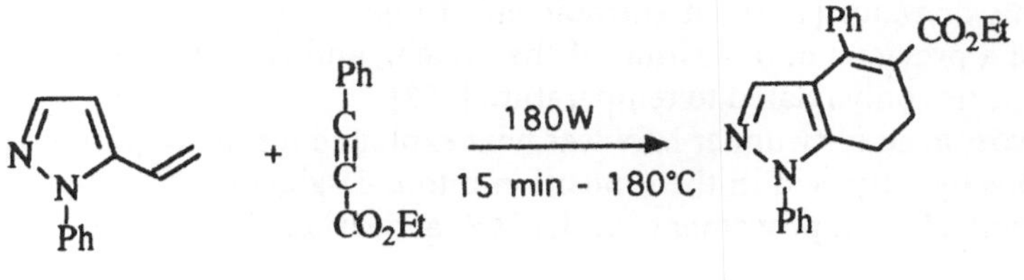

(79)

- retro-Diels–Alder reactions of benzylamino alcohols [153]

Retro-Diels–Alder reactions often require drastic conditions, high temperature and sometimes even flash-vacuum thermolysis (FVT). Such thermolytic procedures have been used to prepare unsaturated amino alcohols from a variety of amino alcohols. Several reactions were performed for a variety of neat liquid adducts and submitted to MW irradiation or to classical heating at the same temperature. The improvements obtained by coupling MW and the solvent-free technique are remarkable if we consider that both classical thermolysis and FVT (leading to decomposition) are poorly productive (Scheme 31).

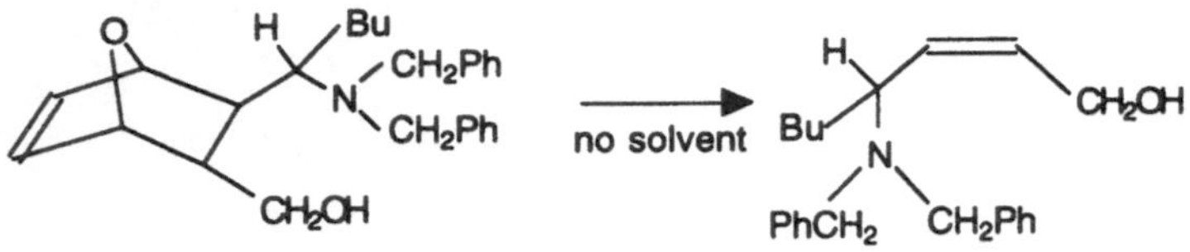

	MW	10 mn	140°C	> 98%
	Δ	10 mn	140°C	< 2%
Scheme 31	Δ	8 h	140°C	84%

6.3.4
Enzymatic Catalysis

It is possible to use enzymes immobilized on solid supports (either mineral or organic) of adequate pH in dry media [154] and, consequently, to operate at higher temperatures than in aqueous or organic media. Two main enzymatic systems including lipases can be used:

(i) *Pseudomonas* lipase (PL) dispersed inside Hyflo Super Cell (HSC) which consists of a diatomaceous silica, and

(ii) SP 435 Novozym, commercialized by Novo (Denmark), which is a *Candida* Antarctica lipase grafted onto an acrylic resin (Accurel). Such systems are thermally stable and exhibit an optimal activity in the range 80–100°C. They can therefore be used either under conventional heating or under MW activation with a monomode reactor to take advantage of a strict control in temperature by concomitant modulation of emitted power.

As an example, racemic 1-phenylethanol was resolved either by transesterification using the PL/HSC system or by esterification with SP 435 Novozym (Scheme 32) [155].

The results show an increased enantioselectivity under MW activation. The origin of such an effect could be manifold: a most efficient removal of light alcohols or water [156], an entropic effect due to dipolar polarization able to induce a previous organization of the system, and conformational changes of proteins not only related to temperature [157].

Novozym activity under MW was next exploited for the regioselective esterifications by fatty acid in the 6-position into α-D-glucose of α-D-glucopyranosides with clear improvements under MW activation (Scheme 33) [158].

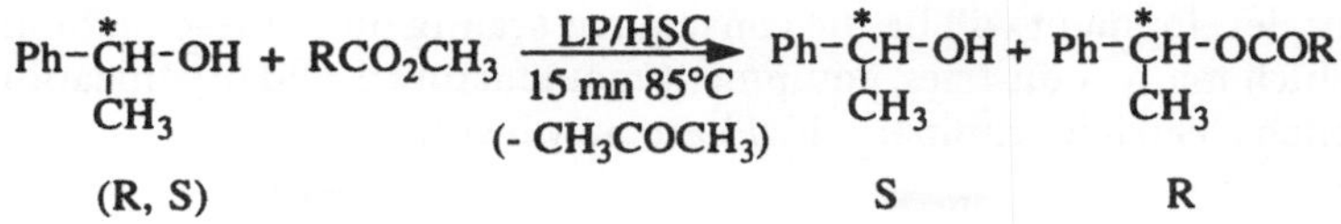

$$\text{Ph-}\overset{*}{\text{C}}\text{H-OH} + \text{RCO}_2\text{CH}_3 \xrightarrow[\substack{15 \text{ mn } 85°C \\ (-\text{ CH}_3\text{COCH}_3)}]{\text{LP/HSC}} \text{Ph-}\overset{*}{\text{C}}\text{H-OH} + \text{Ph-}\overset{*}{\text{C}}\text{H-OCOR}$$

$$\underset{(R, S)}{\overset{|}{\text{CH}_3}} \qquad\qquad \underset{S}{\overset{|}{\text{CH}_3}} \quad \underset{R}{\overset{|}{\text{CH}_3}}$$

$R = \overset{\backslash}{\underset{\diagup}{\text{C}}}=\text{CH}_2$ CH_3	**MW monomode reactor**	conv.: 47 %	ee: 79 %	**E: 42**	
	Δ	oil bath	conv.: 38 %	ee: 50 %	E: 16

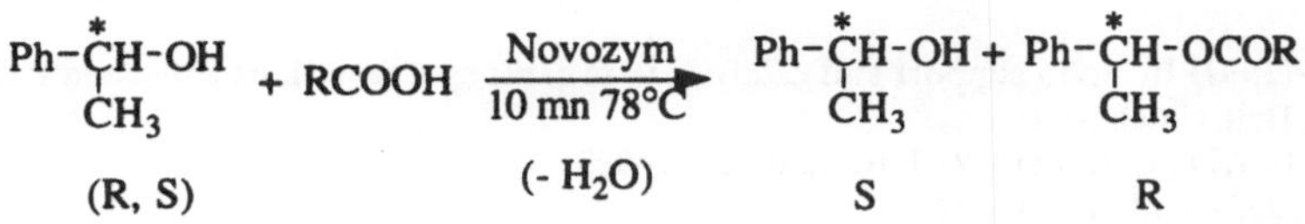

$$\text{Ph-}\overset{*}{\text{C}}\text{H-OH} + \text{RCOOH} \xrightarrow[\substack{10 \text{ mn } 78°C \\ (-\text{ H}_2\text{O})}]{\text{Novozym}} \text{Ph-}\overset{*}{\text{C}}\text{H-OH} + \text{Ph-}\overset{*}{\text{C}}\text{H-OCOR}$$

$$\underset{(R, S)}{\overset{|}{\text{CH}_3}} \qquad\qquad \underset{S}{\overset{|}{\text{CH}_3}} \quad \underset{R}{\overset{|}{\text{CH}_3}}$$

$R = \text{CH}_3\text{-(CH}_2)_6$	**MW monomode reactor**	conv.: 52 %	ee: 93 %	**E: 44**	
	Δ	oil bath	conv.: 48 %	ee: 62 %	E: 9

Scheme 32

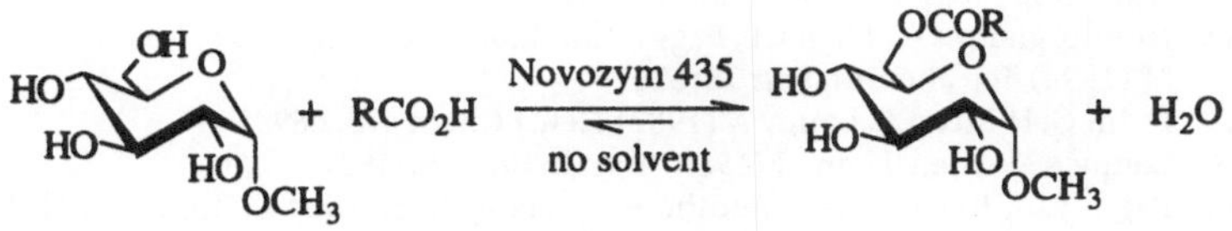

$$\text{HO}\text{-}\overset{\text{OH}}{\diagdown}\text{O} + \text{RCO}_2\text{H} \underset{\text{no solvent}}{\overset{\text{Novozym 435}}{\rightleftharpoons}} \text{HO}\text{-}\overset{\text{OCOR}}{\diagdown}\text{O} + \text{H}_2\text{O}$$

R=CH₃-(CH₂)₁₁

Δ	classical method	reduced pressure	24 h	70°C	53%
MW	**120-40 W**		**5 h**	**95°C**	**95%**
Δ	oil bath		5 h	95°C	55%
MW	120-60 W		5 h	95°C	97%

Scheme 33

7
Perspectives

The choice and use of solvents is both an intuition and a tradition for most organic chemists. The development of solvent-free procedures is a current topic which can harmoniously connect research and the environment. On an industrial scale these methods essentially become unavoidable when one considers the inconvenience involved in solvent use due to handling, cost, toxicity and the safety and pollution problems they generate.

Microwave irradiation is confirmed as a new, more efficient mode of activation when compared to classical heating resulting in rapid reactions with better yields and purer compounds. Coupling microwave and solvent-free procedures is shown to be of great interest and offers attractive potential. More and more, classical conditions should be revisited in this direction thus taking advantage of clean, efficient, safe and economical technology.

One major development will be the consequent scaling up of these methods, an area in which many industries, equipment manufacturers and organizations (e. g. the French electricity company, EDF) are now involved.

8
References

1. Keinan E, Mazur Y (1977) J. Am Chem Soc 99:3861
2. Laszlo P (1987) In: Preparative chemistry using supported reagents. Academic Press, London
3. Smith K (1992) In: Solid support and catalysts in organic synthesis. Ellis Horwood PTR Prentice Hall, Chichester
4. Kabalka GW, Pagni RM (1997) Tetrahedron 53:7999
5. Toda F (1993) Synlett 304
6. Toda F (1995) Acc Chem Res 28:480
7. Starks CM (1971) J Am Chem Soc 93:195
8. Bram. G, Loupy A, Sansoulet J (1985) Israel J Chem 26:291
9. Bram G, Galons H, Labidalle S, Loupy A, Miocque M, Petit A, Pigeon P, Sansoulet J (1989) Bull Soc Chim Fr 247
10. Foucaud A, Bram G, Loupy A (1993) In: Preparative chemistry using supported reagents, Academic Press, London, chap 17, p 317
11. Hondrogiannis G, Than CL, Pagni RM, Kabalka GW, Herold S, Ross E, Green J, Mc Ginnis M (1994) Tetrahedron Lett 35:6211
12. Bram G, D'Incan E, Loupy A (1982) New J Chem 573, 689
13. Loupy A, Seyden-Penne J (1980) Tetrahedron 36:1937
14. Pagni RM, Kabalka GW, Boothe R, Gaetano K, Stewart LJ, Conaway R, Dial C, Gray D, Larson S, Luidhardt T (1988) J Org Chem 53:4477
15. Kropp PJ, Daus KA, Tubergen MW, Kepler KD, Wilson VP, Craig SL, Baillargeon MM, Breton GW (1993) J Am Chem Soc 115:3071
16. Czech B, Quici S, Regen SL (1980) Synthesis 113
17. Tan CL, Pagni RM, Kabalka GW, Hillmyer M, Woosley J (1992) Tetrahedron Lett 33:7709
18. Bodwell GJ, Hougton TJ, Koury HE, Yarlagadda B (1995) Synlett 751
19. Parks GA (1965) Chem Rev 65:177
20. Varma RS, Varma M (1992) Tetrahedron Lett 33:5937
21. Texier-Boullet F (1985) Synthesis 679
22. Rosini G, Ballini R, Sorrenti P (1983) Tetrahedron 39:4127
23. Ranu BC, Bhars S (1992) Tetrahedron 48:1327
24. Loupy A, Tchoubar B (1992) In: Salt effects in organic and organometallic chemistry, VCH, Weinheim, p 37
25. Clark JH (1980) Chem Rev 80:429
26. Yamawaki J, Ando T (1979) Chem Lett 755
27. Villemin D, Ricard M (1984) Tetrahedron Lett 39:1059
28. Oussaïd A, Loupy A unpublished results; Oussaïd A (1997) PhD thesis, Orsay, Paris-South University
29. Ando T, Kawate T, Ichihara J, Hanafusa T (1984) Chem Lett 725
30. Anufriev VP, Norikov V (1995) Tetrahedron Lett 36:2515; Ando T, Kawate T, Ichihara J, Hanafusa T (1983) Bull Soc Chim Jpn 56:1885
31. Ando T, Sumi S, Kawate T, Ichihara J, Hanafusa T (1984) J Chem Soc Chem Commun 439
32. Posner GH, Rogers DZ (1977) J Am Chem Soc 99:8208, 8214
33. Ranu B, Majee A, Das AR (1995) Synth Commun 25:263
34. Dhavale DD, Sindkhedkar MD, Mali RS (1995) J Chem Res S 414
35. Pagni RM, Kabalka GW, Hondrogiannis G, Bains S, Anosike P, Kurt R (1993) Tetrahedron 49:6743
36. Bains S, Pagni RM, Kabalka GW, Palla C (1994) Tetrahedron: Asymmetry 5:2263

37. Kamitori Y, Hojo M, Masuda R, Yoshida T (1985) Tetrahedron Lett 26:4767
38. Kodomari M, Sakamoto T, Yoshitomi S (1990) J Chem Soc Chem Commun 701
39. Ranu BC, Ghosh K, Jana U (1996) J Org Chem 61:9546
40. Breton GW, Kurtz MJ, Kurtz SL (1997) Tetrahedron Lett 38:3825
41. Farwaha R, de Mayo P, Schauble JH, Toong YC (1985) J Org Chem 50:149
42. Slayton RM, Franklin NR, Tro NJ (1996) J Phys Chem 100:15551
43. Santaniello E (1993) In: Preparative chemistry using supported reagents. Academic Press, London, chap 18, p 345
44. Ranu BC, Sarkar A, Saha M, Chakraborty R (1994) Tetrahedron 50:6579; Ranu BC, Majee A, Sarkar A (1998) J Org Chem 63:370
45. Clark JH, Cork DG (1983) J Chem Soc Perkin Trans 1 2253
46. Bram G, Loupy A (1993) In: Preparative chemistry using supported reagents. Academic Press, London, chap 20, p 387
47. Kotushi H, Shimanouchi T, Oshima R, Fujiwara S (1998) Tetrahedron 54:2709
48. Veselovsky VV, Gybin AS, Lozanova AV, Moiseenkov AM, Smit WA, Caple R (1988) Tetrahedron Lett 29:175
49. Patil VJ, Mävers U (1996) Tetrahedron Lett 37:1281
50. Bram G, D'Incan E, Loupy A (1981) J Chem Soc Chem Commun 1066
51. Barry J, Bram G, Decodts G, Loupy A, Pigeon P, Sansoulet J (1983) Tetrahedron 39: 2669
52. Bram J, Loupy A, Roux-Schmitt MC, Sansoulet J, Strzalko T, Seyden-Penne J (1987) Synthesis 56
53. Huang B, Gupton JT, Hansen KC, Idoux JP (1996) Synth Commun 26:165
54. Balogh M, Laszlo P (1993) In: Organic chemistry using clays, reactivity and structure concepts in organic chemistry, vol 29. Springer, Berlin Heidelberg New York
55. Izumi K, Urabe K, Onaka M (1992) In: Zeolite, clay and heteropoly acids in organic reactions, VCH, Weinhein, chap 1, p 21
56. Toshima K, Ishizuka T, Matsuo G, Nakata M (1995) Synlett 306
57. Toshima K, Miyamoto N, Matsuo G, Nakata M, Matsumura S (1996) J Chem Soc Chem Commun 1379
58. Toshima K, Ushiki Y, Matsuo G, Matsumara S (1997) Tetrahedron Lett 38:7375
59. Kellendonk FJA, Heinerman JJL, Van Santen RA (1987) In: Preparative chemistry using supported reagents. Academic Press, London, chap 23, p 455
60. Theng BKG (1979) In: Formation and properties of clay-polymer complexes, Elsevier, Amsterdam
61. Diddams P (1992) In: Solid supports and catalysts in organic synthesis, Ellis Horwood PTR Prentice Hall, Chichester, chap 1, p 23
62. Ben alloum A, Labiad B, Villemin D (1989) J Chem Soc Chem Commun 386
63. Pério B, Dozias MJ, Jacquault P, Hamelin J (1997) Tetrahedron Lett 38:7867
64. Devic M, Schirmann JP, Decarreau A, Bram G, Loupy A, Petit A (1991) New J Chem 15:949
65. Bram G, Loupy A, Majdoub M, Petit A (1991) Chem Ind (London) 396
66. Cornélis A, Dony C, Laszlo P, Nsunda KM (1991) Tetrahedron Lett 32:1423; Clark JH, Kybett AP, Mac Quarrie DJ, Barlow SJ, Landon P (1989) J Chem Soc Chem Commun 1353
67. Cornélis A, Gestermans A, Laszlo P, Mathy A, Zieba I (1990) Catal Lett 6:103
68. Bram G, Loupy A, Villemin D (1992) In: Solid supports and catalysts in organic synthesis, Ellis Horwood PTR Prentice Hall, Chichester, chap 12, p 301
69. Corey EJ, Seebach D (1966) J Org Chem 31:4097
70. Loupy A, Sansoulet J, Harris AR (1989) Synth Commun 19:2939
71. Harris AR (1988) Synth Commun 18:659
72. Toda F, Yagi M, Kiyoshige K (1988) J Chem Soc Chem Commun 958
73. Toda F, Tanaka K, Hamai K (1990) J Chem Soc Perkin Trans 1 3207
74. Kammoun N, Le Bigot Y, Delmas M, Boutevin B (1997) Synth Commun 27:2777
75. Im J, Kim S, Hahn B, Toda F (1997) Tetrahedron Lett 38:451
76. Toda F, Aakai H (1990) J Org Chem 55:3446

77. Rajagopal D, Rajagopalan K, Swaminathan S (1996) Tetrahedron: Asymmetry 7:2189
78. Keller P (1987) Macromolecules 20:462; Majdoub M (1998) PhD thesis, University Tunis II
79. Strzelecka H, Jallabert C, Veber M, Malthête J (1988) Mol Cryst Liq Cryst 136:347; Bayle JP, Bui E, Perez F, Courtieu J (1989) Bull Soc Chim Fr 532
80. Loupy A, Sansoulet J, Vaziri-Zand F (1987) Bull Soc Chim Fr 1027
81. Barry J, Bram G, Decodts G, Loupy A, Orange C, Petit A, Sansoulet J (1985) Synthesis 40
82. Loupy A, Sansoulet J, Pedoussaut M (1986) J Org Chem 51:740
83. Le Ngoc T, Duong-Lieu H, Nguyen-Ba H, Radhakrishna AS, Singh BB, Loupy A (1993) Synth Commun 23:1379
84. Diez-Barra E, De la Hoz A, Diaz-Ortiz A, Prieto P (1992) Synlett 893
85. Diez-Barra E, De la Hoz A, Diaz-Ortiz A, Prieto P (1993) Synth Commun 23:1935
86. Diaz-Ortiz A, Prieto P, Loupy A, Abenhaïm D (1996) Tetrahedron Lett 37:1936
87. Loupy A, Luche JL (1997) In: Handbook of phase transfer catalysis, Chapman and Hall, London, chap 11, p 369
88. Sasson Y, Weiss M, Loupy A, Bram G, Pardo C (1986) J Chem Soc Chem Commun 1250
89. Loupy A, Pardo C (1988) Synth Commun 18:1275
90. Loupy A, Bram G, Sansoulet J, Vaziri-Zand F (1986) New J Chem 10:765
91. Oussaid A, Pentek E, Loupy A (1997) New J Chem 21:1339
92. Kornblum N, Seltzer R, Haberfield P (1963) J Am Chem Soc 85:1148
93. Matsumoto K (1980) Angew Chem 82:1046
94. Bram G, Sansoulet J, Galons H, Miocque M (1988) Synth Commun 18:367
95. Mirza-Aghayan M, Etemad-Moghadam G, Zaparucha A, Berlan J, Loupy A, Koenig M (1995) Tetrahedron: Asymmetry 6:2643
96. Loupy A, Sansoulet J, Zaparucha A, Mérienne C (1989) Tetrahedron Lett 30:333
97. Loupy A, Zaparucha A (1993) Tetrahedron Lett 34:473
98. Jones GB, Chapman BJ (1995) Synthesis 475
99. Conn RSE, Lovell AV, Karady S, Weinstock LM (1986) J Org Chem 51:4710
100. Diez-Barra E, De la Hoz A, Merino S, Sanchez-Verdu P (1997) Tetrahedron Lett 38:2359
101. Diez-Barra E, De la Hoz A, Merino S, Sanchez-Verdu P (1997) In: Halpern ME (ed) Phase transfer catalysis, mechanism and syntheses. ACS Symposium Series 659, chap 14, p 181
102. Diez-Barra E, De la Hoz A, Merino S, Sanchez-Verdu P, Rodriguez A (1998) Tetrahedron 54:1835
103. Van Eldick R, Hubbard CD (1997) In: Chemistry under extreme or non-classical conditions, Wiley, New York
104. Loupy A (1993) Spectra Analyse 175:33
105. Abramovitch RA (1991) Org Prep Proc Int 23:685; Caddick S (1995) Tetrahedron 51:10403
106. Jullien SC, Delmotte M, Loupy A, Jullien H (1991) In: Microwaves and high frequencies, International Congress, Nice (France) 8–10 October, p 654
107. Stuerga D, Gaillard P (1996) Tetrahedron 52:5505
108. Lewis DA, Summers JD, Ward TC, Mc Grath JE (1992) J Polym Sci A 30:1647
109. Binner JGP, Hassine NA, Cross TE (1995) J Mater Sci 30:5389
110. Gedye R, Smith F, Westaway K, Ali H, Baldisera L, Laberge L, Rousell J (1986) Tetrahedron Lett 27:279
111. Giguere RJ, Bray TL, Duncan SM, Majetich G (1986) Tetrahedron Lett 27:4945
112. Loupy A, Petit A, Hamelin J, Texier-Boullet F, Jacquault P, Mathé D (1998) Synthesis, 1213
113. Villemin D, Thibault-Starzyk F (1991) J Chem Educ 346
114. Jacquault P (1992) EU Patent 545995 AI Prolabo Co
115. Guttierez E, Loupy A, Bram G, Ruiz-Hitzky E (1989) Tetrahedron Lett 30:945
116. Trehan IR, Brar JS, Arora AK, Kad GL (1997) J Chem Educ 74:324
117. Singh J, Kaur J, Nayyar S, Kad GL (1998) J Chem Res S 280
118. Perez ER, Marrerro AL, Perez R, Autié MA (1995) Tetrahedron Lett 36:1779

119. Acosta A, De la Cruz P, De Miguel P, Diez-Barra A, De la Hoz A, Langa F, Loupy A, Majdoub M, Martin N, Sanchez C, Seoane C (1995) Tetrahedron Lett 36:2165
120. Csiba M, Cléophax J, Loupy A, Malthête J, Gero S (1993) Tetrahedron Lett 34:1787
121. Bram G, Loupy A, Majdoub M, Gutierrez E, Ruiz-Hitzky E (1990) Tetrahedron 46:5167
122. Van den Eynde JJ, Fromont D (1997) Bull Soc Chim Belg 106:393
123. Ranu BC, Saha M, Bhar S (1997) Synth Commun 27:621
124. Michaud D, Texier-Boullet F, Hamelin J (1997) Tetrahedron Lett 38:7563
125. Varma VS, Chatterjee AK, Varma M (1993) Tetrahedron Lett 34:4603
126. Varma VS, Lamture JB, Varma M (1993) Tetrahedron Lett 34:3029
127. Varma VS, Chatterjee AK, Varma M (1993) Tetrahedron Lett 34:3207
128. Oussaid A, Loupy A (1997) J Chem Res S 342
129. Benhaliliba H, Derchour A, Bazureau JP, Texier-Boullet F, Hamelin J (1998) Tetrahedron Lett 39:541
130. Suarez M, Loupy A, Perez E, Moran T, Gerona G, Morales A, Autié M (1996) Heterocyclic Commun 2:275
131. Cotterill IC, Usyatinsky AY, Arnold JM, Clark DS, Dordick JS, Michels PC, Khmelnitsky YL (1998) Tetrahedron Lett 39:1117
132. Villemin D, Ben Alloum A (1990) Synth Commun 20:3325
133. Villemin D, Ben Alloum A, Thibault-Starzyk (1992) Synth Commun 22:1359
134. Bougrin K, Soufiaoui M, Loupy A, Jacquault P (1995) New J Chem 19:213
135. Limousin C, Cléophax J, Petit A, Loupy A, Lukacs G (1997) J Carbohydr Res 16:327
136. Abenhaïm D, Loupy A, Mahieu C, Séméria D (1994)) Synth Commun 24:1809
137. Jiang Y, Wang Y, Deng R, Mi A (1997) In: Halpern ME (ed) Phase transfer catalysis, mechanisms and syntheses. ACS Symposium Series 659, chap 16, p 203
138. Loupy A, Pigeon P, Ramdani M (1996) Tetrahedron 52:6705
139. Loupy A, Petit A, Ramdani M, Yvanaeff C, Majdoub M, Labiad B, Villemin D (1993) Can J Chem 71:90
140. Abenhaïm D, Loupy A, Ngoc Son CP, Nguyen Ba H (1994) Synth Commun 24:1199
141. Loupy A, Pigeon P, Ramdani M, Jacquault P (1994) Synth Commun 24:159
142. Lewis DA (1992) Mat Res Soc Symp Proc 269:21
143. Barnier JP, Loupy A, Pigeon P, Ramdani M, Jacquault P (1993) J Chem Soc Perkin Trans 1 397
144. Bogdal D, Pielichowski J, Boron A (1996) Synlett 873
145. Bogdal D, Pielichowski J, Jaskot K (1997) Heterocycles 45:715
146. Oussaid A, Le Ngoc T, Loupy A (1997) Tetrahedron Lett 38:2451
147. Diez-Barra E, Abenhaïm D, De la Hoz A, Loupy A, Sanchez-Migallon A (1994) Heterocycles 38:793
148. Jaisinghani HG, Khadilkar BM (1997) Tetrahedron Lett 39:6875
149. Borah HN, Boruah RC, Sandhu JS (1998) J Chem Res S 272
150. Loupy A, Petit A, Bonnet-Delpon D (1995) J Fluorine Chem 75:215
151. Boruah B, Prajapati D, Boruah A, Sandhu JS (1997) Synth Commun 27:2563
152. Diez-Barra E, De la Hoz A, Carrillo-Munoz JR, Diaz-Ortiz A, Gomez-Escalonilla MJ, Moreno A, Langa F (1996) Tetrahedron 52:9237
153. Bortolussi M, Bloch R, Loupy A (1998) J Chem Res S 34
154. Guibé-Jampel E, Rousseau G (1987) Tetrahedron Lett 28:3563
155. Carrillo-Munoz JR, Bouvet D, Guibé-Jampel E, Loupy A, Petit A (1996) J Org Chem 61:7746
156. Chemat F, Poux M, Berlan J (1994) J Chem Soc Perkin Trans 2 2597
157. Porcelli M, Cacciapuoti G, Fusco S, Massa R, d'Ambrosio G, Bertoldo C, De Rosa M, Zappia V (1997) FEBS Lett 402:102
158. Gelo-Pujic M, Guibé-Jampel E, Loupy A, Galema SA, Mathé D (1996) J Chem Soc Perkin Trans 1 2777

Author Index Volume 201–206

The volume numbers are printed in italics